AF691472

Cambridge Imperial and Post-Colonial Studies Series

General Editors: **Megan Vaughan**, Kings' College, Cambridge and **Richard Drayton**, King's College London

This informative series covers the broad span of modern imperial history while also exploring the recent developments in former colonial states where residues of empire can still be found. The books provide in-depth examinations of empires as competing and complementary power structures encouraging the reader to reconsider their understanding of international and world history during recent centuries.

Titles include:

Tony Ballantyne
ORIENTALISM AND RACE
Aryanism in the British Empire

Peter F Bang and C. A. Bayly (*editors*)
TRIBUTARY EMPIRES IN GLOBAL HISTORY

Gregory A. Barton
INFORMAL EMPIRE AND THE RISE OF ONE WORLD CULTURE

James Beattie
EMPIRE AND ENVIRONMENTAL ANXIETY, 1800–1920
Health, Aesthetics and Conservation in South Asia and Australasia

Rachel Berger
AYURVEDA MADE MODERN
Political Histories of Indigenous Medicine in North India, 1900–1955

Robert J. Blyth
THE EMPIRE OF THE RAJ
Eastern Africa and the Middle East, 1858–1947

Rachel Bright
CHINESE LABOUR IN SOUTH AFRICA, 1902–10
Race, Violence, and Global Spectacle

Larry Butler and Sarah Stockwell
THE WIND OF CHANGE
Harold Macmillan and British Decolonization

Kit Candlin
THE LAST CARIBBEAN FRONTIER, 1795–1815

Nandini Chatterjee
THE MAKING OF INDIAN SECULARISM
Empire, Law and Christianity, 1830–1960

Esme Cleall
MISSIONARY DISCOURSE
Negotiating Difference in the British Empire, c.1840–95

T. J. Cribb (*editor*)
IMAGINED COMMONWEALTH
Cambridge Essays on Commonwealth and International Literature in English

Bronwen Everill
ABOLITION AND EMPIRE IN SIERRA LEONE AND LIBERIA

Róisín Healy & Enrico Dal Lago (*editors*)
THE SHADOW OF COLONIALISM IN EUROPE'S MODERN PAST

B.D. Hopkins
THE MAKING OF MODERN AFGHANISTAN

Ronald Hyam
BRITAIN'S IMPERIAL CENTURY, 1815–1914: A STUDY OF EMPIRE AND EXPANSION
Third Edition

Iftekhar Iqbal
THE BENGAL DELTA
Ecology, State and Social Change, 1843–1943

Leslie James
GEORGE PADMORE AND DECOLONIZATION FROM BELOW
Pan-Africanism, the Cold War, and the End of Empire

Robin Jeffrey
POLITICS, WOMEN AND WELL-BEING
How Kerala became a 'Model'

Miguel Bandeira Jerónimo
THE "CIVILISING MISSION" OF PORTUGUESE COLONIALISM, 1870–1930

Miguel Bandeira Jerónimo and António Costa Pinto
THE ENDS OF EUROPEAN COLONIAL EMPIRES
Cases and Comparisons

Gerold Krozewski
MONEY AND THE END OF EMPIRE
British International Economic Policy and the Colonies, 1947–58

Zoë Laidlaw and Alan Lester *(editors)*
INDIGENOUS COMMUNITIES AND SETTLER COLONIALISM
Land Holding, Loss and Survival in an Interconnected World

Javed Majeed
AUTOBIOGRAPHY, TRAVEL AND POST-NATIONAL IDENTITY

Francine McKenzie
THE POLITICS OF PREFERENCE
Redefining the Bonds of Commonwealth 1939–1948

Gabriel Paquette
ENLIGHTENMENT, GOVERNANCE AND REFORM IN SPAIN AND ITS EMPIRE 1759–1808

Sandhya L. Polu
PERCEPTION OF RISK
Policy-Making On Infectious Disease in India 1892–1940

Sophus Reinert, Pernille Røge
THE POLITICAL ECONOMY OF EMPIRE IN THE EARLY MODERN WORLD

Jonathan Saha
LAW, DISORDER AND THE COLONIAL STATE
Corruption in Burma c.1900

John Singleton and Paul Robertson
ECONOMIC RELATIONS BETWEEN BRITAIN AND AUSTRALASIA 1945–1970

Leonard Smith
INSANITY, RACE AND COLONIALISM
Managing Mental Disorder in the Post-Emancipation British Caribbean, 1838–1914

Miguel Suárez Bosa
ATLANTIC PORTS AND THE FIRST GLOBALISATION C. 1850–1930

Jerome Teelucksingh
LABOUR AND THE DECOLONIZATION STRUGGLE IN TRINIDAD AND TOBAGO

Julia Tischler
LIGHT AND POWER FOR A MULTIRACIAL NATION
The Kariba Dam Scheme in the Central African Federation

Erica Wald
VICE IN THE BARRACKS
Medicine, the Military and the Making of Colonial India, 1780–1868

Ryo Ikeda
THE IMPERIALISM OF FRENCH DECOLONIZATION
French Policy and the Anglo-American Response in Tunisia and Morocco

A. B. Leonard and David Pretel
THE CARIBBEAN AND THE ATLANTIC WORLD ECONOMY
Circuits of trade, money and knowledge, 1650–1914

Cambridge Imperial and Post-Colonial Studies Series
Series Standing Order ISBN 978–0–333–91908–8 (Hardback)
978–0–333–91909–5 (Paperback)
(outside North America only)

You can receive future titles in this series as they are published by placing a standing order. Please contact your bookseller or, in case of difficulty, write to us at the address below with your name and address, the title of the series and the ISBN quoted above.

Customer Services Department, Macmillan Distribution Ltd, Houndmills, Basingstoke, Hampshire RG21 6XS, England

The Caribbean and the Atlantic World Economy

Circuits of trade, money and knowledge, 1650–1914

Edited by

A. B. Leonard
Centre for Financial History, University of Cambridge, UK

and

David Pretel
Universitat Pompeu Fabra, Barcelona, Spain

palgrave
macmillan

Selection, introduction and editorial content © A. B. Leonard and David Pretel 2015
Chapters © Contributors 2015

All rights reserved. No reproduction, copy or transmission of this publication may be made without written permission.

No portion of this publication may be reproduced, copied or transmitted save with written permission or in accordance with the provisions of the Copyright, Designs and Patents Act 1988, or under the terms of any licence permitting limited copying issued by the Copyright Licensing Agency, Saffron House, 6–10 Kirby Street, London EC1N 8TS.

Any person who does any unauthorized act in relation to this publication may be liable to criminal prosecution and civil claims for damages.

The authors have asserted their rights to be identified as the authors of this work in accordance with the Copyright, Designs and Patents Act 1988.

First published 2015 by
PALGRAVE MACMILLAN

Palgrave Macmillan in the UK is an imprint of Macmillan Publishers Limited, registered in England, company number 785998, of Houndmills, Basingstoke, Hampshire RG21 6XS.

Palgrave Macmillan in the US is a division of St Martin's Press LLC, 175 Fifth Avenue, New York, NY 10010.

Palgrave Macmillan is the global academic imprint of the above companies and has companies and representatives throughout the world.

Palgrave® and Macmillan® are registered trademarks in the United States, the United Kingdom, Europe and other countries.

ISBN 978–1–137–43271–1

This book is printed on paper suitable for recycling and made from fully managed and sustained forest sources. Logging, pulping and manufacturing processes are expected to conform to the environmental regulations of the country of origin.

A catalogue record for this book is available from the British Library.

A catalog record for this book is available from the Library of Congress.

Typeset by MPS Limited, Chennai, India.

Contents

List of Figures

List of Tables

Notes on Contributors

Manuel Covo is a postdoctoral fellow in the Program of Early American Economy and Society at the Library Company in Philadelphia. In 2013 he was awarded the EHESS PhD Award for the best dissertation in all fields. His dissertation 'Commerce, Empire and Revolutions in the Atlantic World: The French Colony of Saint-Domingue between France and the United States (ca. 1778–ca. 1804)' studied commerce between the French colony of Saint-Domingue and the United States at the end of the eighteenth century. Taking an Atlantic perspective, his thesis connected divergent historiographies: colonial history, the histories of the French and Haitian revolutions, and the history of the early American Republic. Focusing on transnational circulations – smuggling and commercial networks – his study dealt with the role of Atlantic exchanges in the formation of new political entities during the time of revolutions. His research interests are in Atlantic, imperial, and business history.

Martin Daunton is Professor of Economic History and Head of the School of the Humanities and Social Sciences at Cambridge University. He read economic history at the University of Nottingham and completed his PhD at the University of Kent, before taking his first post at the University of Durham. He moved to University College London as Lecturer in economic history, and was promoted to the Astor Professorship of British History before departing for Cambridge in 1997. He has served as President of the Royal Historical Society and a Trustee of the National Maritime Museum. He has written two surveys of British economic history since 1700, *Progress and Poverty: An economic and social history of Britain 1700–1860* and *Wealth and Welfare: an economic and social history of Britain 1851–1951*, and two volumes on the politics of taxation, *Trusting Leviathan: the politics of taxation in Britain 1799–1914* and *Just Taxes: the politics of taxation in Britain 1914–1979*. He is currently completing a book on the economic governance of the world since 1933.

Nadia Fernández-de-Pinedo is a Senior Lecturer in Economic History at the Universidad Autónoma of Madrid. Her principal research interest is Cuban economic history, including several chapters and articles focusing particularly on nineteenth century trade and commercial relations, as well as two books: *El comercio exterior y la fiscalidad de Cuba,*

1790–1864 (2002) and *Las balanzas inéditas del comercio exterior de la Habana de 1803 a 1807* (2000). She was a research scholar at Warwick University, and has been a visiting researcher at universities and research centres in Cuba, the UK, and France. Her work covers Cuba's overseas trade, consumption, transnational networks, and technology. Recently she has embarked on a new project with Patricio Saiz and David Pretel examining Cuban IPRs, technology transfer processes, and institutions. Among previous publications is 'Patents, sugar technology and sub-imperial institutions in nineteenth-century Cuba' (*History of Technology*, 2011).

Knick Harley is a retired Professor of Economic History at the University of Oxford. His research interests revolve around the interactions between the emergence of modern economic growth and the globalisation of the world economy between the eighteenth and twentieth centuries. His published research includes extensive work on trade and industrialisation in the British Industrial Revolution and in the industrialisation of the US. In addition, he has published extensively on the impact of technology on transportation in the late nineteenth century. This research investigates the effects that growing trade had on both the nature of industrial structure and on the rise of international services in the core economies (particularly Britain), and on the geographical expansion of periphery economies into continental interiors. He has served as editor of the *Journal of Economic History*.

Adrian Leonard is a post-doctoral researcher at the Centre for Financial History, University of Cambridge, and Affiliated Lecturer at the university's Faculty of Economics. He has written widely on topics related to marine insurance and the Atlantic World. Current projects include a history of commercial insurance in London in the twentieth century, and editing the Palgrave History of Finance volume *Marine insurance: international development and evolution*. He is co-editor of *The Atlantic World 1400–1850* (2014) and *Questioning Credible Commitment* (2013). Prior to returning to academia to receive Masters and PhD degrees in Economic History from Cambridge, he worked as a writer and commentator on financial issues for publishers including the *Financial Times* and *Reuters*.

Chuck Meide was born and raised in Atlantic Beach, Florida, just north of the oldest port in the US, St. Augustine. He attended Florida State University, receiving Bachelors (1993) and Masters (2001) degrees in anthropology, with a focus on maritime archaeology. He undertook PhD studies in historical archaeology at the College of William & Mary in 2002, where his dissertation research focused on nineteenth-century economic relations on the maritime landscape of Achill Island, Ireland. He has participated in the archaeological investigation of numerous shipwrecks, including a Spanish galleon from Tristan de Luna's 1559 colonisation fleet, the 1588 Spanish Armada vessel *Trinidad Valencera*, and the French explorer La Salle's ship *La Belle* (1686). Since 2006 he has served as the Director of the Lighthouse Archaeological Maritime Program (LAMP), the research arm of the St. Augustine Lighthouse & Museum, where he has directed the excavation of a Loyalist refugee vessel lost in 1782, during the American Revolution.

David Ormrod is Professor of Economic and Cultural History at the University of Kent, and Senior Research Associate in the Centre for Financial History, Cambridge. He has written widely on early modern commercial history and co-convenes the fortnightly research seminar on *The Premodern World* at the Institute of Historical Research. In 2012, he completed an ESRC-funded research project reconstructing urban and agricultural rent movement in London and the South-East since 1580, which will enable scholars to model the growth of the British economy with greater precision. Publications include *The Rise of Commercial Empires: England and the Netherlands in the Age of Mercantilism, 1650–1770* (2003), which entered the American Library Association's list of outstanding academic books of 2003; 'Institutions and the Environment: Shipping Movements in the North Sea/Baltic Zone, 1650–1800', in R.W. Unger (ed.), *Shipping and Economic Growth, 1350–1850* (2011); and 'Agrarian Capitalism and Merchant Capitalism: Tawney, Dobb, Brenner and Beyond', in J. Whittle (ed.), *Landlords and Tenants in Britain, 1440–1660* (2013).

Gert Oostindie is Director of the KITLV/Royal Institute of Southeast Asian and Caribbean Studies and Professor of Caribbean History at Leiden University. His principal areas of research have been Caribbean and Dutch (post)colonial history in a comparative perspective. His books include *Dutch Atlantic Connections, 1680–1800: Linking Empires, Bridging Borders* (ed., with Jessica Vance Roitman, 2014), *Postcolonial Netherlands: Sixty-five Years of Forgetting, Commemorating, Silencing* (2011), *Dutch Colonialism, Migration and Cultural Heritage* (ed., 2008), *Paradise Overseas. The Dutch Caribbean: Colonialism and its Transatlantic Legacies* (Macmillan, 2005), and *Decolonising the Caribbean: Dutch Policies in a Comparative Perspective* (with Inge Klinkers, 2003).

David Pretel is Postdoctoral Research Fellow in Economic History at Universitat Pompeu Fabra in Barcelona, where he is a Lecturer in the

History of Economic Thought. Before coming to Barcelona he was a Max Weber Postdoctoral Fellow in History at the European University Institute in Florence. He has also been Junior Research Fellow at the Institute of Historical Research (University of London), and a research associate in history at Trinity Hall (University of Cambridge). He was educated at Universidad Autónoma Madrid and the University of Cambridge, and during his doctoral studies he was a visiting graduate student at the universities of Bristol, Nottingham Trent, and the UCLA Center for Economic History. His general area of interest lies in nineteenth- and twentieth-century Spanish and Latin American history, especially the history of economic thought and economic history, and the history of science and technology.

Martín Rodrigo y Alharilla is Associate Professor of History at the University Pompeu Fabra (Barcelona), where he is a member of the Research Group on Empires, Metropoles, and Extra-European Societies. His main lines of research focus on the history of modern enterprise in Spain, Catalan naval activity, and in particular the analysis of the Spanish imperial system during the nineteenth century, both in the West Indies and the Philippines, and Spain's political and economic relations with its colonies. He received a BA in Philosophy and Humanities (1992) and a PhD in Economics (2000) from the Universitat Autònoma de Barcelona. He has received scholarships from several institutions and foundations including the Bank of Spain, Maritime Museum of Barcelona, Caja Madrid Foundation, Noguera Foundation, and Gas Natural Foundation. His books include *Los marqueses de Comillas, 1817–1925* (2000, nominated for the Third LID Prize for Entrepreneurial Biography); *La familia Gil: Empresarios catalanes en la Europa del siglo xix* (2010); and as an editor *Cuba: de colonia a República* (2006).

Inés Roldán de Montaud is Full Research Scientist in History at the Spanish National Research Council (CSIC), and has been an Associate Professor at the Faculty of Arts of the Universidad de Alcalá since 1997. She was a research scholar at the London School of Economics, and Ramón y Cajal researcher at the Institute of History of the CSIC. She is the author of several books, monographs, and chapters, and has published numerous research essays in specialised journals both in Spain and abroad. Publications include *La Restauración en Cuba: el fracaso de un proceso reformista* (2000), *La Banca de emisión en Cuba, 1856–1898* (2004), *Las Cajas de Ahorros de las provincias de Ultramar, 1840–1848. Cuba y Puerto Rico* (2010), and as editor *Las Haciendas públicas en el Caribe hispano en el siglo XIX* (2008). Her research work has been focused on the study of Spain in the nineteenth century, particularly in colonial relations. She initially addressed the social structure of slavery, then continued to study elites, political parties, and elections, and lately has turned to financial institutions.

Dale W. Tomich is Professor of Sociology and History and Deputy Directory of the Fernand Braudel Center at Binghamton University (State University of New York). He received his PhD in history from the University of Wisconsin-Madison. He is author of *Slavery in the Circuit of Sugar: Martinique and the World Economy, 1830–1848* (1990) which was awarded the Distinguished Scholarship Award of the American Sociological Association, and *Through the Prism of Slavery: Labor, Capital, and World Economy* (2004). His current work is on Atlantic slavery and world-economy, including a collaborative international project on the making of plantation spaces in nineteenth century, Cuba, Brazil, and the US South. He has been Visiting Professor at various universities in Brazil, the United States, and France.

Nuala Zahedieh is Reader in Economic and Social History at the University of Edinburgh, where she has taught since 1989. A broad range of publications on the Caribbean and the Atlantic economy include *The Capital and Colonies: London and the Atlantic Economy, 1660–1700* (2010). She is currently writing a book about early English Jamaica in an Atlantic context.

1
Experiments in Modernity: the Making of the Atlantic World Economy

A.B. Leonard and David Pretel

'The Atlantic was a European invention', declared David Armitage in his opening chapter of the 2002 edited collection *The British Atlantic World, 1500–1800*. He argued that Europeans were the first to connect the four sides of the Atlantic into a single entity, both as a natural place, and as a system. Echoing Braudel, he explained how they 'integrated' disparate physical parts to 'invent' a geography, one in which most of the action happened on land, but which was bestowed an identity based on the ocean – itself a contemporary unification – which links together its components on *terra firma*.[1]

In much the same way, over the past three decades, historians have invented the Atlantic World. They have drawn together diverse but connected histories, from imperial to maritime, to invent a new field of historical study which includes something for almost everyone. The outcome has been a cascade of academic work in almost all branches of history which can now be identified as Atlantic History. The results have been positive: the Atlantic World as an applied historical creation has led to genuine new insights into a broad range of historiographical subjects, from state formation and the reach of empires, to ideas of place.[2] It has cast a new, sometimes critical light onto some established and widely accepted historiographical principles. Increasingly, Atlanticists have achieved this by adopting the approach advanced by historical sociologists such as Charles Tilly, who have stressed the necessity of rooting out and exposing the unity and interconnection within macrohistorical processes. Such an attitude towards history is particularly fruitful when applied to the study of the development of the Atlantic World economy and its evolving global circuits.[3]

This way of 'doing' Atlantic history is a relatively recent advance, and is only just taking hold. Earlier research into the Atlantic World typically interpreted the interactions between expanding European naval powers and their Atlantic colonies as largely intra-imperial, and often traced their evolution as a simple extension of European imperial rivalries. Developments and exchanges – commercial, social, cultural, or otherwise – between the 'new' areas of the new Atlantic were largely ignored. The Atlantic World was seen as a by-product of European imperialism, to be investigated in a simplistically applied context of metropolitan centres and new-world peripheries. Today, Atlanticists have begun, tentatively, to usurp this approach through comparative, regional, and sectoral Atlantic histories which, like many new Atlanticists themselves, fail to respect imperial boundaries. As a result of this broader analytical approach, the discipline now offers new understandings of what was happening not only on the ground in the American and African Atlantic World, but also in the western European coastal nations which invented it.[4]

Peter Coclanis has declared that Atlantic history is 'now sitting with the grown-ups'. Findings such as those of Martín Rodrigo y Alharilla in this volume are clearly the product of this new, less empire-focussed, Eurocentric treatment. Such work follows the methodological approach demanded by Craig Lockard of world history in the early 1980s, one which takes on board poststructuralist and multicultural criticisms, similarly called for by Bruce Mazlish in the early 1990s.[5] Coclanis has argued repeatedly that Atlantic history ignores too much of the rest of the world, and his trenchant criticism has helped to drive forward a broader Atlantic history, but it could be levelled against any branch of the discipline. If we are 'all Atlanticists now' (another assertion of Armitage, one with which many historians would disagree with strongly), it is perhaps equally valid to say we are all world historians now. Clearly we are not, but we can benefit from the approach they have embraced. In so doing, Atlantic history opens doorways to deeper understandings of broader events, albeit by passing through doors opened much earlier by historical sociologists such as Wallerstein, Tilly, Mintz, and others in the field of world history.[6] For today's Atlanticists, the improved dialogue between theory and empirical evidence is much improved.

What of economic history, which we have bound together in this volume with Atlantic history? The field is perhaps less sexy, but is nonetheless thriving. As a bridge between political and social histories, and a subject area which could and probably should be drawn down in many other genres of enquiry, its importance is increasingly widely acknowledged, even as the regression analyses and formulae of economics employed by cliometricians and others frighten away undergraduates. At the heart of the economic historian's field lies the desire to discover the causes of economic growth, divergence, and inequality. Extending from this question are others which lead to those larger questions. Why did the industrial revolution happen in England? What caused Western Europe to diverge, economically, from the rest? What is the role of institutions in fostering commerce, innovation, and industrialisation?

Many chapters in this volume draw on and build upon existing contributions from economic world history, supported by the theoretical and methodological insights advanced by historical sociology, to provide clues to the answers through the lens of Atlantic economic history. After all, the invention of the Atlantic by early modern Europeans was driven primarily by the quest for financial gain, whether on the part of individual adventurers, or the monarchs (and later, governments) that backed them. Enormous flows of treasure from first the Spanish and then the Brazilian Atlantic had an undeniable impact on European economies. While the contention that this influx was the cause of the great European inflation of the sixteenth century seems logically to be overturned by the reality of the century's robust population growth (David Fischer argues convincingly that 'the price revolution came first, American treasure followed',[7]) the quest for precious metals and their import into the Old World did impact upon local economies, as well as upon foreign relations in a period of great dynastic rivalry. The trade routes, colonies, and plantations which comprised the early Atlantic World were a product of mercantilism, but it was in the Atlantic where the strict proscriptions of the defining early modern economic approach were whittled away, first on a de facto basis, and much later on a de jure basis, too, as the Spanish, French, Dutch, and finally the English abandoned increasingly moribund trading systems.

Thus the distinctly early modern Atlantic World phenomenon was a key element of Europe's transition from a medieval to a modern economy. This development was a product of collaboration between old and new world actors. As evolving European polities made the relatively rapid shift to enlightened modernity, as they first launched into, then ultimately abandoned, imperial mercantilist systems in favour of a more modern, open system of commerce, they discovered in the Atlantic something of a testing ground for new approaches. Unique pressures and risks in distant colonies, and the enormous time-lag which affected every communication with the metropolitan centre, made essential a change of approach. In this way the Atlantic World was a vanguard of modernity, driven not just by geography, technology, demographics, and resources, by also, and critically, by political and economic rivalries, and by human agency.

This is especially true in the arena of Atlantic World trade, commerce, finance, and agricultural and industrial production. The opening of newly discovered lands to Europeans from 1492 increased the land endowment per European capita sixfold.[8] This dramatic change in the supply of land triggered a global economic rebalancing which spawned the first era of globalisation. The completion, by about 1890, of the incorporation of this new land into the Atlantic economy marked the end of that era. The economic exploitation of this massive injection of one of the three factors of production required labour on a massive scale. In combination, new land and immigrant labour (both free and coerced – the latter constituting just one great Atlantic World economic experiment) led to a layered convergence, the moving together of international wages, prices, and national incomes, which was to delineate gainers and losers. O'Rourke and Williamson report a 'really big leap to more globally integrated commodity and factor markets' between 1850 and 1914, as foreign markets influenced local prices. They attribute convergence to the 'open economy forces of trade and mass migration'. US per capita GDP was $2,482.50 in 1870, compared to $3,342.00 in the UK. By 1910 the figures were $5,015.00 and $4,7114.00 respectively.[9]

Economic historians have long discussed the origins of the uneven material development in the Atlantic World, and the economic bifurcation that manifested in the late eighteenth century. Between then and the later nineteenth century, the Atlantic World became divided between economically advanced, industrial territories and so-called 'backward' ones. However, the origins of the late nineteenth century's asymmetric Atlantic economy have remained poorly understood. Any global analysis of the modern, polarised Atlantic economy must seek the answer in a number of complementary explanations, rather than in a single cause (such as slavery) or grand theory (such as entrepreneurial spirit). Demographic and geographic explanations should be conjoined with material explanations which take into account the nature of mercantile cultures and the location of useful knowledge at particular sites. Each of the contributions to this volume, discussed below, have sought such a multifaceted understanding.

Commodity price convergence, easily illustrated through grain prices, was another product of the Atlantic World economy. The Anglo-American wheat price gap fell from 54% in 1870 to nothing in 1913. Intra-European commodity markets also experienced integration, where permitted by trade policies. For example, British barley prices were

42% higher than Danish prices in 1870, but the gap was zero by 1913. Overall, however, as Atlantic World land became increasingly productive, grain markets within Continental Europe became more balkanised, since 'globalization was not a universal phenomenon, even during the comparatively liberal late nineteenth century'.[10] Jeffrey Williamson denies 'commodity price convergence between Asia and Europe over the three centuries following 1492'.[11] However, convergence arising from Atlantic World grain imports was nothing new at the turn of the twentieth century. Recent work by Paul Sharp shows that in the trade in grain between America and Britain, 'US and UK prices almost perfectly follow each other in the long run' since 1700, especially when relative peace made extensive trade possible, even before the 'transport revolution'.[12] In the case of wheat trading, modern convergence was fostered in the Atlantic World economy during the eighteenth century, long before the repeal of Britain's corn laws.

Another clear example of the move through the Atlantic World toward economic modernity lies in the shift away from the preponderance of exclusive trading companies with monopoly rights. Such organisations – whether guilds, companies of chartered adventurers, or early joint-stock entities – had held sway across Europe and the 'known' world for centuries, controlling the production, distribution, import, and export of almost everything, from foodstuffs to playing cards. The jury is out on the question of the efficacy of such controlling institutions, although some recent scholarship is challenging the historiographical notion that guilds encouraged economic growth.[13] However, it is clear that the Atlantic World was a great testing-ground for the operation of commerce outside the context of such restrictive, medieval institutions. The bulk of Atlantic World trade was entirely open to private venturers relatively early on. Exceptions such as the Royal African Company (RAC) often constrained trade; Britain's slaving trade flourished only after the RAC lost its monopoly. The French Mississippi Company offers a similar example, as does Spain's slave *asiento*, which was typically placed in the hands of foreign monopolists.

Meanwhile most of Western Europe's trade to the East was tied to the hands of monopoly East India companies, whether English, Dutch, French, or Swedish. For British consumers, Atlantic World tobacco was produced and sold by competing private traders, while tea was sold by the double monopoly of Canton's *cohong* and the British East India Company. Competition made tobacco almost uneconomically cheap for producers. Tea, in contrast, and despite its widespread consumption in Britain by the end of the eighteenth century, was and remained expensive, even when heavy taxes are discounted, and could be afforded only through adulteration, or when smuggled to avoid both monopoly and taxation: Parliament found in 1745 that for each pound of tea legally imported, three were smuggled.[14] Open trade in all these areas, both Atlantic and Eastern, ultimately smashed the monopolies, following the early Dutch Atlantic example.[15]

Colonisation itself was an experiment in economics on a transoceanic scale, one which was sometimes structured and planned to a greater extent than may initially meet the historian's eye. Historical circumstances such as legal systems and institutional designs always affected the political economies of empires, but in the long run the historical making of the Atlantic economy, in which different nations played distinctive roles, was the consequence of variations in the institutional architecture of the different empires involved, and the divergent commercial policies those empires adopted over the period covered by this volume. As Nuala Zahedieh has pointed out, in 1584 Richard Hakluyt promoted Atlantic World settlement to Elizabeth I as a means of satisfying England's import needs, alleviating population stress, creating a market for her manufactures, and initiating a closed imperial economic system.[16] Hakluyt's remarkably prescient vision could almost have been an historian's analysis of Britain's Atlantic World economic strategy over more than two centuries, and constitutes a grand and successful economic experiment. The English venture which ensued, which appears almost orchestrated in the light of Hakluyt's policy recommendations to the Queen, was an early foray into the capitalistic international trade which characterises the modern world.

Meanwhile Spain attempted to extend her physical territory by creating New Spain from the conquistadors' territories, which were to form an integral part of the united Iberian kingdoms under de jure, if not de facto, central absolutist authority.[17] Often considered a co-venture of the state and church, Spain's Atlantic World empire was also a grand economic experiment in mercantilism, leading to the creation of a currency, the Spanish dollar, founded on plunder but amounting to the most persistent and universal currency the world had and has ever known.[18] And while the ultimate impact of Spain's Atlantic World troves remains a subject of debate, the development of an internationally accepted currency, adopted as a unit of account as far away as Canton in the nineteenth century,[19] and arguably leading directly to its successor, the US dollar, was a great foray into economic modernity.

Experiments in economic modernity continued into the nineteenth and twentieth centuries, when at times their direction was reversed (as is sometimes the case to this day). For example, as Spain's imperial ambitions were fading, her Caribbean colonies became world leaders in sugar production, while her own industrial progress lagged behind that of her metropolitan rivals. Neocolonial conditioning factors, such as informal and free-trade imperialism, which shaped the economic development of the Caribbean peaked in the nineteenth century. At the height of the second imperial era, neo-mercantilist policies, and the growth of international trade and commerce within this context of increasing nationalism and protectionism,[20] affected various features of economic interaction, such as the exchange of commodities and knowledge in the Caribbean.

Indeed, things changed radically in the mid-nineteenth century, with both industrialisation and the increasing impact of multinational corporations, corporate forms of business organisation, and transnational banking. Two connected processes are of particular importance: the expansion of European industrial production and overseas trade, and the development of colonial enterprises in the Caribbean such as plantations to produce genuinely global colonial commodities to be consumed by swelling mass-consumption markets in Europe, as well as in the rising US. Both constitute important components of the transformation of states and empires associated with the end of the imperial era.

Wallerstein's world-system model has often been applied to Atlantic history, identifying (at its simplest, in a teleological way) a metropolitan centre 'core' served by a colonial 'periphery'. Wallerstein says his world-system reflects '*a* world, not *the* world', and that sovereign states are 'to be seen as one kind of organisational structure among others within this single social system'.[21] This model holds particularly well as regards the Caribbean within the Atlantic World. Victor Bulmer-Thomas, for example, declares (without an explanation of world-systems) that 'a small number of states (the core) came to exercise an overwhelming influence on the economic fortunes of the Caribbean countries'.[22] Many of the contributions in this volume, following Wallersteinian insights, adopt a systemic approach that places institutions and politics as well as global political and economic pressures at the centre of the understanding and interpretation of the Atlantic World political economy.

Historically, of course, the boundary between centre and periphery was not easily drawn, during a period of increasing global processes,[23] and some Atlanticists (and even policy-makers) have been guilty of sloppy application of Wallerstein's world-systems approach. As P.J. Cain and A.G. Hopkins observed, 'A primary concern with the underdevelopment of regions outside Europe leads to stereotyped treatment of the "exploiting metropole".'[24] While oversimplification of Wallerstein's model (in both theory and practice) may make it appear naive, its author clearly explained that a specific centre or periphery could change its nature in this context over time. This dynamic can be seen in the evolving historical economic relationships between the Caribbean, the traditional imperial centres, and the US, as the latter shifted from periphery to core. Richard Drayton has observed 'the periphery acting as centre', tracking 'the cultural political and cultural impact [of Caribbean plantations] on the international system'.[25] Contributions throughout this volume illuminate various actions and experiments leading this development, and a contemporary awareness of the shift.

This introduction does not assert that the integrated Atlantic World and its integrated economy led the shift to modernity. Such a claim would be, at best, preposterous. Yet it did act as a crucible in which economic, social, and political experimentation with new ideas and approaches, both imported from the old world and spawned in the new, were allowed to flourish, often unfettered. It may be less preposterous to suggest that together the steady intensification of Atlantic commerce, the free trade which was, in stages, adopted within the region (illicitly or openly, often with state support), the reaction to these developments by various states and the institutions which formed and evolved around them, and the exploitation of the Atlantic World resource windfall marked an important stage in the launch of a modern political economy.

* * *

This collection of essays was conceived alongside a conference panel, *Atlantic Networks and Economic Exchanges Between Europe and the Caribbean,* convened for the Tenth International Congress of the Spanish Association of Economic History. The volume presents linked chapters – many commissioned well after the conference – which together examine the evolving and strengthening interconnections between the changing political economies of Europe and the Caribbean during the 'long' eighteenth and nineteenth centuries. It brings together comfortably research by both well-established authors and early-career historians. Each work transcends national historiography to examine, from a broad, Atlantic World perspective, the transoceanic circuits formed during the period 1650–1914. The volume aims specifically to position the Caribbean within the Atlantic World economy by illustrating circulations of trade capital and credit, knowledge and technology, and economic and commercial exchanges within an interconnected, holistic region. Its main aim is to place the Caribbean economy within a wider Atlantic World and an inter-imperial context, one in which rival empires struggled for supremacy, but also where new approaches adopted outside or in defiance of imperial constructs were tested, and sometimes adopted and embraced.

The pages that follow are not just about the economic history of the Caribbean and European expansion. They are, above all, about the historical formation of the modern political economy to which Europe and its Caribbean territories made a significant contribution. They are also about the exchanges and interconnections which characterised the Atlantic World. The book as a whole is about the Atlantic World's influence on the Caribbean, and the Caribbean's influence on the Atlantic World. Many chapters place major historical generalisations alongside detailed case studies, while acknowledging historical structures and major long-term trends. Through this perspective, the editors seek to contribute to a greater understanding of the interplay between local conditions and long-term macrohistorical dynamics. Chapters in this book also take into account microhistories, balancing the abstraction of systemic approaches with focussed case studies and examples. In this way many chapters provide new historical data, drawing upon the exceptionally rich variety of archival and published sources located in Europe, the Caribbean, and the US.

Studying the long-distance networks that connected the Caribbean and Europe during the 'first age of globalisation' from an Atlantic perspective provides a better understanding of the socioeconomic changes that occurred in both geographical areas. The 'loose' definition of a network provided by Frederick Cooper is useful in understanding the international interactions of the Atlantic World economy. It may be seen as a series of interlocking and overlapping social and commercial networks that were 'less defined than a "structure" but more than just a collection of individuals engaging in transactions'.[26] These networks were cross-national and cross-imperial. However, study of them should focus not only on social networks – on human agency and the people who formed structures of exchange, although this element of research is critical to our understanding of Caribbean economic history – but also on the circuits of exchange that represent the historical reordering of directions of exchange throughout the Atlantic World economy, to garner a more complex picture of the sociotechnical networks involved, effectively ascribing to materiality, environment, and geography an active, non-human agency.[27]

A key mechanism for stabilising long-term commercial exchange and collaboration was mutual trust. Informal institutions such as merchant networks conformed to widely accepted practice, in order to combat risk and reduce transaction and information costs. Reduced risk arose through personal and professional connections with colleagues in foreign cities, and their joint participation in business partnerships and ad hoc ventures. Information was widely shared, for the benefit of all. In a space where enforcement was beyond the reach of official structures, the sanction invoked for nonconformism, or worse, practice deemed dishonest in the eyes of fellow adventurers, was exclusion.[28]

As discussed above, much of earlier Atlantic history has concentrated on the study of national perspectives which focus on a single empire, with its centre located in Europe. Overseas territories in the Americas were considered peripheral. Such studies tend to overlook the examination of the economic interconnections of the Atlantic World economy, and the transnational melting-pot which characterised much of the commerce and development of the Caribbean. At the heart of this volume is the conviction that the dividing lines between the 'centres' and 'peripheries' of empires can be artificial and misleading. Thus, most of the chapters that follow avoid Eurocentric and national histories by removing the frontiers that segment national and colonial histories, substituting an Atlantic World which comprises parts of both the old and new worlds, and which recognise fluid interrelationships between the colonised and the colonisers.

Among the various Atlantic connections, economic historians have recently focused on the study of oceanic trade between different societies, the extension of principles of political economy, long-distance migration, transfers of industrial technologies, and the spread of both diseases and medical advances. The contributors to this volume recognise the necessity for cross-imperial Atlantic history, the approach urged by Armitage in the early years of the twenty-first century, and aim to show how the boundaries between imperial segments of the Atlantic World economy were mutable. Thus, this study of Caribbean economic history focuses deliberately on the analysis of international linkages within the Atlantic World, and places particular emphasis on the essential recognition of the transnational space that constituted the Atlantic World, and the reciprocal influences of the Caribbean, the US, and Europe. In so doing it provides a corrective to the purely national studies of an earlier, less mature Atlantic history.

The essays collected in this volume accomplish this in a variety of ways. David Ormrod does so by looking to the North Sea-Baltic trading system, and its entry into the larger matrix of regional change. Rather than seeking a one-dimensional explanation for this change, he looks at multifaceted causes under two headings: institutions and environment. He argues that a 'balanced view of Atlanticisation' must take into account both institutional and geopolitical dimensions, as well as natural resources and energy supplies (an approach adopted by other contributors). Reflecting upon Kenneth Pomeranz, Ormrod argues that 'abolishing the land constraint' allowed the land-intensive products of the Atlantic to 'bring unique advantages' to Britain, but that these came at a price, a cost partly met by restructuring older trading systems. Thus, the rate of the Atlantic World's rise was closely connected to European 'change and adaptation'.

Shifting patterns of trade also form a core concern of Nuala Zahedieh's chapter, which shows how Britain's mercantilist warfare of the long eighteenth century was not of benefit to trade, but instead damaged the very commerce it was intended to bolster. By exploring inter-imperial trading relationships, she overturns the historiographical argument that the War of the Spanish Succession was self-financing. Knick Harley also overturns an historiographical trope, that of the 'Williams thesis'. Eric Williams proposed that the slave-based Atlantic economy was, through the demand it created for British manufactures, a driver of the industrial revolution in Britain. However, Harley contends that causality did not arise from this single factor. He argues, counterfactually, that the northern mainland colonies would have found alternative export goods to fund their demand for European manufactures.

These alternatives are illustrated by Adrian Leonard, who has shown how trade from Rhode Island to the West Indies was thriving in the eighteenth century, as merchants there began to assume a central role in insuring merchant vessels and cargoes. The contribution adds a cultural dimension to the mix, by exploring the transmission of merchant culture and custom from the old world to the new. The changing trade pattern he illustrates was the beginning of a North American shift from its role as a peripheral nation to one of a core – a shift illustrated by Chuck Meide through a unique combination of marine archaeology and extensive archival research. His very practical illustration of shifting cores and peripheries supports Wallerstein's insistence that relative relationships between centres and metropoles were not immutably fixed. Manuel Covo's contribution shows, uniquely, how the course of history could have been much different, by shedding light on Baltimore's close connections with post-revolutionary France, and how its Caribbean ties influenced this flirtation, and the ultimate outcome.

Martín Rodrigo y Alharilla tracks the movement of capital from the Spanish Caribbean, principally Cuba, back to European centres including Barcelona, Paris, and London. His chapter shows how nineteenth-century capital-flows between the Caribbean and the old world had powerful economic impacts in multiple European 'cores'. In contrast, Gert Oostindie shows how the Dutch Atlantic was thoroughly integrated into the Atlantic World economy, and consciously eschewed stringent imperial connections. The benefits of this integration did not accrue to individual empires, but rather advanced the development of the broader Atlantic World. Meanwhile he contends that the Dutch Atlantic was more important than historiographical credit often grants it.

Dale Tomich looks, rather than to imperial connections, to geographical and physical conditions of sugar production, in connection with the technology deployed to maximise it. Economies operate in specific historical geographical configurations, he argues, with the instruments of production mediating between human labour and nature to form an interdependent unity. These factors must be considered alongside political and institutional ones to garner a true understanding of the sugar island's success. David Pretel and Nadia Fernández-de-Pinedo also consider the dynamics of sugar production, linking human agency and networks to shifting technology. They not only highlight metropolitan Spain's inability to complement Cuba's economic activity, but above all underline the active participation of Cuban Creole elites in the modernisation of sugar plantations. In doing so, they illustrate the key role of local conditions, global developments, and extra-imperial transfers, specifically between non-Spanish Europe, the US, and Cuba. Inés Roldán de Montaud also considers Cuban sugar plantation development, but from the perspective of finance. Her extensive research into Barings' activities in Cuba illustrate the dramatic development of transnational organisations which changed the nature of relationships in the Caribbean. Here, too, Atlantic World connections supersede imperial links in the provision of the capital which underpinned Cuban sugar production. The editors believe that, taken together, these chapters offer an entrance to a new understanding of the development of the Caribbean economy, which had shrugged off many restrictive imperial constraints to evolve and develop within a distinct Atlantic World.

Notes

1. Armitage, D.: 'Three concepts of Atlantic history', in Armitage, D. and Braddick, Michael J.: *The British Atlantic World, 1500–1800*, Basingstoke: Palgrave, 2002, pp. 11, 16–17.

2. For a recent Atlantic World take on state formation, see Chet, Guy: *The ocean is a wilderness: Atlantic piracy and the limits of state authority, 1688–1856*, Amherst: University of Massachusetts Press, 2014. For an Atlantic analysis of historical geography, see Benton, Lauren: *A search for sovereignty: law and geography in European empires, 1400–1900*, Cambridge: Cambridge University Press, 2010.
3. Tilly, Charles: *Big structures, large processes, huge comparisons*, New York, Russell Sage Foundation, 1984, pp. 74–6.
4. For examples of this new Atlanticism, see, for example, Section III of *The Oxford Handbook of the Atlantic World, 1450–1850*, Canny, Nicholas and Morgan, Philip (eds.), Oxford: Oxford University Press, 2011, pp. 289–495. For all Section III's abandonment of imperial approaches, Section II compiles chapters entitled 'The Spanish Atlantic', 'The British Atlantic', and so on. For a further discussion of non-imperial, non-Eurocentric Atlantic history, see Greene, Jack P, 'Hemispheric history and Atlantic history', in Greene, Jack P. and Morgan, Philip D., *Atlantic history: a critical appraisal*, Oxford: Oxford University Press, 2009, pp. 299–315. For an example of social, regional Atlantic history, see Candlin, Kit: *The last Caribbean frontier, 1795–1815*, this series, Basingstoke: Palgrave Macmillan, 2012.
5. Lockard, Craig A.: 'Global history, modernization and the World-System approach: a critique', *History Teacher* Vol. 14, No. 4 (1981), pp. 489–515; Mazlish, Bruce: 'Global history in a postmodern era', in Mazlish, Bruce and Buultjens, Ralph (eds): *Conceptualizing Global History*, Boulder: Westview, 1993, pp. 113–27.
6. See, for example, Fredrikson, G.M.: 'From exceptionalism to variability: recent developments in cross-national comparative history', *Journal of American History*, Vol. 82, No. 2 (1995), pp. 587–604; Halperin, C. J.: 'Comparative history in theory and practice: a discussion', *American History Review* Vol. 87, No. 1 (1982), pp. 123–43; Burke, Peter: *History and social theory*, second edition, Cambridge: Polity, 2005, pp. 21–6.
7. Fischer, D.H.: *The great wave: price revolutions and the rhythm of history*, Oxford: Oxford University Press, 1996, p. 82.
8. Webb, Walter Prescott: *The great frontier*, London: Secker & Warburg, 1953, p. 17.
9. Boyer, George: 'The convergence of living standards in the Atlantic Economy, 1870–1930', in *The new comparative economic history: essays in honor of Jeffrey Williamson*. London: MIT Press, 2007, p. 320.
10. O'Rourke, Kevin: 'The European grain invasion, 1870–1913'. *Journal of Economic History*, Vol. 57, No. 4 (Dec., 1997), pp. 781–4.
11. Williamson, Jeffrey: *Globalisation and the poor periphery before 1950*, Cambridge, MA: MIT Press, 2006, p. 25.
12. Sharp, Paul: 'The long grain invasion of Britain: market integration and the wheat trade between North America and Britain from the eighteenth century'. Working Paper, Department of Economics, University of Copenhagen. www.econ.ku.dk/english/research/publications/wp/2008/0820.pdf, viewed 30 November 2011.
13. On the negative impacts on growth of restrictive institutions, see Ogilvie, Sheilagh: *Institutions and European trade: merchant guilds 1000–1800*, Cambridge: Cambridge University Press, 2011. For the positive role of guilds

in fostering innovation, see Epstein, S.R. and Prak, Maarten: *Guilds, Innovation and the European Economy, 1400–1800*, Cambridge: Cambridge University Press, 2008.
14. Parkinson, C.N.: *Trade in the eastern seas, 1793–1813*, Cambridge: Cambridge University Press, 1937, pp. 94–5; Chet, *The ocean is a wilderness*, p. 70.
15. See Oostindie, this volume.
16. Cf. Zahedieh, Nuala: '*Economy*', in Armitage, D. And Braddick, Michael J.: *The British Atlantic World, 1500–1800*, Basingstoke: Palgrave, 2002, pp. 51–68.
17. Cañeque, A: *The king's living image: the culture and politics of viceregal power in colonial Mexico*, Routledge, New York, 2004, p. 122.
18. Andrew, A.P.: 'The end of the Mexican dollar', *Quarterly Journal of Economics*, Vol. 18, No. 3 (May, 1904), p. 322.
19. See, for example, the turn-of-the century accounts of Jardine Matheson, held at Cambridge University Library.
20. See Hobsbawm, Eric J.: *Nations and nationalism since 1780: programme, myth, reality*, Cambridge: Cambridge University Press, 1990.
21. Wallerstein, Immanuel: *The modern world-system I: capitalist agriculture and the origins of the European world-economy in the sixteenth century*, revised edition, London: University of California Press, 2011 (1974), p. xviii.
22. Bulmer-Thomas, V.: *The economic history of the Caribbean sine the Napoleonic Wars*, Cambridge: Cambridge University Press, 2012, p. 21.
23. See Drayton, R.: 'The collaboration of labour', and Bayly, C.A.: '"Archaic" and "modern" globalisation in the Eurasian and African Arena, c. 1750–1850', in Hopkins, A.G.: *Globalization in world history*, London: Pimlico, 2002, pp. 103, 50; Burke, *History and social theory*, pp. 82–8.
24. Cain, P.J. and Hopkins, A.J.: 'The political economy of British expansion overseas, 1750–1914', *Economic History Review*, Vol. 33, No. 4 (Nov., 1980), p. 464.
25. Drayton, R.: 'The collaboration of labour: slaves, empires, and globalizations in the Atlantic World, c. 1600–1850', in Hopkins, A. G. (ed.): *Globalization in World History*. London: Pimlico, 2002, p. 99.
26. Cooper, Frederick: 'Networks, moral discourse and history', in Callaghy, Thomas M., Kassimir, Ronald, and Latham, Robert (eds.): *Intervention and transnationalism in Africa: global–local networks of power*, Cambridge: Cambridge University Press, 2001, p. 23.
27. On the materialist approach that takes into account geography and nature see Drayton, R.: 'Maritime networks and the making of knowledge', in Cannadine, David (ed.): *Empire, the sea, and global history: Britain's maritime world, c. 1763–c. 1840*, New York: Palgrave Macmillan, 2007.
28. See, for example, Milgrom, P.R, North, D.C. and Weingast, B.R.: 'The role of institutions in the revival of trade: the Law Merchant, private judges, and the Champagne fairs', *Economics and Politics*, vol. 2, no. 1 (1990), pp. 1–23.

The editors of this book gratefully acknowledge financial and institutional support from the University of Cambridge, the European University Institute, the Economic History Society, and the Spanish Ministry of Economy and Competitiveness (ref. HAR2013-40760-R).

2

From Seas to Ocean: Interpreting the Shift from the North Sea-Baltic World to the Atlantic, 1650–1800

David Ormrod

Expansion of long-distance trade was a key component of Europe's commercial revolution, but changes *within* the trading world of the northern seas – the North Sea and the Baltic – have been substantially underestimated in explanations of the rise of the world economy from *c*.1500–1800. Historians of the northern Netherlands are necessarily exempt from the strictures which follow, conscious that for centuries Dutch-Baltic trade represented 'what was literally a lifeline' and the 'mother of all trades'.[1] American and British historians, on the other hand, have allowed the emergence of the Atlantic economy, and especially of plantation economies and swelling Atlantic World populations, to dominate narratives of growth to a degree that has obscured understanding of the restructuring of intra-European trade. Still, agreement has been absent amongst the latter group about the degree to which the expansion of trade beyond Europe acted as an engine of growth, particularly through the connections between trade, colonisation, and industrialisation.[2] These debates have been summarised elsewhere, and are too well known to require further comment here, but it may prove useful briefly to scrutinise some influential recent arguments which have given 'Atlanticisation' a prominent position in the literature in order to identify the pathways and mechanisms which are alleged to underlie the 'first great divergence', along with some of the flaws in the formulation of the problem.[3] This chapter will then consider critical shifts within the North Sea-Baltic trading system which entered the larger matrix of regional change under two headings: institutions and environment. The view expressed here is not that the Atlanticists are necessarily wrong or misguided, but simply that that their narratives are incomplete and limited.

Historiography and historical statistics

The influential article by Acemoglu, Johnson, and Robinson on the rise of Europe and Atlantic trade provides a useful starting point.[4] Their argument suggests that the rise of Western Europe originated in the centuries from 1500 to 1800, an 'historically unprecedented period of sustained growth' which was closely linked with the expansion of Atlantic trade. 'Atlantic' is interpreted widely to include not only trade with imperial Atlantic plantation and settler economies, but at times also Asian commerce, as well as trade with Africa. The linking of maritime expansion with economic growth in the centuries following the geographical discoveries is hardly a novel claim, but more so is the suggestion that it was specifically Atlantic expansion that induced fundamental institutional changes protecting private property and commercial interests, particularly those of merchant groups previously existing 'outside the royal circle' in monarchical states. Thus, the rise of Europe 'reflects not only the direct effects of Atlantic trade and colonialism, but also a major social transformation induced by these opportunities.'[5] In the case of Britain, the struggle for the liberalisation of overseas trade is told in true Whiggish mode, culminating in the dismantling of commercial monopolies in the wake of the Glorious Revolution. But state power in general is assumed to be antithetical to sustained economic growth, leaving little place for mercantilist regulation in shaping trade patterns, encouraging strategic industries, supporting grain markets, and so on – something which successive Whig governments would have found baffling.[6]

Acemoglu, Johnson, and Robinson's methods involve running a series of regressions which eliminate some common explanations for the rise of Western Europe such as Protestantism, war-making, inter-state rivalry, and the strength of a country's classical heritage, while testing the remaining hypothesis that rates of urban growth were highest in countries with the greatest potential for Atlantic trade. Urbanisation is used throughout as a proxy for GNP per capita, and the 'Atlantic Trade Potential' is defined as the ratio of Atlantic sea coast to land area.[7] The maths is secondary to the main argument, however, since the authors accept that the contribution of profits from intercontinental trade was only modest. The really important variable is institutional change, but this is neither measured or tested: it is assumed or picked up from current trends in the literature, representing a 'marriage between the Marxist thesis linking the rise of the bourgeoisie and the development of the world economy (e.g. among others, Williams, Frank, and

Wallerstein) and the neoclassical emphasis on the development of political institutions and secure property rights in Western Europe (e.g. North and Thomas, E. L. Jones, North.... &c)'.[8] The most robust findings, those derived from statistical tests, relate to the experience of Atlantic ports and cities compared with that of other cities in Western and Eastern Europe. Comparisons are also made between the volume of Atlantic and Mediterranean voyages. Throughout all of this, trade, shipping, and the expansion of ports and cities in the North Sea-Baltic zone remain invisible. London and Amsterdam, confusingly, are described and analysed solely as 'Atlantic cities'.

A more balanced view of the Atlantic dimension is found in Kenneth Pomeranz's account of the 'Great Divergence' between Europe and China.[9] Pomeranz also emphasises the importance of a responsive and representative institutional structure in promoting European commercial growth, but accords less significance to commerce with the new world, the slave trade, and overseas coercion, suggesting that 'these arguments cannot be dismissed, but neither are they compelling'.[10] In the case of Britain, however, its access to the land-intensive products of the Atlantic world brought unique advantages, together with its capacity to exploit its own reserves of fossil fuels found in coal. Both these advantages can be seen in one sense, and especially in comparison with Asia, as windfall gains, yet they originated in a purposeful search to find ways of lifting resource constraints, of 'abolishing the land constraint'. The exploitation of new world resources – extensive growth on the periphery – is nicely balanced by an emphasis on intensification within the metropolitan core.

The view that the most pressing problems of diminishing returns were met principally by home-grown solutions has been most clearly expressed by Tony Wrigley. The character of the English industrial revolution, according to Wrigley, is defined by the switch from organic materials to coal and iron, to a mineral-based economy which released a growing population from the land constraint.[11] In his 1999 Prothero Lecture, Wrigley took this argument several stages further and explained the divergence of England from the rest of Europe as the consequence of 'an intensification rather than an extensification of her territory'. Like the seventeenth-century Dutch economy, rapidly rising agricultural productivity and the increased use of fossil fuels permitted an abnormally high rate of urbanisation and major increases in output per head.[12] Empire, trade, and overseas expansion are assigned a minor role in the story.

In these contributions are three possible types of approach to the question of the Great Divergence, the question of what made the development path of northwest Europe unique, despite surprising similarities with eastern Asia.[13] They are differentiated by the relative importance attached to endogenous and exogenous processes influencing growth, whether in terms of potential for long-distance trade, resource endowments, or institutional arrangements. Yet the internal/external issue depends largely upon economic geography and the spatial limits of the region which we decide to adopt for our enquiries. I have suggested elsewhere that the divergence of England in the long eighteenth century is best understood in its larger regional context, one of growing dominance within the North Sea-Baltic economy, itself a macro-region within the emerging world economy.[14] Like the Mediterranean region, it was a subsystem of the larger European world-economy (*economie-monde*, in Braudellian language), and of equal weight. The term 'macro-region' is used here to describe a region complete in itself, with its own spatial hierarchy, including its own core and peripheral areas. Each has the capacity to merge with or incorporate others. Conversely, as Braudel pointed out, during periods of crisis or contraction – phase B movements – world-economies tend to fragment into their smaller component parts. The shift from Northern Europe to the Atlantic can thus be mapped as shown in Figure 2.1.

The rise of the Dutch Republic and the opening of transoceanic trade routes began during a long upward phase of a movement which stretched from the 1540s until the mid-seventeenth century. England's commercial revolution, on the other hand, is usually dated from the 1660s and 1670s, by which time the decline of the Italian states was well advanced. The shift in Europe's centre of gravity from the Mediterranean to the North Sea thus occurred as economic expansion gave way to contraction. England's displacement of the northern Netherlands, accelerating after the Dutch collapse of 1672, was based not on a buoyant growth of exports, but on a long phase of import-led growth and intense commercial rivalry. The development of northwest Europe's Atlantic trades was fuelled initially by a demand for imports, especially of Caribbean sugar and Virginia tobacco, and not until the later seventeenth century did commercial profits accruing from these trades become significant for England.[15] For the northern Netherlands, however, recent research by Klooster and Enthoven reveals that trade with the Caribbean, Brazil, and Spanish America was significantly larger and more profitable than was previously recognised, from the truce of 1609 to the loss of Brazil in 1654.[16] The following tables show the distribution of imports between the Americas, Asia, and European markets for England and the Northern Netherlands. The import figures bulk much larger than exports until the post-1750 decades for England; exports of timber and groceries from the British North American mainland to the Caribbean

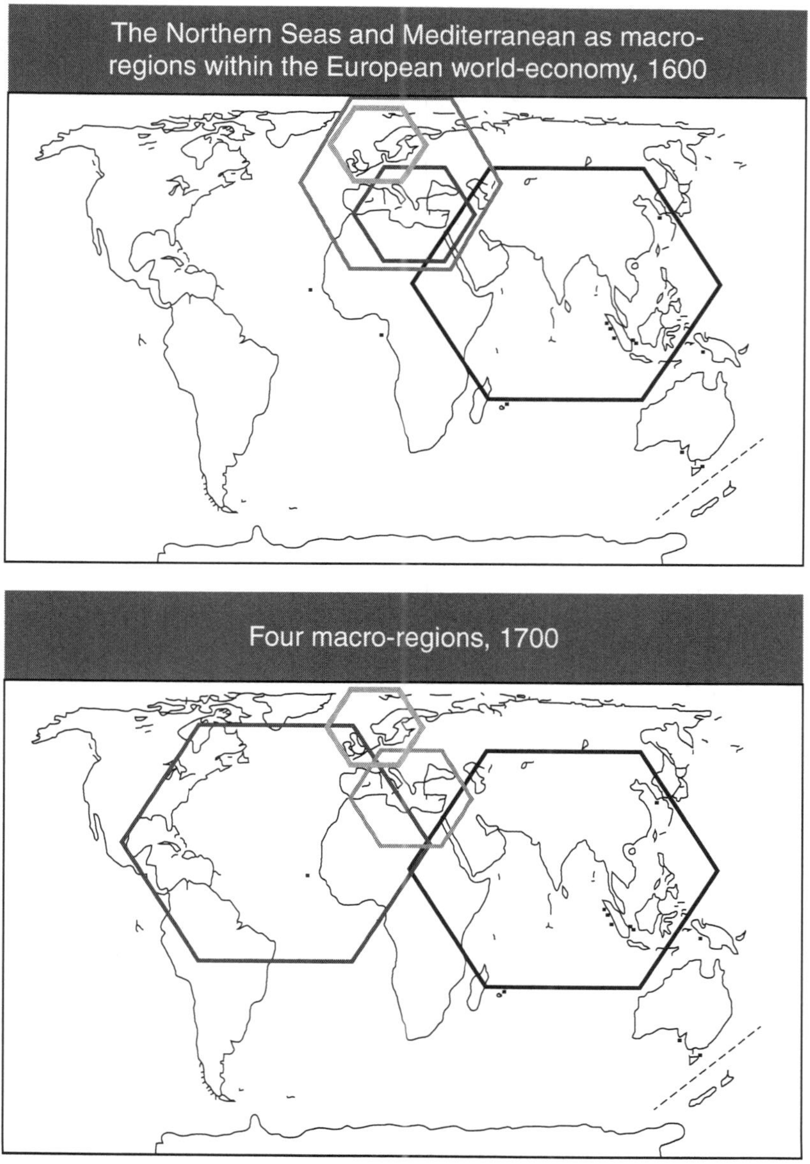

Figure 2.1 Macro-regions, 1600 and 1700

Table 2.1 English overseas trade flows, 1699–1774

Annual average % value (official values, 1696–1709)

	Atlantic	*Asia*	*N. Sea/Baltic*	*Mediterranean*
1699–1701				
imports	18.9	12.9	34.2	26.6
exports	12.2	2.8	47.7	33.5
re-exports	15.7	0.7	62.6	11.3
1722–4				
imports	24.8	14.3	28.9	26.4
exports	15.0	1.9	34.7	42.5
re-exports	17.9	–	66.1	6.5
1752–4				
imports	32.7	13.2	27.0	19.5
exports	20.3	7.9	28.9	34.2
re-exports	18.0	2.3	57.6	8.2
1772–4				
imports	37.4	15.2	22.3	14.4
exports	42.4	7.3	18.0	22.4
re-exports	16.7	1.1	55.4	7.8

Annual average value, £'000 (official values, 1696–1709)

	Atlantic	*Asia*	*N. Sea/Baltic*	*Mediterranean*
1699–1701				
imports	1,107	756	2,001	1,555
exports	539	122	2,114	1,484
re-exports	312	14	1,243	224
1722–4				
imports	1,679	966	1,950	1,783
exports	758	93	1,750	2,141
re-exports	487	19	1,794	176
1752–4				
imports	2,684	1086	2,215	1,597
exports	1,707	667	2,423	2,879
re-exports	627	81	2,012	285
1772–4				
imports	4,769	1,929	2,841	1,829
exports	4,176	717	1,769	2,211
re-exports	972	63	3,223	453

Source: see note 18.

plantation colonies exacerbated this imbalance by meeting Caribbean demand. By the early eighteenth century the Dutch had arrived at a more even import/export balance for a smaller volume of trade.[17]

It is at once clear that intra-European trade was much more important in value terms than transatlantic trade for both England and the Netherlands up to 1750. For England, Atlantic trade entered a new phase of export-led growth after 1750 with the take-off of the mainland American market for manufactured goods. During the previous century the Americas had been important chiefly as a source of re-exports for redistribution to nearby Europe. Sugar, tobacco, and Asian textiles, together with exports of English manufactured goods, provided the purchasing power to pay for essential Baltic imports of timber, naval stores, and Swedish and Russian iron. Across the Atlantic the burgeoning provisioning trade between American merchants and Caribbean consumers relieved Britain of the responsibility for supplying these input requirements necessary for the production of the commodities yielded by the Caribbean plantation economies.

At the same time a dense network of bilateral and triangular voyages across the North Sea linked London and the east coast ports of England and Scotland with those of the Netherlands, Germany, and Scandinavia. They involved the exchange of bulk cargoes (British grain and coal for Norwegian timber) and high-value goods between Britain, the Netherlands, and Germany. The transhipment of goods along the

Table 2.2 Estimated imports into the Dutch Republic, 1636 (£'000 at 1696–1709, constant prices)

Trade area	***£'000***	*%*
Atlantic	2,059	31.3
Asia	941	14.3
Europe	3,577	54.4
Total	6,578	100.0

Table 2.3 Estimated value of Dutch overseas trade, 1780 (£'000 at 1696–1709, constant prices)

Trade area	***£'000***	*%*
Atlantic	6,474	22.9
Asia	3,505	12.4
Europe	12,486	44.2
Not specified	5,799	20.5
Total	28,264	100.0

Source: For notes to Tables 2.2 and 2.3, see endnote.[18]

great river systems of northern Europe, especially the Rhine and the Elbe, drew large areas of the German hinterland into these extensive circuits of commerce.[19] Because much of this trade involved relatively short hauls, the ratio of utilised to available shipping tonnage was very high compared to long-distance trades to the Americas or Asia.[20] Table 2.4 shows that the bulk trades of the North Sea–Baltic zone handled a much greater volume of cargo than the transatlantic trades before 1750, with a smaller volume of available shipping. Around 1750 almost half the British merchant fleet was committed to transatlantic traffic, but conducted a considerably smaller volume of higher-value trade.[21] Spare capacity was reduced by the practice of chartering vessels as tramp ships, where the risks of trade were shared between merchants and masters. In these respects, the nearby and northern European trades were characterised by a high degree of shipping efficiency compared with long hauls across the Atlantic. It seems that no major decline in freight rates and transaction costs occurred across the North Sea region before 1800. The productivity of Dutch shipping was already high in 1600, with England making up lost ground in the decades after 1660. Rather, it was improved information exchange which played a key role in bringing northern European markets closer together in the early modern period, through price synchronisation.[22]

Table 2.4 Tonnage of shipping required for English bulk trades

		Coal exports	*Grain exports*	*Salt exports*	*Timber imports*	*Sugar imports*	*Tobacco imports*
1700	Northern Europe	4	**2** 2	1	178		
	Nearby Europe	108	**17** 23	7	10		
	Caribbean					23	
	North America				2		15
	All markets	**83** 113	**23** 31	8	190	23	15
1753	Northern Europe	21	**2** 2	1	263		
	Nearby Europe	293	**100** 140	18	10		
	Caribbean				10	47	
	North America				20		28
	All markets	**246** 328	**144** 203	19	303	47	28

Sources: Davis (1962); **Ormrod (2003)**; 3-year averages are shown centred on 1700 & 1753.[23]

These figures suggest that England in 1700 was, in Brinley Thomas's words, 'very much a member of the then European economic community'.[24] The 'Americanization' of English foreign trade lay in the future, and the contribution of the colonial periphery was limited to providing a flow of goods for re-export during a long period of import-led growth. Of greater strategic value to England were imports from the Baltic, and a series of projects to develop the linen industries of Scotland and Ireland along lines which would rule out competition with England's woollen industry, her leading export sector.[25] The contribution of these local peripheries to English interests was clearly substantial, as were the terms on which England's colonial re-exports were disposed of in Europe. These terms were essentially laid down in the Navigation Acts of 1651, 1660, and 1663, restricting the use of foreign shipping and creating the structures of a national entrepôt system centred on London. The terms of trade would be shaped by an increasingly complex system of import duties, bounties, drawbacks, and prohibitions which aimed to influence the development of commodity markets and to meet the revenue requirements of the emerging fiscal-military state.[26]

At this point it is important to underline the centrality of politics to these developments. The consolidation of state power from the mercantilist experiments of the Cromwellian regime to the Anglo-French wars of 1689–1713 was decisive in reshaping the contours of the North Sea-Baltic trading world to England's advantage, and to the disadvantage of Holland, France, and Sweden. Indeed the struggle for supremacy at the core of the emerging world-economy was fought initially in the North Sea zone itself, rather than on distant colonial peripheries. The Navigation Acts were initially intended to limit Dutch control of England's Baltic and Norwegian trade, and could be enforced only with difficulty on the colonial periphery, in spite of the English capture of New Amsterdam in 1664 and the introduction of more stringent procedures in 1673 and 1696.[27] The customs figures and the early history of English mercantilism are consistent with the view that until the third quarter of the eighteenth century income and employment generated by the North Sea-Baltic zone were much more substantial than those produced by transatlantic trade.

The links between colonial trade and European development were complex, but it is clear that they were mediated though the rise of the North Sea trading system to a much greater extent than Acemoglu, Johnson, and Robinson, and even than Pomeranz suggest. It seems unrealistic to present the Atlantic frontier as a kind of *deus ex machina* which transformed Europe's prospects.[28] Indeed, the evidence suggests a reverse kind of logic in which northwest Europe was able, from its own resources, to initiate a long phase of westward expansion which ultimately incorporated the Atlantic frontier into its own commercial orbit. Furthermore, it can be argued that it was England's aggressive mercantilist strategies which channelled this development, moving it from a long phase of import-led growth dependent on new consumption habits towards a new pattern of export-led industrialisation in the European core. During the earlier, import-led phase, import substitution strategies proved notably successful in several sectors of British industry, including linen manufacture, cotton, silk, paper, and ceramics.[29] Thus, the key question is less 'What did the Atlantic periphery contribute to the rise of Europe?' than 'How was Northern Europe able to initiate the "Atlantic thrust" and mobilise the immense capital requirements of the commercial revolution?' During the last forty years of the seventeenth century in England, the rate of new investment in commerce, as distinct from industry, was, in Ralph Davis's words, 'abnormally high', and the volume of merchant capital involved grew much faster than the value of trade itself.[30] Where did it come from?

Institutional change and commercial reorganisation

To some extent the capital required to operate transatlantic trade came from the process of colonial development itself. As the eighteenth century progressed, a growing proportion of British American trade came to be controlled and financed by merchants in colonial ports, as they moved from commission trade to become exporters on their own accounts, building on the experience they had gained in the Caribbean provisioning trade.[31] However, planters and colonial merchants in the early colonial period relied on the merchants of London, Bristol, and other ports to finance colonial exports to Europe. That trading capital arose from European sales of colonial goods and home manufactures. The critical change which facilitated the Atlanticisation of British overseas trade was the refinancing of the North Sea economy in such a way as to release merchant capital for deployment elsewhere, including Atlantic trade.[32] Mainland trade to the Caribbean also freed capital in Europe, and facilitated North American imports of British manufactures by resolving the settler colonies' balance of payments challenge.[33]

The Navigation Acts aimed to exclude Dutch and other foreign shipping from imperial trade, which in practice placed foreign merchants in a position of dependence upon British intermediaries and commercial networks. But more stringent external regulation and protection were balanced by a new freedom to engage in trade, as internal corporate monopolies were dismantled in the 1690s. Foreign merchants and their agents were now free to settle in London and to participate fully in the export and re-export trades to nearby Europe, especially in woollen textiles. A growing proportion of Anglo-Dutch and Anglo-German trade was henceforth financed on the accounts of foreign merchants, leaving British merchants to act either as commission agents, or to diversify their trade outside Europe. By 1695 foreign capital tied up in the nearby European trades accounted for over £300,000, or nearly one-third of England's total domestic exports. This movement of European capital into the British commodity trade paved the way for the well-known influx of Dutch and Huguenot capital into the newly created national debt and East India stocks.[34]

Up to a point Acemoglu, Johnson, and Robinson are right to borrow North and Weingast's familiar claim that the establishment of representative government following the Glorious Revolution helped to secure property rights in England for landowners and merchants,[35] but the liberalisation of trade and its impact on different interest groups were more interesting and complex than the former allow. The notion that trade was simply widened to include merchants 'outside the royal circle' including colonial merchants, planters, and slave traders is an enormous understatement.[36] Atlantic traders had already played a radical role in the opposition to the Crown during the civil wars, and in the framing of Cromwell's Navigation Act. The reforms of the 1690s, however, introduced a new degree of inclusiveness, leading to the internationalisation of London's commodity and financial markets. These were the years when London became a truly cosmopolitan world city, preceded during the 1680s by an enormous influx of Huguenot refugees – of between 40,000 and 50,000 – and the creation of a new Anglo-Dutch and Huguenot establishment at the core of the city's governance and financial life. At least 25 directors of the Bank of England came from immigrant Huguenot circles from 1719 to 1785, together with several directors of the East India Company.[37]

As foreign capital moved into England's nearby European trades, British merchants found the import trades from mainland America and the Caribbean increasingly attractive, not least because the Navigation Acts gave them leverage over their Dutch and other foreign rivals. The shift is easier to describe than to quantify, and I have provided several examples elsewhere. Prominent merchants active in the Dutch and North Sea trades who transferred some of their interests to transatlantic ventures during the first half of the eighteenth century included Robert Hackshaw; Isaac, William, and Daniel Minet; Ralph Carr; Thomas

Hall; Henry Gambier; Henry Hope; and George Aufrere. Hackshaw, for example, was trading on a large scale in the 1690s to Germany, the Netherlands, southern Europe, New England, and Barbados. Thirty years later, he was specialising wholly in commission trade for New York and New England merchants, and left total assets of £24,000 at his death.[38] Several merchants combined commission business with trade on their own accounts, and the majority of those listed above were able to provide essential shipping services for their colonial correspondents. These tendencies became stronger as the eighteenth century wore on, as R.C. Nash has shown, as indigenous American traders emancipated themselves from dependence on British merchant capital. In their transatlantic trade with New England, New York, Pennsylvania, the Caribbean, and to a lesser extent with the Chesapeake, British merchants took on the role of commission agents, handing colonial goods for the accounts of American correspondents.[39] The collapse of the Amsterdam re-export market for tea and Asian textiles in the 1720s and 1730s seems to have been a major factor in encouraging some to increase their transatlantic commitments, particularly in the tobacco trade, which retained its buoyancy throughout.[40] Here again the Huguenot connection was important at an early stage in establishing new transatlantic networks linking London, La Rochelle, South Carolina, Massachusetts, and New York.[41]

The new balance between internal free trade and external protection in England after 1689 meant that the nationality and status of a merchant mattered much less than the integrity of the imperial monopoly created by the Navigation Acts. Regulated trading along the lines of the Company of Merchant Adventures had proved to be incompatible with such mercantilist measures, but the sharply rising level of import duties instigated during the 1690s meant that the British-dominated North Sea-Baltic of the eighteenth century lost the free-trade character that it possessed under the domination of Amsterdam a century earlier. As Leos Muller has suggested, two broadly different situations in the history of the region in the early modern period stand in contrast: trades centred on Amsterdam, Hamburg, Danzig, and Antwerp had been connected much more with location (as gateway systems) and rather loose political structures, while trades centred on London, Copenhagen, and Stockholm depended more on states' clearly formulated mercantilist policies.[42]

The energy crisis and the environment

Political and institutional changes clearly played a decisive role in shaping the shift from the North Sea-Baltic world to the Atlantic. But what of environment, geography, and resources? Acemoglu, Johnson, and Robinson clearly place their emphasis on the former, while others such as Pomeranz give prominence to the lifting of resource constraints inherent in the Atlantic contribution to Europe, together with the exploitation of home-produced fossil fuels. The exploitation of the Americas undoubtedly helped to release Europeans from a regime of diminishing returns, and to sustain an expansive process of Smithian growth. But when and to what extent did these effects being to make their impact felt? The evidence produced above suggests that colonial development, far from representing a windfall gain, depended on resource-saving both on the colonial periphery, and within the European core itself. Further, it suggests that the Atlanticisation of European trade was functionally related to the restructuring of the commercial economy of Northern Europe and its Baltic periphery.

Pomeranz, on the other hand, views European access to American resources as a windfall comprising foodstuffs, raw materials, and energy, and one which was unavailable to Asia. Taken together, these resources provided Europe with a measure of ecological relief which indeed can be quantified.[43] If the annual energy output of the British coal industry in 1815 represented the equivalent of 15 million acres of forest, the land area required to produce the cotton, sugar, and timber imported from North America in 1830 would have amounted to between 25 and 30 million acres. The great bulk of this acreage, however – 23 million acres – is highly counterfactual, not to say virtual, since it represents the assumed sheep acreage which would have been needed to replace real cotton imports with home grown woollen yarn. In fact, these real cotton imports required only half a million American acres. It is of course possible that more intensive sheep farming, new breeds, and the production of heavier fleeces might have provided much of the surplus.

Given a total arable acreage of only 17 million acres in the England of the 1860s, a net addition of even three or four million acres – the amount needed to produce West Indian sugar and American timber imports in the 1830s – would be of major significance. So too would be the acres required to feed and fuel the Caribbean plantation economies, which were vast consumers of timber, livestock, and foodstuffs produced on the American continental mainland. But Pomeranz is making a much larger claim: that extensive gains arising from new world primary production were *twice* as significant as the benefits of European intensification derived from coal.[44] Here he differs substantially from Tony Wrigley who, as we have already noted, has raised the transition from an organic- to a mineral-based economy to a position of critical significance. Wrigley's argument provides an interesting counterpoint to that of Pomeranz, and stresses that England's relative advance (or divergence) was in train long before the industrial revolution. The issue of timing is one which Pomeranz acknowledges, but does not engage with.[45] Apart from his discussion of new world silver supplies, the main thrust of his logic is applied to the early nineteenth century and the industrial revolution, particularly the growth of cotton imports. The possible role of commercial or Smithian growth remains unexplored, so it is not yet understood whether the costs of Atlanticisation were outweighed by its benefits, by making available new, better, or cheaper raw materials and foodstuffs than were available from European sources during the long eighteenth century.

In the early seventeenth century the Netherlands, followed by England, began to follow a divergent path from the rest of Europe. Both countries pursued strategies of commercial growth and imperial expansion, and both, significantly, were limited by shortages of land and energy supplies. According to Wrigley, England gained the edge over the Netherlands because of the availability of coal.[46] Peat extraction, beyond a certain point, produced negative environmental consequences for Holland, including the build-up of inland lakes and the destruction of farmland. The Dutch depended increasingly on British coal imports for their processing industries, to a much larger degree than historians have supposed.[47] By the early eighteenth century Holland also suffered from the pollution of water supplies due to rising sea levels, polderisation, and river improvements, which together restricted the rate at which surface water was dispersed. This must have raised costs and lowered the quality of those industries which depended on good-quality water supplies, notably the brewing, distilling, sugar refining, bleaching, dyeing, and tanning industries.[48]

England suffered none of these environmental problems, and in the age of the sailing ship, transatlantic commerce was an energy-efficient form of long-distance trade. Nevertheless, the costs of maintaining the imperial state and the Atlantic trading system were considerable, and were not fully balanced by the inward flow of primary products entering Britain. Brinley Thomas identified two moments in the history of Atlantic expansion when the strain on organic energy resources was especially significant: first, from the 1630s to the 1680s, coinciding with the earlier phase of Atlantic expansion, and secondly, from the end of the Seven Years' War, that is, from 1763 through the 1770s and beyond.[49] For Thomas, Britain's Atlantic commitments, together with population pressure from the mid-eighteenth century, forced the metallurgical industries to solve their earlier technological problems, and triggered the industrial revolution.

Although the extent of Britain's energy crisis continues to be debated, its outlines are clear enough.[50] The price of charcoal rose by 150% during the first subperiod, compared with an increase of only two per cent in the general price level at a time of population stagnation. Cromwell's rearmament programme and the commercial revolution triggered a major advance in progress in the energy sector: of timber, charcoal, and bar iron. Pig iron output failed to keep pace, and the 1640s and 1650s saw a massive increase in the import of Swedish bar iron. The commercial revolution also placed new demands on the shipbuilding industry. Imports of Baltic timber soared during the 1670s and 1680s, multiplying tenfold in volume, supplemented by Norwegian supplies, along with an important group of forest products: wood ash, potash, pitch, and tar.[51] In a real sense the solution to the energy crisis in Britain involved the import of surplus energy from abroad. By 1700 as much as half the shipping tonnage entering English ports was carrying timber imports from Norway and the Baltic, a level sustained until the early 1750s. The North American contribution to the timber trade was disappointingly small during the first half of the eighteenth century. It amounted to one per cent by volume in 1700, rising to 6.6% in 1752–4.[52] The colonies were initially much more successful in supplying high-value woods, especially tropical hardwoods and walnut, and in developing a successful shipbuilding industry for both intercolonial trade and the export market. By the 1770s the cost differential between British and American shipbuilders had widened considerably, with British costs reaching almost twice the colonial level. In 1774 nearly a third of British-owned ships were American built.[53] The slow introduction of the sawmill in Britain, delayed until about 1800, exacerbated the situation.[54]

Pomeranz admits that American timber exports to Britain were trivial before 1800, and focuses mainly on sugar and cotton as providing Britain's main sources of energy and ecological relief. The importance of cotton was real enough, substituting for an equivalent (counterfactual) British flax acreage of 200,000 acres in 1815. But the benefits derived from the high calorific value of sugar are much less clear from a dietician's point of view, particularly in the long term.[55] Much more persuasive is the emphasis on Atlantic re-exports as paying for Baltic timber and other primary products – which returns the argument to the strategic role of the Baltic periphery. By the early 1770s – Brinley Thomas' second phase – Britain still depended on northern Europe for imports of charcoal, iron, timber, and naval stores, which amounted to about two-thirds of total consumption. In a revealing reinterpretation of the historiography of the British eighteenth-century iron trade Evans, Jackson, and Rydén have shown how the diffusion of technological progress was frustratingly slow, so that the British market was dominated by imported bar iron from Sweden and Russia until the early 1790s. Both were able to cater for a range of specialised needs. The connection with the Atlantic economy, as these authors see it, involved bringing 'semi-processed materials from the Baltic and disgorging manufactured metalwares into the Atlantic basin'.[56] Alongside these commercial perspectives the political/strategic dimension deserves emphasis: the inescapable fact that the struggle for a British Atlantic provided a major stimulus to innovation in the metallurgical industries. Thomas goes as far as to say that 'the strain on organic energy resources due to the blue water strategy and the Westernization of foreign trade was a major reason why Britain lost the American War'.[57]

For most of the eighteenth century the new world was important primarily as a source of subtropical groceries and semi-luxury goods for Europeans, rather than as a provider of basic energy supplies. Historians have been strangely silent about the long-term impact of the two main staple trades, tobacco and sugar, on health and well-being, and their hidden or external costs. During the third quarter of the eighteenth century, however, British exports to the Americas surged forward. By 1773 the export surplus approached around £1 million, as population growth in the northern mainland colonies increased demand for British manufactured goods. O'Brien and Engerman estimate that during the period 1784–6 to 1804–6, Americans purchased about 60% of the addition to British exports, and Europeans about one-third.[58] Deirdre McCloskey has argued that in the absence of expanding colonial markets the home market would have provided compensatory demand.[59] To argue otherwise, she suggests, is to repeat the mercantilist fallacy of static markets. Pomeranz follows McCloskey's argument, but the evidence to support it is actually very thin. In a low-wage economy afflicted by rising food prices it is not immediately obvious that domestic demand could have advanced more rapidly than it did. Simon Smith has subjected McCloskey's anti-mercantilist model to careful scrutiny, and showed that 'without colonies, British revenues from exporting would have certainly been reduced, but it is inconceivable that revenues would have fallen by an amount commensurate with the colonial market's actual share'.[60]

Pomeranz is broadly correct in his identification of the exploitation of coal reserves and colonies as critical to explanations of the contours of European growth. However, for the two centuries before 1830 the significance or weight of these two elements should be reversed to give priority to the switch to a mineral-based economy, underpinned by essential supplies of Baltic iron and primary products. Greater prominence should be allocated to inter-state rivalries within the core of the emerging world system: the North Sea-Baltic zone. If Britain's coal reserves represented a kind of windfall, an accident of geography, the mercantilist state was an institutional construct which accelerated the integration of North America into the world economy. The British variant of mercantilism depended on naval power rather than standing armies, as in France, and this gave Britain unique advantages. Armies represented a heavy drain of energy and resources in the early modern period, whereas the sailing ship was probably the most efficient high-energy converter of the age, capable of deployment in trade as well as war. For these reasons a balanced view of Atlanticisation must involve, on the one hand, the institutional and geopolitical dimension, and on the other, questions relating to the availability of natural resources and energy supplies. 'Abolishing the land constraint' came at a price, part of which was met by the restructuring of the North Sea-Baltic trading system. The momentum behind the rise of the Atlantic World depended to a large degree upon change and adaptation within Europe itself.

Notes

1. Van Tielhof, M.: *'The mother of all trades': The Baltic grain trade in Amsterdam from the late 16th to the early 19th century*, Leiden: Brill, 2002, p. 1.
2. Emmer, Petre-Grenouilleau and Roitman, J. V. (eds.): *A deus ex machina revisited: Atlantic colonial trade and European economic development*, Leiden: Brill, 2006, pp. xiii–xxix. See also the chapter by Harley, this volume.
3. Zahedieh, N. provides a useful summary of this literature, *The capital and the colonies: London and the Atlantic Economy, 1660–1700*, Cambridge: Cambridge University Press, 2010, pp. 3–5.
4. Acemoglu, D., Johnson, S. and Robinson, J.: 'The rise of Europe: Atlantic trade, institutional change and economic growth', *American Economic Review*, No. 95, (June 2005), pp. 546–79.
5. Acemoglu *et al.*, 'Rise of Europe', p. 562.
6. Gauci, P. (ed.): *Regulating the British economy, 1660–1859*, Farnham: Ashgate, 2011, especially the editor's 'Introduction', pp. 6, 19; J. Hoppit, *idem.*, 'Bounties, the economy and the state in Britain, 1689–1800', pp. 139–60; Daunton, M.: *Progress and poverty: An economic and social history of Britain, 1700–1850*, Oxford: Oxford University Press, 1995, pp. 533–57.
7. It isn't clear why length of coastline matters more than one or two gateway ports with good harbours and access to river systems.

8. Acemoglu *et al*, 'Rise of Europe', p. 551.
9. Pomeranz, K.: *The great divergence: China, Europe and the making of the modern world economy*, Princeton, NJ: Princeton University Press, 2000.
10. Pomeranz, *Great divergence*, p. 185.
11. Wrigley, E. A.: *Continuity, chance and change: the character of the industrial revolution in England,* Cambridge: Cambridge University Press, 1988.
12. Wrigley, E. A.: 'The divergence of England: the growth of the English economy in the seventeenth and eighteenth centuries', *Transactions of the Royal Historical Society*, 6th ser. (10), 2000.
13. Pomeranz, *Great divergence,* pp. 3–10.
14. Ormrod, D.: *The rise of commercial empires: England and the Netherlands in the age of mercantilism, 1650–1770,* Cambridge: Cambridge University Press, 2003, pp. xiii, pp. 334–5; Ormrod, D.: 'The North Sea as a core region in the early modern world: the shift from Amsterdam to London', XIVth IEHA Congress, Helsinki 2006, Session 36; Ormrod, D.: 'Institutions and the environment: shipping movements in the North Sea/Baltic zone, 1650–1800', in Unger, R. W. (ed.): *Shipping and economic growth, 1350–1850,* Leiden: Brill, 2011, pp. 135–66.
15. O'Brien, P. K.: 'European industrialization: from the voyages of discovery to the industrial revolution', in Pohl, H. (ed.): *The European discovery of the world and its economic effects on pre-industrial society, 1500–1800,* Stuttgart: Koch, Neff & Oetinger & Co., 1990, pp. 156–7; Ormrod, D.: 'Northern Europe and the expanding world-economy: the transformation of commercial organization, 1500–1800', in Cavaciocchi, S. (ed.), *Prodotti e techniche d'Oltremare nelle economie Europee, secc. XIII-XVIII,* Prato 1998, Firenze: Le Monnier, p. 672. Zahedieh provides the following figures showing the rising value of Caribbean sugar imports into London, compared with Virginia and other plantation tobacco between the years 1663/9 and 1701 (£000, rated and official values): 1663/9: sugar £200.3, tobacco £204.9; 1701: sugar £501.6, tobacco £154.5 (*Capital and the colonies,* p. 189). British Caribbean sugar imports continued their inexorable rise during the eighteenth century, almost quadrupling in value between 1699/1701 and 1772/74, while tobacco imports roughly doubled in value during the same period. Davis, R.: 'English foreign trade, 1700–1774', *Econonomic History Review*, No. 15 (1962), pp. 300–1.
16. Klooster, W.: 'An overview of Dutch trade with the Americas, 1600–1800', in Postma, J. and Enthoven, V. (eds), *Riches from Atlantic commerce: Dutch transatlantic trade and shipping, 1585–1817,* Leiden: Brill 2003, pp. 365–83; *idem*, Enthoven, V.: 'An assessment of Dutch transatlantic commerce, 1585–1817', pp. 385–445. A more pessimistic view emerges from the writings of Emmer, Pieter: 'The economic impact of the Dutch expansion overseas', in O'Brien, P. K. and Prados de la Escosura, l. (eds): *The costs and benefits of European Imperialism from the Conquest of Ceuta, 1415, to the Treaty of Lusaka*, 1974, *Revista de Historia Económica,* 16 (1998), pp. 157–74; and *idem*, 'The West India Company, 1621–1791: Dutch or Atlantic?', in Emmer, Pieter: *The Dutch in the Atlantic economy, 1580–1880: trade, slavery, and emancipation,* Aldershot: Ashgate Variorum, 1998, pp. 66–89.
17. Steensgaard calculated the contribution of Dutch extra-European trade at 20–25% of the Republic's total trade for the years 1752–4, including re-exports, which compares with a figure of roughly 50% for England and 33%

for France, Steensgaard, N.: 'The growth and composition of the long-distance trade of England and the Dutch Republic before 1750', in Tracy, J. (ed.): *The rise of merchant empires: long-distance trade in the early modern world, 1350–1750*, Cambridge: Cambridge University Press, 1990, pp. 102–52.

18. *Notes to Tables 2.1–2.3:* The trade values in Table 2.1 are derived from the Ledgers of the Inspector-General of Imports and Exports (National Archives, Customs 3), assembled by Davis, R. *English Foreign Trade*, and Ormrod, *Commercial Empires*, 2003, Appendix 1, p. 355. They are expressed not in current prices, but in formalised unit values ('official values') which closely match the level of wholesale prices prevailing during the years 1696–1709, when the official list became formalised. So as to facilitate comparison with estimates of the value of Dutch overseas trade shown in Tables 2.2 and 2.3, the Dutch statistics have been converted into constant sterling prices prevailing from 1696 to 1709, derived from the database complied by van Zanden, Jan Luiten: *Prices and wages and the cost of living in the western part of the Netherlands, 1450–1800*, International Institute of Social History, Amsterdam, www.iisg.nl/hpw/brenv.php. Van Zanden based his price data on the institutional price series collected by Posthumus, Wilhelmus: *Inquiry into the history of prices in Holland, 1943–64*, Leiden: Brill, 1946, supplemented by new material collected by Tine de Moor. While these prices do not correspond exactly to merchants' wholesale prices, they must represent a closer approximation than retail prices. The Dutch data are taken from Enthoven, *Dutch Transatlantic commerce*, tables 14.13 and 14.19, which include European data compiled by Jonker, J., Sluyterman, K., and Bruijn, J.R.
19. Ormrod, *Commercial Empires*, pp. 287–91.
20. Ormrod, 'Institutions and the Environment', pp. 149–58; Zahedieh, N., 'Productivity in English Atlantic shipping in the seventeenth century: evidence from the Navigation Acts', in Unger, *Shipping and economic growth*, p. 118.
21. Davis, R.: *The rise of the English shipping industry in the seventeenth and eighteenth centuries*, London: Macmillan, 1962, p. 42.
22. Ormrod, *Commercial empires*, pp. 273–6; van Bochove, C.: *The economic consequences of the Dutch: economic integration around the North Sea, 1500–1800*, published PhD dissertation, Utrecht 2008, ch. 2.
23. Figures in bold show the author's revised estimates for shipping capacity required for coal exports, arising from contemporary inconsistencies in measuring the outports' coal trade, especially to Ireland, together with false entries of 'vessels forced overseas' for tax reasons; grain export figures incorporate an adjustment for malt and barley, which occupied less shipping space than Davis realised. Ormrod: *Commercial empires*, pp. 253–5, 259–62, 292–4.
24. Thomas, B.: 'The first Atlantic economy, 1700–76', in Thomas, B.: *The industrial revolution and the Atlantic economy: selected essays*, London: Routledge, 1993, p. 36.
25. Ormrod, *Commercial empires*, pp. 168–74.
26. Ormrod, *Commercial empires*, pp. 317–21; O'Brien, P. K. and Hunt, P. H.: 'The rise of a fiscal state in England, 1485–1815', *Historical Research*, 66 (1993).
27. The passage of the *Imperial Act of Frauds* of 1696 coincided with the permanent establishment of the Commission for Trade and Plantations: Harper, L.: *The English navigation laws*, New York: Columbia University Press, 1939, ch. 4; Ormrod, *Commercial Empires*, pp. 45–6, 313.

28. O'Brien, P. K. and Prados de la Escosura, L. (eds), 'The costs and benefits for Europeans from their empires overseas', pp. 29–86 in O'Brien and Escosura, *Revista de Historia Económica*, 16 (1998), and note 15 above.
29. Inikori, J.: *Africans and the industrial revolution in England: a study in international trade and economic development*, Cambridge: Cambridge University Press, 2002, ch. 9; Ormrod, *Commercial empires*, ch. 5.
30. Davis, *English Foreign Trade*; Davis, R: *A commercial revolution: English overseas trade in the seventeenth and eighteenth centuries*, London: Historical Association, 1967, p. 12.
31. McCusker, J.J. and Menard, R. R.: *The economy of British America, 1607–1789*, Chapel Hill: University of North Carolina Press, 1985, pp. 79–80.
32. Ormrod, *Commercial empires*, pp. 322–33.
33. McCusker and Menard, *Economy of British America*, p. 155. See also Leonard, this volume.
34. Ormrod, *Commercial empires*, pp. 89–99, 126–7, 327; Jones, D. W.: *War and economy in the age of William and Marlborough*, Oxford: Oxford University Press, 1988, p. 256; Chapman, S. D.: *Merchant enterprise in Britain: from the industrial revolution to World War I*, Cambridge: Cambridge University Press, 1992, pp. 29–35.
35. North, D. C. and Weingast, B. R.: 'Constitutions and commitment: the evolution of institutions governing public choice in seventeenth century England', *Journal of Economic History*, 49 (1989).
36. Acemoglu *et al.*, *Rise of Europe*, p. 562.
37. Carter, A.M.C.: 'The Huguenot contribution to the early years of the funded debt, 1694–1714', Carter, A.M.C.(ed.): *Getting, spending and investing in early modern times: essays on Dutch, English and Huguenot economic history*, Assen: Van Gorcum, 1975, p. 87; Ormrod, *Commercial empires*, p. 92.
38. Ormrod, *Commercial empires*, pp. 330–2.
39. Nash, R.C.: 'English transatlantic trade, 1660–1730: a quantitative study', University of Cambridge Ph.D. thesis, 1982; *idem*, 'The organisation of trade and finance in the British Atlantic economy, 1600–1830', unpublished conference paper, 1999; Ormrod, *Commercial mpires*, p. 329.
40. Glamann, K.: *Dutch Asiatic trade, 1620–1740*, The Hague: Nijhoff, 1958, pp. 216–18; Ormrod, *Commercial empires*, pp. 195–202.
41. Bosher, J.F.: 'Huguenot merchants and the Protestant International in the seventeenth century', *William and Mary Quarterly*, 3rd ser., LII (1995), pp. 77–102; Nash, R.C.: 'The Huguenot Diaspora and the development of the Atlantic economy: Huguenots and the growth of the South Carolina economy, 1680-1775', pp. 75–105, in Janzen, O.U. (ed.), *Merchant organization and maritime trade in the North Atlantic, 1660–1815*, St John's: International Maritime Economic History Association, 1998; Ormrod, D.: 'The Atlantic economy and the "Protestant Capitalist International", 1651–1775', *Historical Research*, LXVI (1993), pp. 197–208. Nash shows that the years from the 1680s to the 1720s were notable for their distinctive presence in Carolina's mercantile community; thereafter, the dominance of English firms increased (p. 89).
42. Personal correspondence, 18 February 2007.
43. Pomeranz, *Great divergence*, ch. 6, *passim*, and Appendix D, pp. 313–15.
44. Ibid., p. 276.

45. Ibid., *Great divergence*, p. 6.
46. Wrigley, *Continuity and change*, pp. 57–60; *idem*, *Divergence of England*, pp. 125–6.
47. Ormrod, *Commercial empires*, pp. 247–55.
48. Faber, J.A., Diederiks, H., and Hart, S.: 'Urbanisering, industrialisering en milieuaantasting in Nederland in de periode van 1500 tot 1800, *AAG Bijdragen*, 12 (1965), pp. 251–71.
49. Thomas, B.: 'Britain's energy crisis in the seventeenth century', pp. 1–33, and *idem*, 'The end of the charcoal iron age', pp. 60–80, both in Thomas, B.: *The Industrial Revolution.*
50. Allen, B.: 'Was there a timber crisis in early modern Europe?', in Cavaciocchi, S.: (ed.), *Economia e energia, secc. XIII-XVIII*, Prato, Firenze: Le Monnier, 2003; Warde, P.: *Energy consumption in England and Wales, 1560–2000*, CNR 2007, p. 15.
51. Fedorowicz, J. K.: *England's Baltic trade in the early seventeenth century: a study in Anglo-Polish commercial diplomacy*, Cambridge: Cambridge University Press, 1980, p. 124; North, M.: 'The export of timber and timber by-products from the Baltic region to Western Europe, 1575–1775', in North, M.: *From the North Sea to the Baltic: essays in commercial, monetary and agrarian history, 1500–1800*, Variorum, 1996, pp. II/1–14.
52. Davis, *Shipping industry*, p. 184.
53. Ibid., p. 68; Zahedieh, N.: 'New World resources and the expansion of England's merchant marine, 1660–1775', in Cavaciocchi, S. (ed.), *Ricchezza del mare, richezza dal mare, Secc. XIII–XVIII*, Prato, Firenze: Le Monnier, 2006, pp. 511–22.
54. van Bochove, *Economic consequences*, chs 6 and 7.
55. Pomeranz himself points out that sugar is today often derided as a source of junk calories, although 'valuable' in poorer diets, *Great divergence*, p. 275.
56. Evans, C., Jackson, O. and Rydén, G.: 'Baltic iron and the British iron industry in the eighteenth century', *Economic History Review*, LV (2002), p. 663; Evans, C., and Rydén, G.: *Baltic iron in the eighteenth century Atlantic World*, Leiden: Brill, 2007; Harris, J.R.: *The British iron industry, 1700–1850*, London: Macmillan, 1988, pp. 50–1.
57. Thomas, *First Atlantic economy*, p. 46.
58. O'Brien, P.K. and Engerman, S.L.: 'Exports and the growth of the British economy from the Glorious Revolution to the Peace of Amiens', in Solow, B.L. (ed.): *Slavery and the rise of the Atlantic system*, Cambridge: Cambridge University Press, 1991, p. 186.
59. McCloskey, D.N. (ed.): *Enterprise and trade in Victorian Britain: essays in historical economics*, 1981; Thomas, R.P. and McCloskey, D.N.: 'Overseas trade and empire', in Floud, R. and McCloskey, D.N. (eds), *The economic history of Britain since 1700, I: 1700–1860*, Cambridge: Cambridge University Press, 1981, pp. 100–1.
60. Smith, S.D.: 'British exports to colonial North America and the mercantilist fallacy', *Business History*, 37 (1995), p. 58.

3

On the Rocks: a New Approach to Atlantic World Trade, 1520–1890

Chuck Meide

The natural flows of air and sea had a profound effect on European colonisation and exploitation of the Caribbean. Early European mariners quickly discovered that the clockwise rotation of wind and currents in the North Atlantic basin, most notably that of the Gulf Stream, greatly facilitated navigation to and from the new world. It was also learned that within the Caribbean, the prevailing currents and winds flowed to the west, sometimes reaching speeds of up to five knots. This made it easy to enter, but difficult to exit the Caribbean through the Lesser Antilles. Wooden-hulled sailing ships could leave with some difficulty via the Anegada, Mona, and Windward Passages in the Greater Antilles, though by the early sixteenth century a more desirable route was discovered through the Leeward Passage.[1] This involved sailing through the Yucatan Channel to enter the Gulf Stream in the Gulf of Mexico, and to coast with it through the Straits of Florida and out into the Atlantic Ocean. Thus, virtually all ships destined to or from the Gulf of Mexico, the Caribbean, or the Atlantic side of Central and South America eventually came through Florida waters. Given the treacherous nature of the reefs and storms along this route Florida's coastline acted as a kind of cultural sieve, culling a representative sample of this shipping in the form of thousands of sunken vessels which lie preserved on the seafloor, awaiting discovery and excavation by modern archaeologists.

This phenomenon makes the study of Florida shipwrecks ideally suited to gaining a broader understanding of interregional trade and the evolution of and interaction between economic core and peripheral areas during the development of global capitalism in the sixteenth through twentieth centuries. Keith Muckelroy was the first maritime archaeologist to propose analysing shipwrecks as components within larger military and economic systems.[2] Since he wrote, and especially in recent years, increasing numbers of archaeologists have used data from shipwrecks for such research.[3]

The present study is different in that instead of focusing on the material remains of wrecks in the archaeological record, it relies on shipping losses extant in the documentary record. This chapter comprises an analysis of a dataset of 1,431 shipwrecks lost in Florida waters between 1520 and 1890, modified from a larger database originally compiled by state underwater archaeologists in the Florida Bureau of Archaeological Research (BAR). At least one study of similar scope has been undertaken to date, which utilised a database of 4,000 shipwrecks in the Gulf of Mexico.[4] Evan Garrison's valuable and innovative study focused primarily on spatial patterning, in order to generate a predictive model of likely shipwreck locations, based on factors such as shipping routes, winds and currents, port locations, historic hurricane paths, and the location of shoals, bars, barrier islands, reefs, and other hazards. The present analysis, however, is less concerned with where or how the shipwrecks took place than with where they were going to or coming from, and what commodities they were carrying. The quantification of these types of variables exposes patterns reflecting the ongoing processes of economic restructuring that took place throughout Europe and the new world between the sixteenth and twentieth centuries.

Cores, peripheries, and the expansion of global capitalism

Immanuel Wallerstein proposed that in the development of a capitalist world economy, the global division of labour consists of core and peripheral areas.[5] Core regions are those with political, economic, and technological power, and their sophisticated infrastructures – including shipbuilding industries – are used to extract and transport commodities from remote or peripheral areas. Wallerstein maintained that prestige or sumptuary goods alone could not perpetuate what he termed the Modern World-System, which instead required the production of bulk goods, those which were seen as necessary by increasingly consumer-driven societies. One notable example is that of sugar, whose transformation from a rarity to a luxury, and finally to a household necessity, has been well chronicled by Sidney Mintz.[6] The global division of cores and peripheries was therefore driven by the impetus to open new markets for such manufactured or processed goods, and to obtain exploitable resources such as raw materials, cash crop produce, and labour. Wallerstein also discussed intermediate areas and semi-peripheral societies which exhibit features of both cores and peripheries, and serve as links or buffers between them. As the system evolved during the spread of modern capitalism starting in the sixteenth century, it displayed a certain degree of fluidity: some areas, notably North America, transformed from a periphery to semi-periphery to a core power in their own right. Wallerstein's World Systems Theory has been used by many archaeologists studying both pre-capitalist and modern societies, including a number of maritime archaeologists.[7]

Ships were the primary means of linking core and peripheral areas, and maintaining their interdependence in this political and economic system. To paraphrase Braudel, ships and capitalism were basically the same thing, for without ships there could have been no movement of people or commodities across the world's oceans, and therefore no colonialism, capitalism, or widespread consumerism.[8] The records of shipping losses in Florida, whose waters fronted the major shipping lanes for the Caribbean and Gulf of Mexico, will, as a representative sample of sixteenth- through nineteenth-century shipping, reflect the evolution of the mercantile-capitalist trading system in these and surrounding regions. Patterns discernible in these records, many of which denote the goods carried and the routes of particular vessels, allows a diachronic assessment of the production, distribution, and consumption of material goods, and of the transition of various localities from peripheral to core regions, and vice versa.

The dataset

The BAR Shipwreck Database is a listing of 2,560 shipwrecks, dating between 1521 and 1949, all in or believed to be in Florida waters. While a summary of the geographical distribution of the wrecks in this database has been presented, for the most part this valuable resource has not been subject to any significant analysis.[9] The shipwrecks included in the BAR database are ultimately known from two sources of information: the documentary record, which includes Admiralty Court and other government records, newspaper accounts, and insurance claims; and the archaeological record, which consists of the actual vessel and cargo remains on the seafloor.

This analysis focuses on vessels whose year of sinking is known, and which were lost from 1521 through the end of 1880s. This excludes some 320 wrecks in the BAR database whose date of sinking remains unknown (mostly archaeological sites whose locations are known but which have not yet been identified by name), and another 803 which were lost from 1890 onward. Six duplicate entries were also eliminated.

Thus, the entire body of Florida shipwrecks was narrowed down to a more manageable sample of 1,431, dating between 1520 and 1890. The period represents the full transition of Florida from an extra-peripheral region at the start of European colonisation through its culmination as a polity and arguably an economic core (or at least semi-core) in its own right, with modern patterns of intra- and inter-regional shipping from fully developed ports, at the dawn of the twentieth century.[10]

The basic unit of analysis is the individual shipwreck, with up to 30 variables for each. In some cases, such as ships of the 1622, 1715, and 1733 Spanish treasure *flotas* wrecked by hurricanes, both archaeological data and a rich documentary record exist. These are, however, the exception rather than the norm. The majority of the vessels in the database are known solely from archival sources. This has implications which must be kept in mind when analysing the distribution of certain variables. The cargoes or trade items in this analysis, for example, are known only (for the very most part) from the documentary record, which excludes items not considered worthy of mention, or illicit goods being smuggled. In 77.8% of the selected sample, cargo was not mentioned. Much of this is due to an unavoidable problem with the historical record: the lack of detail accompanying earlier shipwreck accounts, and the lower probability of earlier losses being recorded. There is obviously potential for archaeological excavation to help flesh out this paucity in the historical record. On the other hand, while an excavated shipwreck site may provide hard evidence of cargo items, the vessel's identity, age, and ports of origin and destination may remain a mystery. Regardless of these problems, the selected sample includes 319 ships with one or more known cargoes, and 519 whose routes were known, and thus represents a sizable sample from which to make inferences about the larger population of ships sailing through Florida waters. The remainder of this chapter presents data related to the overall characteristics of the 1520–1890 shipwreck sample, patterns of broad interregional exchange, the flow of goods between specific commercial centres or ports, and characteristics of some of the more important commodity trades.

Shipwreck frequency, type, nationality, and function

Table 3.1 shows the total number of wrecked vessels by decade. It is readily apparent that these numbers are not randomly distributed across time, but instead show a more or less steady increase as Florida made the gradual transition from remote backwater to commercial crossroads.

Table 3.1 Frequency of shipwreck by decade

Decade	***Frequency***	*%*
1520s	3	0.2
1530s	2	0.1
1540s	5	0.3
1550s	18	1.3
1560s	13	0.9
1570s	16	1.1
1580s	13	0.9
1590s	8	0.6
1600s	2	0.1
1610s	10	0.7
1620s	17	1.2
1630s	14	1.0
1640s	4	0.3
1650s	3	0.2
1660s	3	0.2
1670s	3	0.2
1680s	6	0.4
1690s	5	0.3
1700s	5	0.3
1710s	14	1.0
1720s	2	0.1
1730s	29	2.0
1740s	15	1.0
1750s	22	1.5
1760s	38	2.7
1770s	43	3.0
1780s	34	2.4
1790s	39	2.7
1800s	28	2.0
1810s	107	7.5
1820s	63	4.4
1830s	104	7.3
1840s	135	9.4
1850s	116	8.1
1860s	144	10.1
1870s	205	14.3
1880s	143	10.0
Total, sixteenth century	*78*	*5.5*
Total, seventeenth century	*67*	*4.6*
Total, eighteenth century	*241*	*16.7*
Total, nineteenth century	*1,045*	*73.1*
Total all shipwrecks	**1,431**	**100**

Closer attention reveals a more distinct pattern. After a sharp initial rise in numbers of shipwrecks in the 1540s and 1550s, shipping (reflected by number of ship losses) appears to decrease at the end of the sixteenth century, and then to rise sharply in the 1620s, only to decline and level off, with only incremental increases through the end of the seventeenth century. This distribution can probably be explained by a number of factors: during the period in question, nearly all of the shipping in Florida waters was Spanish. Spain enjoyed an economic boom during the first half of the sixteenth century, as vast amounts of riches from her conquests flowed back to the metropole. During the second half of the century, however, Spain experienced an unprecedented rate of inflation (probably related to the massive influx of specie) which caused an economic depression, particularly affecting the shipbuilding industry.[11] The loss of the Crown's famed Armada in 1588 certainly contributed to the crisis. The sharp rise in shipwrecks in the 1620s is not, however, related to global events, but to the fact that Spain lost an entire fleet in the Florida Keys as a result of the hurricane of 1622. This event has skewed the data, implying an artificially high increase in shipping. If the 1620s anomaly is ignored the recovery from this depressed period of shipping appears gradual, and apparently the numbers of Spanish ships plying the trade of the Indies never surpassed those of the heyday of the early to mid-sixteenth century.

The period after the seventeenth century shows much more dramatic changes in the frequency of shipwrecks in Florida waters. After two initial spikes in the first half of the eighteenth century the number of shipwrecks increases rapidly, and, despite some decline at the turn of the century and a few minor dips thereafter, skyrockets throughout the remainder of the nineteenth century. Again, much of this can be explained by geopolitical events and artificial spiking due to hurricane events. The sharp increases in the 1710s and 1730s are due to the inflated numbers of lost ships because of two more Spanish *flota* losses by hurricane in 1715 and 1733. The rapid increase thereafter is most likely due to the arrival onto the scene of England, a new colonial power. English shipping surpassed that of the Spanish by the 1740s, and in 1763 England gained the territory of Florida from Spain, prompting further English activity in the region. Following the defeat of England after the American Revolution, and her loss of Florida, the overall number of ships in the 1780s dips briefly, but after a low point in the 1800s numbers rise sharply until the end of the nineteenth century. Indeed, the 1,045 ships lost in the nineteenth century comprise almost three-quarters of the entire sample. This apparent increase of shipping is again likely related to the development of a new regional power, this time the US. Its ship losses were almost equal to British losses by the 1820s, when Florida became a US territory, and US ships dominate the dataset from the 1830s onward.

Table 3.3 displays the variation in vessel type of the 874 vessels whose type is known, 61.1% of the total sample. Type refers to distinctions of rigging or hull form, such as the galleon, schooner, brig, or side-wheel steamer, as opposed to differences in function, like warship, merchantman, packet vessel, etc. For the most part vessel type does not play a significant role in this analysis, though it is certain that the variation in type changes over time. Distinctive Spanish types, such as galleons, caravels, *naos*, and *falachos*, appear in the earlier periods dominated by Spanish shipping. Innovations in rigging systems intended to reduce crew sizes led to types such as brigs, barks, and schooners, which required smaller capital investments. Schooners, which use a rig allowing an especially small crew, were used extensively in the coastal trade, and are the most popular vessel type in the sample (220 vessels, or 25.2% of all known vessel types). Finally, starting in the nineteenth century, steam-powered vessels made their appearance, and rapidly proved

Table 3.2 Vessel nationality

Vessel Nationality	*Frequency*	*%*	*Valid %*
American	310	21.7	38.3
Austrian	1	0.1	0.1
Belgian	1	0.1	0.1
Canadian	1	0.1	0.1
Confederate	37	2.6	4.6
Dutch	4	0.3	0.5
English	197	13.8	24.4
English or American	1	0.1	0.1
French	20	1.4	2.5
German	5	0.3	0.6
Irish	2	0.1	0.2
Italian	3	0.2	0.4
Norwegian	8	0.6	1.0
Portuguese	2	0.1	0.2
Scottish	4	0.3	0.5
Spanish	211	14.7	26.1
Swedish	2	0.1	0.2
Total, known nationality	809	56.5	100
Total, unknown nationality	622	43.5	
Total, unknown shipwrecks	1,431	100	

Table 3.3 Vessel type

Vessel type	*Frequency*	*%*	*Valid %*
4th Rate (warship)	3	0.2	0.3
5th Rate (warship)	2	0.1	0.2
Aviso	1	0.1	0.1
Barge	1	0.1	0.1
Bark	78	5.5	8.9
Barkentine	19	1.3	2.2
Brig	84	5.9	9.6
Brig Sloop	2	0.1	0.2
Brigantine	45	3.1	5.1
Caravel	1	0.1	0.1
Clipper	1	0.1	0.1
Dory	1	0.1	0.1
Falacho	1	0.1	0.1
Floating Dry Dock	1	0.1	0.1
Frigate	26	1.8	3.0
Galleon	41	2.9	4.7
Goleta (Spanish Schooner)	4	0.3	0.5
Gunboat	3	0.2	0.3
Hermaphrodite Brig	1	0.1	0.1
Longboat	1	0.1	0.1
Nao (Spanish Merchantman)	51	3.6	5.8
Patache	8	0.6	0.9
Pink	1	0.1	0.1
Refuerzo	1	0.1	0.1
Sail Steamer	1	0.1	0.1
Schooner	220	15.4	25.2
Ship	83	5.8	9.5
Ship Sloop	2	0.1	0.2
Sidewheeler	51	3.6	5.8
Sidewheeler, River	2	0.1	0.2
Sloop-Of-War	1	0.1	0.1
Sloop	32	2.2	3.7
Snow	3	0.2	0.3
Steam Screw	11	0.8	1.3
Steam Screw Schooner	1	0.1	0.1
Steamboat, River	10	0.7	1.1
Steamer	72	5.0	8.2
Sternwheeler	7	0.5	0.8
Sternwheeler, River	1	0.1	0.1
Total, Known Vessel Type	**874**	**61.1**	**100**
Total, Steam-Powered Vessels	156	11.0	17.8
Total, Unknown Type	557	38.9	
Total, All Shipwrecks	**1,431**	**100**	

their efficiency, though many merchants of lesser means continued to man sailing vessels well into the era of steam.[12] In all, 156 vessels were powered by steam, 17.8% of the total sample of known types.

Spanish shipwrecks dominate the sample until the final quarter of the seventeenth century. Before that time only the occasional Portuguese, Dutch, French, or English vessel appears in the dataset to challenge Spanish hegemony. The situation changes rapidly after the 1680s, however. More English ships appear at that time, and English shipping begins in earnest in the 1730s, surpassing that of Spain by the following decade. The numbers of English shipwrecks around Florida increase when Britain maintained control of the territory from 1763 to 1783. With the loss of the Revolutionary War, England ceded Florida back to her rival, Spain. While British shipping suffered a brief setback at this time, Spain's economic woes would not be overcome, and by the end of the eighteenth century Spanish shipping had slowed to a trickle, despite her repossession of Florida. Unable to control her mainland colony, Spain transferred Florida to the newly established United States in 1821, and the territory saw an influx of settlers and the establishment of a number of commercial ports, including Fernandina, Jacksonville, Tampa, St Marks, St Joe, and Apalachicola. By the 1830s American ship losses surpass those of Britain, and the overall number of shipwrecks increases dramatically as Florida continued its economic development through statehood in 1845, through to 1890.

The final factor is vessel function. Table 3.4 reveals a distribution of functions which reflects broader economic and political systems, and regional and global events. The first decade, the 1520s, contains the only exploratory vessel in the entire sample. It was around this time, shortly after the conquest of Mexico, that Spain had successfully charted most of the coastline of her new world holdings, and would thereafter concentrate on colonisation and the extraction of resources. In keeping with the mercantilist principles of the day Spain maintained a strict trading monopoly with her colonies, keeping them dependent on the metropole by making trade with any other nation illegal. Needless to say, enterprising individuals from nations such as England, France, and the Netherlands also craved colonial commodities, and when they couldn't conduct illicit trade for them they often resorted to outright piracy. To protect this lucrative trade, and especially the flow of mineral wealth from the colonies to the mother country, Spain passed legislation calling for the periodic sailing of *flotas* (fleets) from Spain to the Caribbean, which ordinarily took place twice a year. These convoys consisted of both merchantmen and the warships charged with their

Table 3.4 Vessel function

Vessel Function	***Frequency***	*%*	***Valid %***
Blockade runner	18	1.3	3.6
Exploratory vessel	1	0.1	0.2
Fishing	6	0.4	1.2
Merchantman	354	24.7	70.4
Military transport	16	1.1	3.2
Packet	6	0.4	1.2
Passenger ship	1	0.1	0.2
Pilot boat	3	0.2	0.6
Pirate vessel	1	0.1	0.2
Privateer	3	0.2	0.6
Slaver	5	0.3	0.8
Survey vessel	1	0.1	0.2
Tender	5	0.3	1.0
Utilitarian	1	0.1	0.2
Warship	81	5.7	16.1
Wrecker	2	0.1	0.4
Total, known function	**503**	**35.2**	**100**
Total, unknown function	928	64.8	
Total, all shipwrecks	**1,431**	**100**	

protection, along with a number of tender and supply vessels. Other than the single exploratory vessel already mentioned, the portion of the dataset dating from 1520 to 1700 consists entirely of small numbers of tenders and military transports, and large numbers of merchantmen and warships.[13] This pattern is to be expected, given the monopoly of the Spanish *flota* system in the sixteenth and seventeenth centuries. The spike of warship losses in the 1620s is caused by the abnormal proportion of galleons (the term galleon, as opposed to *nao*, always indicated a warship) wrecked by the hurricane of 1622.

The pattern of vessel function changes after the start of the eighteenth century. The first slaver in the database (*Henrietta Marie*, lost on the return trip from Jamaica to London, and excavated by archaeologists in the 1980s) makes its appearance in 1700, reflecting the growing dependence on slave labour in the Caribbean and North American colonies.[14] In the following decade merchant ships begin to outnumber military ones, and by the 1730s merchantmen outnumber the combined figure of warships and military transports by five to one. Three major vessel function trends are readily discernible in the eighteenth and nineteenth centuries: a significant decrease in the proportion of warships to merchant vessels, especially after 1800; a sharp increase in the numbers of merchant vessels, consistent through the entire period, but especially evident after the end of the Napoleonic Wars in 1815; and an overall increase in the diversity of vessel functions. While merchant ships dominate the nineteenth century, slavers, privateers, military transports, packets, fishing vessels, utilitarian or work boats, wreckers or salvage vessels, and pilot boats also appear. Warship numbers remained high through much of the 1700s due to almost constant warfare. The considerable decline of warships and the concurrent rise in merchant ships after the first decade of the nineteenth century is probably a direct result of the 1805 Battle of Trafalgar, in which the British fleet utterly destroyed the combined Franco-Spanish fleets. While the Napoleonic Wars raged on for another decade, the theatre was focused on European combat, and the need for merchant fleet protection in the Caribbean was curtailed significantly. An unprecedented century of international peace followed from 1815, which greatly stimulated merchant shipping.[15] Warships almost disappear from the dataset, other than during a brief resurgence (along with a rise in privateers and military transports) in the 1860s related to the US Civil War, which saw blockade-running to and from Confederate ports.

Major trade routes and patterns of interregional exchange

One of the primary goals of this chapter is to delineate patterns of exchange between regions, and from port to port. On a broad scale this has been accomplished by defining major trade routes between the following five geographical areas: Europe; Africa; the Caribbean and Latin America;[16] the South (comprising the southern American colonies); and the North, including Canada. The major trade route (both origin and destination in terms of the regions listed above) is known for a total of 517 vessels, or 36.1% of the entire sample. This known sample is broken down by trade route in Table 3.5.

It is immediately apparent that some routes were much more heavily travelled than others. Seven trade routes have six or fewer representative shipwrecks (Africa to Caribbean/Latin America, Africa to the South, Caribbean/Latin America to Africa, Caribbean/Latin America to Caribbean/Latin America, Europe to Caribbean/Latin America, Europe to the North, and the South to Africa). Only nine ships wrecked en route from the Caribbean/Latin America to the North. Three trade routes, Europe to the South, South to the Caribbean/Latin America, and the South to the North, fall into a middle range, with 18, 20, and 22

Table 3.5 Frequency of major trade routes

Major route	***Frequency***	*%*	***Valid %***
Africa to Caribbean/Latin America	1	0.1	0.2
Africa to South	1	0.1	0.2
Caribbean/Latin America to Africa	5	0.3	1.0
Caribbean/Latin America to Caribbean/Latin America	6	0.4	1.2
Caribbean/Latin America to Europe	168	11.7	32.5
Caribbean/Latin America to North	54	3.8	10.4
Caribbean/Latin America to South	34	2.4	6.6
Europe to Caribbean/Latin America	5	0.3	1.0
Europe to North	1	0.1	0.2
Europe to South	18	1.3	3.5
North to Caribbean/Latin America	9	0.6	1.7
North to South	46	3.2	8.9
South to Africa	2	0.1	0.4
South to Caribbean/Latin America	20	1.4	3.9
South to Europe	66	4.6	12.8
South to North	22	1.5	4.3
South to South	59	4.1	11.4
Total, major route known	**517**	**36.1**	**100**
Total, route unknown	914	63.9	
Total, all shipwrecks	**1,431**	**100**	

ship losses, respectively. At the other end of the spectrum the six most heavily travelled routes in the sample are Caribbean/Latin America to the South (34 vessels), the North to the South (46), the Caribbean/Latin America to the North (54), transhipment from the South to the South (59), the South to Europe (66), and, by far the most significant route, the Caribbean/Latin America to Europe (168).

While these numbers certainly reflect the traffic of various trade routes, other factors are at play. It is clear from the historical record, for example, that the trade route from Africa to the Caribbean was well established. That only one such vessel shows up in this sample does not mean this route was not heavily used, but rather that, since it terminates in the Caribbean before entering Florida waters, only in very rare instances would a distressed ship using this route end up wrecking anywhere near Florida. The sample works best with – or is most representative of – vessels leaving the Caribbean or Latin America, or going to or leaving the South (though southern states above Georgia are probably also underrepresented). Other routes may produce artificially low numbers in the database simply because of their geographic location.

This explains why ostensibly few or no ships appear travelling from the North to Europe or its reciprocal route (both of which are situated nowhere near Florida waters), and relatively few travelling from the North to the Caribbean (a route adjacent only to Florida's east coast). On the other hand the very low numbers of vessels bound from Africa to the South, and vice versa, may suggest that these two routes were indeed less frequently travelled. This assumption seems safe in that significantly higher numbers of vessels were plying other routes to or from the South.

Examination of the frequencies of all 17 routes by decade reveals how the importance of specific routes waxed and waned over time. Aspects of the distribution are familiar: shipping from the Caribbean/Latin America dominates from the sixteenth through the early eighteenth century. This again reflects the Spanish trade monopoly and transport of extracted resources from periphery to core by the *flota* system. Other activity took place during this period, however. Between the 1550s and 1580s shipping is seen from the Caribbean/Latin America region to elsewhere in the same region, and to the South, and at the same time shipping from the South to elsewhere in the South, and also back to the Caribbean/ Latin America. This pattern makes it clear that the export by *flotas* of new world commodities was not the only economic activity taking place in the colonies. Inter-island trade was important for the growing needs of colonists in the Spanish Caribbean, and between the islands and New Spain (Mexico).[17]

Also at this time, Spain attempted seriously to maintain a military presence in Florida, which had been explored but was inhabited only by Natives. One of the primary reasons was to protect the vital shipping lanes that connected Spain's core and peripheral regions. A 1559 attempt to colonise present-day Pensacola failed after a hurricane wrecked Tristan de Luna's fleet. In 1565 Spain successfully founded St Augustine in Florida, in order to oust the French from their recently established foothold at nearby Fort Caroline (present-day Jacksonville). To support this military outpost regular shipments were made between it and Spain's established colonies in the Caribbean (mainly Cuba) and New Spain. In time other regions of Florida were also colonised through a series of Franciscan missions, and their agricultural output helped to feed St Augustine and to victual the *flota* gathered in Havana for the homeward voyage. The single South to South route wreck in the 1560s actually represents the war fleet commanded by Jean Ribault on route from Fort Caroline for a pre-emptive attack on St Augustine. It was wrecked by a storm, and thus doomed the nascent French colony to failure. These various shipping patterns illustrate the growing complexity of the world system perpetuated by the Spanish Crown and coveted by her rivals; in order to maintain the connection between the metropole and colonial empire numerous smaller links were formed between peripheral areas, and ruthlessly defended against competing colonial powers.

The pattern changes dramatically in the middle of the eighteenth century. Before that time the only route differing from those related to Spanish colonial activity was that of two ships bound from the Caribbean/Latin America to the North in the 1690s. As Spain had no colonies in the northern portion of the continent, this voyage (*Burroughs* and *Reformation,* sailing together from Jamaica to Philadelphia, in 1696) reflects England's growing foothold in the West Indies, and its ties to her colonies in North America. Starting around 1750, however, many of these new trade routes appear in the dataset. The more important trends include an increase in shipping from North to South, from the Caribbean and Latin America to both the North and South, in transhipments between Southern ports, and from the South to Europe. These new flows of goods came, to a degree, at the expense of shipping from the Caribbean and Latin America to Europe, though this remains one of the primary routes. In many ways this explosion of trade route diversity mirrors similar changes seen in vessel function and nationality at about the same time. The patterns suggest that the colonial periphery increased in complexity as various core nations successfully challenged Spain's hegemony in the new world. New trade routes represent new links between, for example, England's colonies of Jamaica and Barbados in the Caribbean, Bermuda in the Atlantic, and New England and South Carolina in the northern and southern portions of North America. At the same time the well-established trade artery from the Caribbean to Europe was – while no longer dominant – still of fundamental importance, as not only Spain, but also her successful rivals, continued to transport goods from the most lucrative tropical colonies to the metropole.

Comparing routes from peripheries to cores

If the North was the first peripheral region in the new world to undertake the transformation from a peripheral to a core region, its imports from longer-lasting peripheries such as the Caribbean and the South should increase, at the expense of its rival core power in Europe. This can be shown by considering four of the major trade routes. The volume of Caribbean/Latin America trade to Europe experienced a critical period of change, beginning in the 1750s. The trade to Europe had dominated until that decade, with only a few blips foreshadowing the upcoming importance of trade between the West Indies and the North. In the decades following 1750, shipping from the Caribbean to the North comes to rival that to Europe, and in some decades – including the last in the study period, the 1880s – surpasses it. Caribbean exports to Europe remain robust, however, and Southern exports to Europe increase quite dramatically over the nineteenth century, often double the quantity of those to the North. The ties between rising Southern exports and the trade to Europe in the nineteenth century are likely due to the widespread cultivation of cotton in the former region, and the burgeoning textile industry in the latter. While Europe still clearly retains a position of economic power, a large share of exports from these two regions was shunted to the North. This division of exports does not significantly reduce those flowing to the original European core, because of the intensification and development of the various peripheral regions in the new world and the related overall increase in shipping. The South's exports to Europe in the 1850s, for example, outnumber those of the Caribbean to Europe at any point in the entire sixteenth and seventeenth centuries, and it is just one of the heavily trafficked routes in that decade. This pattern is quite different from that seen in earlier centuries.

Another way of analysing shipping volume to the four most important core/periphery regions is to consider the numbers of ships bringing goods from all other regions bound for Europe, the North, the South, and Caribbean/Latin America. The latter has consistently low numbers of incoming ships; it is not until the 1850s that more than six vessels from the sample were headed there. Both the South and North have minimal numbers until around the middle and late eighteenth centuries, respectively. Shipping to Europe, as we have already seen, is the most consistently robust, and increases significantly after the start of the eighteenth century. The data suggest that numbers of incoming ships reflect the changing economic status of a region, though the basic nature of the dataset – for example, its tendency to underrepresent shipping to, as opposed to shipping from, the Caribbean – must be kept in mind.

In order better to understand the economic implications of these findings, the same data (numbers of incoming ships to major geographical areas over time) can be considered by region of origin, cargo category, and ship nationality. In Table 3.6 individual cargoes have been classified into 13 distinct categories. The categories are used to simplify the task of understanding the flow of goods between geographical

Table 3.6 Frequency of cargo categories, all shipwrecks

Cargo category	*Frequency*	*%*	*Valid %*
Agricultural produce	6	0.4	2.0
Building materials	9	0.6	3.0
Cash crop produce	96	6.7	32.4
Coal	4	0.3	1.4
Foodstuff	19	1.3	6.4
Forestry resources	45	3.1	15.2
Manufactured goods	18	1.3	6.1
Maritime resources	2	0.1	0.7
Military supplies	13	0.9	4.4
Raw material, metal	3	0.2	1.0
Slaves	2	0.2	0.7
Sumptuary goods	54	3.8	18.2
Tropical produce	26	1.8	8.8
Total, all categories	**297**	**20.7**	**100**
Total, unknown or no cargoes	1,134	79.3	
Total, all shipwrecks	**1,431**	**100**	

areas. In the following, the term region is used in a narrower sense than the geographical areas defined by major trade routes (for example, Louisiana or Florida as opposed to the South).

Shipping to Europe

Analysing shipping to Europe by region of origin, cargo category, and vessel nationality shows, perhaps as expected, that the Caribbean region was the most important region of origin at least until the 1810s. In earlier periods, although the bulk of shipping originated in the Spanish Caribbean, isolated vessels set out for Spain from Mexico and South America. These voyages took place in small numbers between the 1550s and 1680s, which suggests that the *flota* system – which by law left from Havana – was not universally followed by Spanish seafarers. Most of the shipping at this time was in sumptuary goods (including gold, silver, spices, porcelain, and other riches of Spain's new world and Asian colonies), though lower numbers of cash crop produce (most likely sugar, though possibly indigo, coffee, or tobacco) also figure. The first English ships appear in the record around the turn of the eighteenth century, but very rapidly come to dominate all shipping to Europe in the 1740s through 1750s. With the English ascendancy comes a major change in cargo categories. Sumptuary goods disappear between the 1740s and 1760s, replaced by increased volumes of tropical produce (typically sugar). Like the Spanish example all the initial English shipping originated in the Caribbean. It was not until the latter half of the eighteenth century that other regions of origin, including Florida, Alabama, Central America, Louisiana, and South Carolina, appear; not until the 1840s do these ships (most from Florida and Louisiana) outnumber those from the Caribbean. English ships continue to make up a significant portion of the aggregate, but after 1810 American, Austrian, Dutch, French, German, Irish, Portuguese, Spanish, and Swedish vessels were also plying the transatlantic trade. While this cultural diversity is not unexpected, the number of American ships bringing goods to Europe does seem surprisingly low, certainly when compared to the overall number of American ships in the dataset for this period. This may support the assertion that by this time the US had developed into an economic core, and would therefore be bringing in raw materials and bulk cargoes to its own industrial centres, instead of those in Europe. Plantation products are dominant for most of the period between 1810 and 1860. Imports of cash crops level off and then decline after the 1840s, however, and are surpassed in the 1870s by forestry resources (lumber and naval stores) and tropical products (including commodities such as coconuts, fruit, tropical wood, and guano).

Shipping to the North and South

Shipping to the North, mainly from the Caribbean, is evident from as early as the 1690s, but does not really start in earnest until the 1750s. The earliest ships were English (and colonial American), though after the 13 North American colonies gained their independence they are almost exclusively American vessels (with a few English or other European exceptions). Earlier vessels arrived almost exclusively from Britain's West Indies colonies, including some traffic from British Honduras in the middle of the eighteenth century. Trade with Spanish and then American New Orleans is seen in the decades before and after the turn of the nineteenth century. While the Caribbean remained the principal region of origin, starting in the 1830s the North saw ships arriving from an increasing variety of places, including Florida, Alabama, Mississippi, Louisiana, Texas, Mexico, Central America, and Britain. Nothing is known of the goods imported until the 1820s, when growing imports of plantation produce were of prime importance until their peak in the 1850s. In the second half of the nineteenth century, imports of plantation products were replaced by increasing amounts of forestry resources and tropical produce.

Regular maritime traffic into the South begins much earlier than in the North; consistent, if relatively low, numbers of vessels plied this route throughout much of the sixteenth and the first decade of the seventeenth centuries. In contrast to that of the North this pattern is explained by Spain's initial claim to much of the North American continent, and her attempts to maintain and supply strongholds along the Gulf and southern Atlantic coasts. Early shipping – all to outposts in Florida – was about evenly split from Mexico and the Caribbean. The former region was more suited to supplying Pensacola, and the latter to St Augustine. Shipping to Southern ports begins in earnest around the 1760s, and experiences an overall increase in numbers through a period of dramatic rises and falls of traffic. A number of other differences between Northern and Southern shipping patterns are also apparent. Southern ports received a greater variety of goods from a greater variety of regions. Shipping from the Caribbean, while at times substantial, was not nearly as important as it was to either Europe or the North. Considerable shipping reached the South directly from Britain, and after the 1820s even more from the North, mainly from New York and Massachusetts. Other traffic to Southern ports included ships from Central and South America, Africa, Portugal, Pennsylvania, Maryland, and Canada. The South shows a three-part pattern in terms of vessel nationality: at first the traffic is mainly Spanish, then English, and finally American. In the transitional 1760s, though, Spanish, English, and French shipwrecks occurred. This was the decade when Florida changed hands from Spanish to English rule, and the single French wreck (*Le Tigre*, on route from Saint-Domingue to New Orleans in 1766) brings to mind that France also maintained colonial centres in the South: Mobile and New Orleans.

Southern imports show a much greater diversity than the three cargo categories shipped to the North. During the initial Spanish period through the transfer of Florida in 1763 imports consisted mainly of military supplies, plus foodstuff and plantation produce. With the English takeover of Florida efforts were made to introduce the plantation system, which had proved so lucrative in the West Indies.[18] The implementation of this policy is reflected in a cargo of 100 slaves bound for St Augustine on the *Dove* from Africa in the 1770s. Shipments of military supplies remained important through the end of British rule twenty years later, but would dwindle during the relative security of the nineteenth century. Florida became a US territory in 1821. Around that time population and cotton agriculture were rapidly expanding both there and elsewhere in the South. Imported goods in the dataset reflect changes in demography and economy. Starting in the 1820s manufactured goods are the most important import, but foodstuffs, metal, coal, building materials, and some tropical and plantation produce were also shipped.

Shipping to the Caribbean and Latin America

Shipping to the Caribbean and Latin America created another distinctive pattern. The area had the lowest number of incoming ships, and was the only region examined where shipping does not increase significantly over time (primarily because vessels bound for the Caribbean would not tend to wreck off Florida). Regions of origin are noticeably diverse during the entire study period, unlike the other areas. Florida is probably the most consistent source of incoming traffic. Spain and Spanish Florida made up the most common source of inbound ships before the 1780s, but afterwards ships from many areas appear, including Canada, New York, Massachusetts, Maine, Connecticut, Pennsylvania, South Carolina, Georgia, Florida, Louisiana, Africa, Britain, and the southern United Netherlands. Patterns of vessel nationality are similar, though not identical to those in other areas. Spanish ship losses alone dominate until the 1770s, when English shipwrecks become commonplace. Unlike the other regions, however, Spain retained her Latin American colonies until the nineteenth century, and some of its Caribbean colonies through the end of the period, so it is not surprising to see at least one Spanish shipwreck as late as the 1820s. A similar cultural overlap is observed during the shift from English to American maritime primacy: between 1820 and 1890 American ships are the most frequent in the database, but English vessels were still sailing to the Caribbean and Latin America as late as the 1870s and 1880s. One Dutch shipwreck, *Marianne*, en route from Antwerp to Havana, was lost in 1818, a decade straddling the twilight of Dutch economic activity in colonies such as St Eustatius and Sint Maarten.[19]

Commodities shipped to the Caribbean and Latin America included manufactured goods, forestry resources, foodstuffs, agricultural produce, coal, and slaves. The sugar monoculture of the Caribbean required a constant influx of agricultural produce, foodstuffs, and slave labour, and many of the islands faced problems of deforestation, necessitating additional imports of lumber. Coal appears after the 1850s. As peripheral areas for the duration of the period, the Caribbean and Latin America did not have manufacturing centres of their own, and therefore all manufactured goods would have been imported from elsewhere, initially from Europe, but later also from the northern US and certain industrial centres in the southern states (notably New Orleans). One anomalous cargo category was that of sumptuary goods, represented by a single shipment of specie on *El Nauva Victoriosa* bound from Spain to Vera Cruz in 1771. This seems unusual given the almost continuous shipments of metallic riches along the reciprocal route, from Mexico to Spain. One possible explanation is that by the 1770s, just prior to the dismantling of the *flota* system, so little silver resources remained in Mexico that specie for troop payments or other needs had to be imported from the stockpiles in the metropole.

Exports from the Caribbean: a three-phased political economy

The following analysis of exports from these various regions narrows the focus from cargo category to actual cargo, and from region of destination to actual port of destination. Sumptuary goods dominate the trade from the Caribbean/Latin America for Europe through to the 1730s, with only a few instances of plantation produce, which is the most important category until the middle of the nineteenth century, when it is supplanted by tropical produce, and, to a lesser degree, forestry products. Thus three major phases can be seen in the political economy of the Caribbean and Latin America: the sumptuary goods trade (1520–*c*.1750), the sugar trade (prior to 1740–1850), and the diversified tropical products trade (1850–90).

More can be learned by breaking down the categories. The first phase, between 1520 and around 1750, is characterised by Spanish shipping of sumptuary goods from the Caribbean to Spain. These goods (comprising 35 primary cargoes) consist of specie or ingots of gold and silver (68.6%), unspecified luxury goods (14.3%), hides (8.6%), spices (5.7%), and porcelain (2.9%). The second economic phase is dominated by English exports of sugar, rum, and molasses, although such exports appear sporadic until the early to mid-eighteenth century. Of only three plantation crop cargoes shipped before 1740, two were sugar, while the third (the earliest) was indigo.

Significant sugar production actually started earlier, so that the first and second phases of this proposed Caribbean political economy scheme overlap during at least part of the seventeenth century. It has been well established that the English and French sugar industries developed in the West Indies during the last three quarters of the seventeenth century.[20] Several reasons could influence an inaccurate reflection of this activity in the shipwreck database before 1740. One is simply that fewer cargoes from this period are identifiable than in later years. Between 1700 and 1850, some 37 vessels bound for London were wrecked in Florida waters. Of these, only four had known cargoes (all were sugar or rum), but only two had sailed from unknown regions. Twenty-eight of the 37, some 75.7% of all ships bound for London between 1700 and 1850, were sailing from Jamaica, including the vast majority of those with unknown cargoes. Given that Jamaica's economy was dominated by the sugar sector, chances are that at least most of these ships carried sugar, rum, or molasses.[21] The data also help to clarify the duration and volume of the Caribbean sugar trade: it appears that the middle or sugar phase of the proposed three-phased political economy of the Caribbean can be defined as lasting from at least 1700 through around 1850.

Two cargoes dominate the eight primary cargoes of plantation produce exported after 1760. Some 75% was sugar and rum, and the remaining quarter was coffee, the export of which was limited to the 1830s and 1840s, the final two decades of the sugar phase, constituting a transitional period when British West Indian sugar was suffering from a price collapse due to overproduction and the introduction of non-Caribbean sources of sugar. The problem was exacerbated by the emancipation of slavery in 1834. In the face of the rapid devaluation of sugar, British West Indian planters experimented with alternative crops, particularly coffee.[22] This search for economic diversification led to a third Caribbean economic phase, from 1850 until at least 1890, characterised by the export of a wide variety of goods, in contrast to the homogenous exports of the earlier phases. Commodities shipped to Europe during the third phase include tropical woods (36.4%), lumber (18.2%), rum, sugar, hides, furniture (presumably crafted from tropical hardwood), and corn (9.1% each).

Shipping to and from the South and North

Most of the pre-1830 cargoes from the South to Europe are unknown, yet it is clear that the two most important products of the South were cotton before the US Civil War, and lumber and naval stores after it. These and other southern goods were shipped to a diverse selection of European ports throughout the period. Southern merchants had close ties to those in Liverpool, especially between 1810 and 1890. Twenty-eight ships were wrecked en route to this city, as opposed to five each lost en route to London and Bordeaux, and two each bound for Amsterdam, Le Havre, Bremen, and the Clyde. The extensive trading between the South and Liverpool consisted primarily of cotton exports for England's growing textile industry.

Shipping from the South to the Caribbean/Latin America reflects the latter region's traditional sugar monoculture and deforested islands. It is unsurprising that the main products imported from southern ports include food, in the form of rice, and lumber. The greatest trade by far took place with Havana, with its long-standing and sustained status as a principal Caribbean port, and other Cuban ports (13 out of 20 vessels, or 65%). Southern shipments were also despatched for various ports in the British and French Caribbean from 1780–1850 (20% and 10%, respectively), and in South America after 1860 (20%).

Relatively little trade between the northern and southern portions of North America is evident until the nineteenth century. Only two North-to-South routes are found in the database before 1800, both of which were vessels wrecked on their way to St Augustine in the 1760s. Early voyages such as these were likely state-sponsored endeavours in support of England's new acquisition of Florida. Indeed, one of the two wrecks is that of the *Industry*, which was bringing specie, munitions, carpenter's tools, flour, and other supplies to the newly established garrisons at St Augustine, St Marks, and Pensacola. The flow of specie has thus appeared in two economic contexts: as the transfer of mineral wealth from periphery to core (as in the Spanish *flota* system), and representing military subjugation of strategic peripheral regions when transferred (in processed or specie form) from a core or semi-periphery, to fund soldiers' pay.

After the 1810s and the end of the Napoleonic Wars a prolonged peace mitigated the need for further military support along the southern frontier.[23] Maritime trade began to prosper, peaking in the 1840s and 1850s, after which the outbreak of the US Civil War brought shipping between the North to the South to a near standstill. Prior to the war a variety of goods were shipped south from northern industrial centres such as New York and Boston, assorted cargo being the most frequent import. While this category is vague, it probably refers to various manufactured household and hardware goods such as cutlery, clocks, ceramics, furniture, musical instruments, mattresses, carpets, candles, soaps, books, medicine, hardware, firearms, tools, sheet metal, millstones, stock iron, ship's equipage, steam engines, and other machinery. Other antebellum exports include dry goods (clothing and textile products) and foodstuffs such as hams, bacon, dried fruit, coffee, tea, lard, nuts, fresh and salted mackerel, onions, flour, meal, wines, liquors, whiskey, bread, and corn.[24] More or less similar kinds of imports resume after the war: assorted cargo, general merchandise, furniture, coal, iron, bagging, bricks, and ice.

While the cargoes of Northern ships do not seem to undergo a significant post-war change, their ports of destination do. Before the 1860s (excluding the two eighteenth-century shipwrecks) the most important southern port is clearly New Orleans, with 13 shipwrecks or 43.3% of the antebellum subsample. Mobile, with seven losses (23.3%), is also a major destination, followed by Florida's Gulf coast ports, Pensacola and Apalachicola (two wrecks each). After the war and a decade of almost no trade between the regions, the traditional southern shipping centres changed. In general the shift was away from the major industrial centres, Mobile and New Orleans, to more diversified, smaller localities. Galveston emerges as the most important post-war port, with three wrecked vessels or 23.1% of the postbellum subsample. Mobile retains a significant, but reduced share of imports, with two wrecks (15.4%). Jacksonville and Mosquito Inlet, on Florida's east coast, each saw identical levels of trade (two wrecks each); Jacksonville was the only port present in both subsamples which experienced increased commerce in the two decades after the war. Only one wrecked vessel was bound for New Orleans, a significant decline from its former supremacy. This setback must have been temporary – New Orleans and Mobile are the two most important shipping centres along the Gulf coast today – so for at least the initial decades after the Civil War an economic restructuring among the southern semi-peripheries was clearly at play. A likely explanation is that these southern industrial centres, relative economic cores in an ambient peripheral region, were seen as strategic targets by the invading Union army, and suffered a much greater loss of infrastructure. Smaller ports whose facilities were relatively intact would have prospered temporarily under such conditions, and some, like Jacksonville, would be propelled into fully developed centres of shipping, cores in their own right.

The data provide a variety of information related to shipping and the exchange of goods from the South to the North. Little trade is evident between these two regions before the nineteenth century, with only one example in the dataset of a wrecked ship bound from the South to the North before 1800. This was the *Noah's Ark*, an American merchantman bound from New Orleans with a cargo for Philadelphia, which was wrecked on the Florida Keys in 1795. *Noah's Ark* was a precursor to a burgeoning maritime commerce between the two regions in the following century, when southern shipping would expand rapidly, then suffer a collapse with the outbreak of the Civil War and the Union blockade. Still, the number of outgoing ships in the 1860s is the same as that recorded in the previous decade. Of the three shipwrecks that occurred while en route from Southern to Northern ports during the 1860s, one, the *Tiger* en route from New Orleans to Baltimore, took place in February 1860 before the war started. The other two, the *Waltham* and *Ringgold*, en route from New Orleans to Boston and Chasahowika, Florida to New York respectively, were lost in October 1865, after its end.

The effect of the Civil War on post-war southern shipping appears to have been the opposite of that on post-war northern shipping. Northern vessels called upon different southern ports after the war, but carried the same types of export goods. Southern vessels, in contrast, visited the same northern ports they had before the war, but with different types of cargoes. While imports into southern ports increased only gradually after the 1860s, shipments to northern ports appear to have regained their former level rapidly.

The most important Northern ports visited by ships inbound from the South were Boston and New York. Philadelphia appears to have been more active around the turn of the century, and sporadic vessels were bound for Baltimore between the 1810s and 1860. However, Boston and New York together accounted for 75% of all Southern imports during the nineteenth century before the Civil War. After it, only one example appears, a vessel bound for Nova Scotia, and although New York gained importance over Boston, the pair continued to command an aggregate 88.9% of all southern imports. Thus the post-war restructuring of viable commercial centres which took place in the South did not take place in the North.

Changing political economies in the South: exports before and after the Civil War

If sugar reigned in the West Indies, then surely cotton was king in the antebellum South. The data proffer details of many aspects of the cotton trade. Of 38 vessels whose routes are known and were carrying cotton, the most frequent cargo in the entire dataset, all but one originated in the South. According to the database – which almost certainly underemphasises the exports of the southern states on the Atlantic coast above Florida – more than half of this cotton came from Louisiana (55.3%), followed by Alabama and Florida (15.8% each), Texas (7.9%), and Georgia. Britain imported half of all this cotton (88.2% of this amount, or 44.1% of all cotton, was shipped to Liverpool), with 20.6% going to the Northern states, 11.8% to France, and 5.9% each to the Netherlands, Belgium, and Germany. By 1860 half of the world's output of cotton was produced by the southern US, while a quarter of Britain's population was employed in the textile industry. With the outbreak of the US Civil War valid concern emerged that that thousands of British mill labourers might actually starve to death.[25]

The rise and fall of Southern cotton imports to the North, Britain's industrial rival, is clear. Of all known exports from the South to the North prior to the outbreak of war, fully 85.7% were cotton, compared to only 12.5% of all known Southern exports after it, when 25% were sugar and 62.5% lumber. The ascendancy and decline of the South's cotton trade occurred on an international scale. Its exports continued through the end of the nineteenth century, but never again would production equal that of the bumper crops from the late 1850s into the spring of 1860.[26] After the war southern fields and infrastructure lay in ruin. European textile barons had already secured alternate sources of raw cotton from India and Egypt, but the most devastating blow to the cotton economy was the change to a wage-labour system. Like sugar in the Caribbean, cotton could not survive the transition brought on by emancipation.

As in the Caribbean, Southern entrepreneurs sought to diversify their economy by pursuing activities other than cotton cultivation. Limited sugar production occurred in the southern states such as Louisiana, Texas, and Florida. The burgeoning lumber industry, however, was the most important commercial activity during the Reconstruction era. Lumber mills had been established throughout the heavily forested regions of the South since the eighteenth century, though timber milling did not make a significant economic impact until the advent of steam-powered mills capable of producing high-grade boards in the 1840s.[27] The data show that forestry resources output exceeds that of plantation produce in the 1870s, immediately after the war, and that the timber industry was clearly the most significant thereafter. Florida rapidly became one of the world's most important sources of lumber, and its industry was centred on Pensacola, though other ports such as Jacksonville shipped out considerable quantities.

Brick-making was one of the earliest true industries introduced to the South. Evidence exists of a brickyard at Pensacola as early as the British period in 1767, and the long-lasting Bonifay brickworks were established in 1807, during the second Spanish period.[28] By the early 1820s an estimated 100,000 bricks were being produced annually in Pensacola; by the 1850s production was over two million.[29] Southern brick-making industries look to have come to a halt during the Civil War, but exports resumed the following decade.

Other aspects of Reconstruction-era economic diversification visible in the data include a post-war rise in the export of non-plantation agricultural produce. With the advent of ice machines in 1851 commercial fishing developed along the Gulf. By 1869 a thriving market for snapper, oysters, and other seafood had emerged; refrigeration allowed the fish to be transported north and west by rail from ports such as Pensacola.[30] The Florida citrus industry also began to develop in the late nineteenth century. Other exports of the 1870s–1880s include further foodstuffs and manufactured goods.

As these various diversification efforts led to economic recovery and increased shipping, the interior of Florida and the western territories opened for development and expansion, with the spread of inland steam navigation by river, and rail construction on land. Jacksonville became Florida's premier deepwater port, establishing a regular pattern of exchange with domestic and international commercial centres via steamship and railroad. Federal government subsidies of harbour and channel dredging and maintenance, and improvements in cartography and navigational aids, lead to growth of established ports such as Pensacola, and promising new ones like Tampa.[31] By the end of the nineteenth century Florida, along with much of the South, had completed a transition from a periphery to a core in its own right.

A wreck model of evolving political economies

Like that of the Caribbean and Latin America, the evolution of the political economies of the North and South can be viewed as a multi-phased process. A salient difference between the regions is that while the Caribbean and Latin America remained peripheral in economic significance, the South and especially the North shifted from peripheries to cores. Table 3.7 summarises the overall results of this analysis, describing the changes in economic status for each of the key geographic areas and for world trade as a whole. The temporal and economic divisions in the table are based on the patterns of shipwrecks in the dataset. It shows clearly that economic restructuring played out quite differently in the Caribbean and the northern and southern portion of North America.

The first phase of regional economic transition encompasses the period of European contact and initial colonisation. This phase in the Caribbean and most of Latin America was already coming to a close by 1520, hence only the single example of an exploratory caravel in the dataset (and only three as opposed to four economic phases). The initial phase began in the South, as it was colonised by the Spanish

Table 3.7 Evolution of political and economic structures for major regions

<table>
<tr>
<td>Global Trade (all geographic areas)</td>
<td colspan="2">1520–c.1750
Trade between old and new worlds is carried out mostly in Spanish ships; Cargoes are homogenous, mainly sumptuary goods shipped from the Caribbean and Latin America (some transported from East Asia) to Europe; More or less even proportions of warships and merchant vessels in dataset; Rise in English shipping towards end of period.</td>
<td>c.1750–c.1810
Maritime trade abruptly becomes diversified in regards to cargoes carried and regions of origin. English ships from the metropole and colonies in the West Indies and North America now dominate the seas; Spanish shipping dwindles.</td>
<td>c.1810–1890
Rapidly increasing trade on a worldwide scale. Number of merchant ships far surpasses that of warships; US ships outnumber British by 1830s in new world trade; End of Napoleonic Wars in 1815 heralds a prolonged period of peace, global industrialisation, and full development of world capital markets; The 'international economic system [becomes] the axis of the material existence of the [human] race'.[32]</td>
</tr>
<tr>
<td>Caribbean and Latin America</td>
<td>1520–c.1740
Periphery; After initial periods of exploration and conquest, Spain consolidates hold on much of the new world; Exploits Native and imported African slave labour to extract mineral wealth and ship it along with other sumptuary goods back to Spain in flota; Havana, as a centre of shipping, shipbuilding, and commerce, approaches semi-periphery status.</td>
<td colspan="2">pre-1740 (c.1700)–c.1850
Periphery; Political economy dominated by English and French sugar production and export, reliant on imported African slave labour; Protracted colonial warfare to seize or destroy rival sugar islands and merchant shipping ends in 1815; Emancipation in 1834 leads to end of sugar monoculture and experimentation with alternate cash crops such as coffee.</td>
<td>c.1850–1890
Periphery; Post-emancipation diversified economy based on export of diverse tropical goods produced through wage-labour system.</td>
</tr>
<tr>
<td>North</td>
<td>1520–1680s
Super-periphery; No activity represented in the dataset (due in part to natural biases in the shipwreck sample).</td>
<td>1690–1750
Periphery; English colonists maintain some trade with Caribbean; Little shipping elsewhere represented by dataset.</td>
<td>1750–1810
Semi-periphery to semi-core; Develops a number of commercial centres and various industrial activities; Most trade still with Caribbean, though some with southern ports; Commerce intensifies after gaining autonomy from the English metropole in 1783.</td>
<td>1810–1890
Core; Numerous commercial centres such as Boston, New York, Baltimore, and Philadelphia; Rapidly expanding industrialisation rivals that of England and the rest of Europe; Fully developed shipbuilding, manufacturing, and textile industries; Significant increase in exports of manufactured goods and imports of raw materials to and from peripheral and semi-peripheral regions; Prevents the autonomy of southern states in the Civil War and subsequently abolishes slave labour system; Its own political economy suffers little change after the war.</td>
</tr>
</table>

(*continued*)

Table 3.7 Continued

South (Florida and the Gulf Coast)	**1520–1760** Super-periphery; Very remote area characterised by Spanish military outposts (by 1560s) subsidised by supplies shipped from semi-peripheral regions in the Caribbean and Mexico; By seventeenth century in parts of Florida a Franciscan mission system in place which uses Native labour to produce food surplus to supply St. Augustine and *flotas* in Havana.	**1760–1810** Periphery; Still provided with military support by core; Develops slave-based plantations to produce agricultural resources for export to North and Europe; Exports steadily increase; Some developed ports	**1810–1865** Semi-periphery to semi-core; Dramatic rise in shipping to Europe; Extensive slave-based plantation system; Cotton exports dominate export trade; Imports mainly manufactured goods; Some industrial centres	**1865–1890** Semi-core to core (or at least sporadic core centres); Diversified economy (lumber, seafood, and manufacturing industries, etc.) replaces slave-based cotton monoculture; Reorganisation of ports, limited areas of industrialisation

in the sixteenth century, but its economic transformation progressed more slowly. The earliest decades of colonisation in any region typically witnessed little commercial activity. Instead, subsidies in the form of specie, munitions, and other supplies were provided by the metropole to secure and develop what were hoped to be lucrative or strategic peripheries. This pattern persisted later in some regions of the South; Florida remained a frontier region in the 1760s. Its military supplies in this period came only indirectly from the English metropole, via the more developed semi-peripheral colonies to the north. The pattern had remained virtually unchanged from that in place two hundred years before, when supply ships came instead from the Spanish semi-peripheries of Havana and Vera Cruz.

The *flota* system dominates interregional trade in the Caribbean and Latin American region until the 1740s. This well-organised system was designed to transport valuable resources from the peripheral regions of empire to the metropole of Spain safely and efficiently, and was in place while other regional peripheries remained underdeveloped. It was dependent on a complex web of relationships between various other peripheral regions in the Spanish empire for defence and support (for example, military outposts along shipping lanes, and agricultural produce for the transatlantic voyage). Rival colonial powers attempting to disrupt or appropriate this trade would often attack these peripheral links: France tried to wrest control of Florida by establishing Fort

Caroline in 1564, and Cromwell's grandiose Western Design succeeded in the English conquest of Jamaica in 1655.

Little if any European contact was made in the North before the seventeenth century. The first activity in the dataset is in the 1690s, after England had established a number of colonies with economies based on agriculture and maritime commerce. Trade from them generally increased until the beginning of the next phase, around 1750, and it grew more quickly after that. Around this time a global, fundamental change occurs in the number of basic trade routes, the nationality of international shipping, the proportions of merchant and naval shipping, and the nature of commodities being shipped. This broad pattern is due in part to the roughly simultaneous evolution of political economies in the North, the South, and the Caribbean.

In the Caribbean, the *flota* system had all but exhausted the mineral wealth of the new world by the 1740s. It was replaced by widespread plantation agriculture. The system had been developing since the second quarter of the seventeenth century, and evidence in the dataset suggests it was flourishing by the start of the eighteenth, though it does not become readily apparent in the database until the decline of the *flota*.

In the South, similar plantation production increased in importance during the latter half of the eighteenth century. While the South at this time remained a periphery, localised areas were beginning to develop into semi-peripheral economies. Such commercial ports as Charleston, Savannah, and New Orleans began to make considerable contributions to the wealth of their respective empires. Even the relative backwater of Florida experienced the stirrings of future industrial production, such as at the early brickyards in Pensacola. In the North commercial centres such as Philadelphia, Baltimore, New York, and Boston were more fully developed and industrialised. Colonial shipbuilding industries contributed to the further intensification of maritime trade, mainly with British colonies in the Caribbean. Even before the 13 North American colonies achieved autonomy the North had made the transition from periphery to semi-periphery or semi-core.

During the next phase, from 1810, the North had made the transition to a fully developed core. No longer dependent on the core industries of Europe, Northern textile mills and factories produced a wide range of manufactured products which were shipped to ports in the Caribbean and the South in exchange for raw materials. Two cores henceforth shared peripheral regions under a dual system made possible by the rapidly expanding merchant shipping industry and worldwide consumption of bulk commodities. Southern shipping patterns also indicate a change around 1810. Slave-labour cotton production has pervaded the entire region and dominated outgoing cargoes, supplying the looms of two world cores. Cotton was the premier Southern commodity, but further industrial development in specific areas led to cargoes of Southern-made manufactured goods, which were transported mainly between semi-peripheries and peripheries within the South. In the Caribbean, sugar was still the main export, and was shipped mainly to Europe for further processing and consumption. Little industry emerged in the Caribbean other than sugar manufacturing facilities; plantations' increasingly mechanised equipment was supplied by European, Northern, and, to a lesser degree, Southern industrial centres.[33]

The final economic transition took place in the middle of the nineteenth century, slightly earlier in the Caribbean than in the South, and not at all in the core regions of Europe or the North. The last shift is characterised by a collapse of the slave-based system of plantation monoculture, followed by a period of recovery and stabilisation produced by a diversification of economic activities in the new wage labour system. For each peripheral or semi-peripheral region this shift is apparent in the dataset by the disappearance of sugar or cotton cargoes, and their replacement with a wide variety of alternate products. In the Caribbean this occurred in the 1850s, and was precipitated by a global collapse in the price of sugar and the emancipation of slavery in British colonies, while in the South it took place in the 1860s–1870s and was a result of the physical and commercial devastation of the Civil War and subsequent end of slavery, but the results were remarkably similar in many respects. One major difference was that the South continued to develop centres of industry and make the full transition to a core or at least semi-core economy, while the Caribbean remained marginalised.

A neatly compartmentalised schema such as that presented in Table 3.7 is certainly a simplification of almost 400 years of economic history. Numerous exceptions to the posited generalisations could be cited, many of them due to the inherit biases of the dataset, but the model remains useful for explaining the complex changes which took place during the spread of colonialism and capitalism throughout the Atlantic World. The increasingly interrelated network of exchange between cores and peripheries, driven by mass consumerism on both sides of the Atlantic, was critical to the formative stages of modernity. Ships, as the vectors of capitalism and the carriers of peoples, goods, and ideas, were integral to the process of modernisation, and to our better understanding of how it took place.

Notes

1. Smith, Roger C.: 'Treasure ships of the Spanish Main: the Iberian–American maritime empires', in Bass, G.F. (ed.): *Ships and shipwrecks of the Americas*, London: Thames & Hudson, 1988, p. 85.
2. Muckelroy, Keith: *Maritime archaeology*, Cambridge: Cambridge University Press, 1978.
3. See, for example, Gould, Richard A.: *Archaeology and the social history of ships*, Cambridge: Cambridge University Press, 2000; Souza, Donna J.: *The persistence of sail in the age of steam: underwater archaeological evidence from the Dry Tortugas*, New York: Plenum Publishing, 1998; and Staniforth, Mark: *Material culture and consumer society: dependent colonies in colonial Australia*, New York: Kluwer Academic/Plenum Publishing, 2003.
4. Garrison, Ervan G.: 'A diachronic study of some historical and natural factors linked to shipwreck patterns in the northern Gulf of Mexico', in Barto, J. (ed.): *Underwater archaeology: proceedings from the Society for Historical Archaeology Conference*, Baltimore: Arnold, 1989, pp. 12–18; see also Garrison, Ervan G. *et al*: *Historic shipwrecks and magnetic anomalies of the Northern Gulf of Mexico: Re-evaluation of Archaeological Resource Management Zone 1*, 3 vols, New Orleans: US Department of the Interior, Minerals Management Service, 1989.
5. Wallerstein, Immanuel: *The modern world-system*, 3 vols, New York: Academic Press, 1974, 1980, 1989.
6. Mintz, Sidney Wilfred: 'Time, sugar, and sweetness', *Marxist Perspectives*, Winter 79/80, pp. 56–72; *Sweetness and power: the place of sugar in modern history*, New York: Penguin, 1986.
7. Champion, T. C.: *Centre and periphery: comparative studies in archaeology*, London: Routledge, 1995; Sanderson, Stephen K.: *Civilizations and world systems: studying world-historical change*, Walnut Creek: Altamira Press, 1995; Staniforth, *Material Culture*, p. 35.
8. Braudel, Fernand: *Civilization and capitalism, 15th–18th century, vol. 1: the structures of everyday life*, New York: Harper & Row, 1981, p. 514.
9. Smith, Roger C. *et al*: *An Atlas of Maritime Florida*, Tallahassee: University Press of Florida, 1997, pp. 38–9.
10. Garrison, *A diachronic study*, Table 1.
11. Meide, Chuck: *A plague of ships: Spanish ships and shipbuilding in the Atlantic Colonies, sixteenth and seventeenth centuries*, manuscript on file, Department of History, College of William and Mary, Williamsburg, Virginia, 2002, p. 22.
12. Souza, *The persistence of sail*.
13. For the purposes of this study, military transports refer to vessels carrying either military supplies (specie is one of the more common such cargoes present in this study) or troops, rather than the more conventional definition of vessels carrying only the latter.
14. While slave ships are, by definition, merchantmen, I have classified them separately to help track the slave trade in the dataset. There is some justification for this functional distinction, as (especially later-period) slavers were often purpose-built for the trade, to meet the specific needs of vessels carrying captive human cargoes. Blockade runners, another specialised variant

of the merchant vessel, also comprise a separate category, and they too were often custom designed and built for their trade.

15. Polanyi, Karl: *The great transformation: the political and economic origins of our time*, Boston, MA: Beacon Press, 1957, pp. 3–19.
16. These two areas were combined due to the relatively few incidents of shipwrecks to or from South or Central America, and the fact that most of the shipping from these regions and Mexico passes through the Caribbean regardless, in addition to the cultural ties between the Caribbean and Latin America throughout the colonial period and beyond.
17. Turner, Samuel P.: 'Inter-island trade and Spanish colonial expansion from Española 1512–1517', unpublished PhD dissertation, King's College London, 1998.
18. Smith *et al*, *Atlas of Maritime Florida*, pp. 10–11.
19. Eastman, John Arnold: 'An archaeological assessment of St Eustatius, Netherlands Antilles', unpublished Master's thesis, College of William and Mary, Williamsburg, Virginia, 1996, p. 33.
20. Dunn, Richard S.: *Sugar and slaves: the rise of the planter class in the English West Indies, 1624–1713*, Chapel Hill: University of North Carolina Press, 1972; Mintz, *Sugar and sweetness*, pp. 36–45.
21. Delle, James A.: *An archaeology of social space: analyzing coffee plantations in Jamaica's Blue Mountains*, New York: Plenum Press, 1998, p. 46.
22. Ibid., pp. 50–1.
23. A shipment of military supplies was dispatched to Florida as late as the 1830s, sent (in this case from Mobile to Tampa on the steamer *Convoy*, wrecked 1836) to support the US Army's efforts forcibly to expulse the Seminole Indians from Florida during the Second Seminole War (1835–42).
24. Meide, Chuck: 'The archaeology of economic transition: the excavation of the San Marcos shipwreck, a mid-nineteenth century merchant sailing ship lost at St Marks, Florida', unpublished Master's thesis, Florida State University, Tallahassee, 2001, pp. 43–4.
25. Murphy, R.W. (ed.): *The blockade: runners and raiders*, Alexandria: Time-Life Books, 1983, p. 14.
26. Ibid., p. 15.
27. Burns, Jason M.: *The life and times of a merchant sailor: the archaeology and history of the Norwegian ship Catharine*, New York: Kluwer Academic/Plenum Publishers, 2002, p. 21; Smith *et al.*, *Atlas of Maritime Florida*, p. 18.
28. Scott-Ireton, Della A.: 'Brickmaking in Hispanic West Florida', unpublished manuscript on file at the Department of History, University of West Florida, Pensacola, p. 3.
29. Polk, James Knox: 'Pensacola commerce and industry, 1821–1860', unpublished Master's thesis, University of West Florida, Pensacola, 1971.
30. Smith *et al.*, *Atlas of Maritime Florida*, p. 20.
31. Smith *et al.*, *Atlas of Maritime Florida*, p. 12.
32. Polanyi, *Great transformation*, p. 18.
33. See Pretel and Fernández-Pinedo, this volume.

4

Commerce and Conflict: Jamaica and the War of the Spanish Succession

Nuala Zahedieh

In the mercantilist world of the long eighteenth century the state was expected to use its resources, including its military strength, in the struggle for economic supremacy in Europe. In Richard Pares's words, 'the navy was a branch of business'.[1] Powerful groups such as the Jamaica lobby repeatedly sought to solve economic problems with military action, and they frequently got their way. There has been surprisingly little systematic effort to assess the real economic gains of this undoubtedly economic war, but most historians have felt that, in general terms, the policy paid off.[2] According to Curtis Nettels,

> the Peace [of Utrecht] brought the advantages for which England had gone to war. Both the Dutch and the French had been crowded out of the favoured position in Spanish colonial trade... by the end of the war, markets so long closed or partially closed had been forced open.[3]

Yet a closer look at the impact of the war on Jamaica, which played a central role in Spanish American trade, suggests that Nettels was unduly triumphalist, and although certain groups did gain, the Peace fell far short of bringing 'the advantages for which England had gone to war'.

West India merchants, especially those trading with Jamaica, had an especially close interest in the questions of control and access to Spain's American empire, questions which lay at the heart of the wars of 1689–1713. Although the history of the British Caribbean is generally written in terms of sugar production, this was not the whole story. The region was valued at least as much, if not more, as a gateway to Iberian America and the fabled treasure of its bullion mines. The image of Spanish America as a source of almost limitless wealth – an El Dorado – went back to the first discoveries. By the late seventeenth century, Spain was described as a 'dead carcass upon whom all the rest do prey', but the carcass continued to provide rich pickings.[4] Despite some decline after 1600 the mines of Peru and Mexico were still producing a vast treasure, and a marked revival occurred at the end of the century. Between 1660 and 1700 annual silver output averaged 10 million pesos (£2.5 million), and far exceeded America's total sugar production, which Noel Deerr valued at not much over £1 million in 1700, of which British colonies accounted for 40% to 50%.[5] England's early Caribbean colonies were settled on the eastern periphery where, on account of the wind system, they were out of the way of Spanish harassment, but also inconvenient for trade. By contrast, Jamaica, which was captured in 1655, was located in the heart of the Spanish Indies, with easy access to major Spanish ports and trade routes. Although, despite its size and fertile soils, Jamaica did not become Britain's leading sugar producer until well into the eighteenth century, it did quickly prove itself strategically placed to gain access to Spanish wealth: a 'silver mine without the labour and expense of mining'.[6]

Although Spain (or rather the Kingdom of Castile) maintained its claim to a monopoly of the American treasure trade, exclusivity became impossible to enforce. Others could gain access to American wealth in a number of ways. Plunder was the simplest strategy, and as Jamaica was an ideal base for predatory activity, privateering dominated the island's economy in its early years in English hands. The trade required little start-up capital, and offered quick returns in both physical wealth and a store of useful knowledge about the region.[7] The privateers helped to push back the commercial frontier, but their own trade faced diminishing returns as they stripped the coastal regions of easy plunder and encouraged the Spaniards to raise their defence capabilities.[8] Plunder was not a strategy for long-term growth, and peaceful trade offered far better returns over the long run. Spanish America had an estimated population of between six and eight million in 1700 (twenty times as big as that of British America), including a number of large cities such as Lima, whose population peaked at 80,000 before the earthquake of the 1680s.[9] Strong regional economies provided a market for slaves and capital goods, while a growing number of high-spending consumers demanded food, drink, and a range of manufactured goods. Even slaves were seen wearing elegant shawls and silk stockings at the bullfight in Lima.[10]

Like other imperial powers Spain maintained its claim to regulate access to its American markets. Given the scale of the task it established clear priorities with precedence given to protecting the bullion trade, which was organised in supposedly annual fleets from Seville, and, after

1680, Cadiz: the *flota* which collected the Mexican silver from Vera Cruz and the *galeones* which collected the Peruvian silver at Portobello.[11] In theory, all trade in manufactured goods was confined to the outward fleets and occasional licensed ships, and carried on Spanish account. In practice by 1680 around 95% of the manufactured goods on board the fleets was of foreign production, and a high, although unknown, proportion was loaded on foreign account with, or without, cover of a Spanish merchant (a straw man).[12] Despite high transaction costs (above all, fees and bribes) the fleet trade could be very profitable, but was subject to growing delays and slow turnover as the supposedly annual fleets became increasingly irregular.[13] Between 1660 and 1700, on average, only one *flota* sailed every two years, and one *galeones* every three.[14] Furthermore, the English were losing market share to the French, who used their growing influence at the Spanish court to secure commercial concessions, and who also supplied the most desirable goods. Cargoes were dominated by textiles (accounting for almost 90% of non-Spanish goods in 1686), and lightweight fabrics were in especially high demand. The French excelled in the production of both linens and silks, while England struggled to compete in both. England did have a long-standing competitive advantage in woollen cloths, but by the late seventeenth century France was challenging here too. In 1686 an estimated 40% of non-Spanish goods on the monopoly fleets were French, whereas just under 20% were English.[15]

As English merchants faced increasing costs and competition in the Cadiz trade, the acquisition of Jamaica provided a new opportunity to develop a more direct route to Spanish colonial markets, as the Dutch had done at Curacao.[16] Transactions costs in direct trade were reported to be half what they were with the fleets, and trade proved easy outside the tightly regulated bullion routes, although markets were small and easily glutted.[17] It was more difficult and dangerous to penetrate the most commercially attractive monopoly staging posts at Cartagena, Portobello, Vera Cruz, and Havana, but even this could be done under cover of the slave trade. On account of the agreement at Tordesillas in 1494, Africa lay in Portugal's half of the globe. Spain allowed this one trade to fall into foreign hands, and never attempted itself to supply its colonies, even with slaves.[18] Licenses were granted to foreigners, making a mockery of the attempt to maintain a commercial monopoly, as the legal access to major ports provided easy cover for a spectacularly lucrative illicit trade in manufactured goods. As a result the slave *asiento* was seen as the richest trade in the Indies, and as soon as Jamaica fell into English hands the conquerors took steps to secure a share of this coveted commerce.[19] In 1662 the slave contract was renewed and its holders permitted to collect slaves from any nation not at war with Spain. Thus the *asientists* became 'straw men' for the Dutch and other suppliers. In 1663 England's newly formed Royal African Company obtained a contract to supply the *asiento* with 3,000 slaves per year from Jamaica, and after various difficulties the trade was placed on a solid footing in the 1670s. By the 1680s the holders of the *asiento* maintained a permanent agent in Jamaica who shipped out between 1,000 and 2,000 slaves a year, or half the island's deliveries.[20] Prices in the Spanish colonies were two or three times as high as those on the island, and the small group of investors who controlled the trade saw it as a 'much better way of making money than making sugar'.[21] By 1686 the island was reputed to have bullion exports amounting to between £100,000 and £150,000, more or less matching sugar exports valued at £172,000.[22]

As Jamaica underwent three decades of rapid and largely uncontested growth, in which it secured the 'governing share' of illicit commerce in the Caribbean, some established traders with Old Spain resented what Defoe later called a 'thieving, rogueing trade' which merely took money out of one pocket to put in another.[23] Others had high hopes that the new route would expand England's flagging share of Spanish American commerce. However, in the 1690s the island began to face unwelcome competition.[24] Until that decade the French settlers who had established control of western Hispaniola in the 1660s had largely concentrated on plunder and hunting, but as the Cadiz trade was disrupted by war the buccaneer colony was encouraged to take a new interest in more peaceful business.[25] Although the peace of 1697 secured an undertaking that Charles would not leave his crown to the French heir, the treaty recognised the French claim to hold western Hispaniola, or Saint-Domingue. With growing French influence at the Spanish court, fears arose of further commercial concessions.[26]

The fears were soon realised. From 1699 the *Company de Cacheu*, the Portuguese firm which had held the *asiento*, turned to subcontractors in Saint-Domingue in addition to its established suppliers in Jamaica and Curacao.[27] Equally if not more alarming, French merchants, above all those of St Malo with capital to spare after the peace ended privateering, began to develop a new trade route in competition with the Caribbean. In August 1698 a leading merchant of St Malo joined with a Paris merchant in forming a company for trade in the Pacific, or the South Seas. In December Jacques de Beauschesne embarked on a voyage to sail through the Straits of Magellan and into the South Seas, previously little-explored by north Europeans other than pirates.[28] De Beauschesne used his country's friendship with Spain as cover for a new and spectacularly profitable trade. As there had been no fleet at Portobello since 1696 Peru was starved of goods, and merchants welcomed direct supplies from Europe. De Beauschesne was reported to have made a gross mark up of 750%, but even so his prices were a third of those demanded by merchants in the Caribbean, and clearly threatened to undermine their business. Tales that de Beauschesne returned to France in 1701 with goods worth £25 million were no doubt fanciful, but certainly raised English envy and emulation, and completely changed the shape of what was seen as possible in Spanish American trade.[29]

It emerged after the death of Charles II in 1700 that he had defaulted on his promise to divide his empire, and had nominated Louis XIV's grandson, Philip of Anjou, as his heir. The commercial consequences were soon apparent.[30] In 1701 the new king ended the slave contract held by the Portuguese *Company de Cacheu*, which had spread its custom between Jamaica, Curacao, and Saint-Domingue, and awarded an exclusive ten-year contract to a French Guinea company newly formed under the leadership of Ducasse, a former privateer. Furthermore, the new company was given permission to despatch returns direct to France – a concession never before allowed to any foreigner.[31] In August that year three French ships were permitted to sail into the South Seas under pretence of providing naval protection. In fact they went to build on the earlier steps to establish a new direct trade to Peru, which would undermine traditional Caribbean commerce.[32] It was clear to all observers, especially the leading Jamaica merchants whose own interests were threatened, that 'a nation [the French] whom we least expected in trade had become apprized of the importance and advantages of the Spanish West Indian commerce [and] used all their skill and interest to ingross it mostly to themselves'. France seemed poised to monopolise the inexhaustible treasure of the Spanish Indies, which would put it in a position to effect 'the conquest of the world'.[33]

In 1700 William III had been disappointed to find that he faced strong resistance to taking action against France. However, France's provocative behaviour pushed commercial interests to combine forces with those who had dynastic grounds for discontent over Philip's succession. Ten days after the French slave *asiento* was signed, London and Amsterdam formed an alliance for joint military and naval operations. It was intended to counter the threat that France and Spain, 'so closely united and cemented', would destroy the 'free intercourse of navigation and commerce which England and Holland have in the Indies and other parts'.[34] War began in 1702 with wholehearted support from

English traders to Old and New Spain, who united behind the grand aim of removing the Bourbon from the Spanish throne and installing the Habsburg pretender, and so preventing Spain's American empire falling into France's lap. They also articulated narrower commercial aims: to divert the Cadiz trade from French to English merchants; to evict the French from Hispaniola and so reduce competition in the contraband trade; to prevent the French trading in the South Seas; and to remove the *asiento* from French hands. The material ambitions underpinning the mercantilist drive to capture overseas markets were, of course, cloaked in the language of liberty for all of Europe.

England fully expected that war against Spain in the West Indies would more than pay for itself. War with Spain would offer England 'ample recompense, price and reward' for its heroic defence of its 'own liberty and the liberty of Europe in general', as it would be able to grab Spanish treasure, and in this Jamaica's strategic location could be put to profitable use.[35] Despite repeated disappointments, the expectation lingered from the days of Elizabeth, fuelled by tales of the audacious exploits of Drake and other national heroes.[36] In fact war in the Indies fell far short of paying for itself. Apart from the massive expense of convoying merchant ships across the Atlantic, the naval presence maintained at Jamaica cost around £100,000 a year.[37] This was a source of great profit to certain private individuals, above all contractors including Gilbert Heathcote, the island agent who remitted £12,000 a year for the use of the navy and charged 18% for his services.[38] However, there were no great prizes for the state. In 1701 Benbow, with ten Ships of the Line, tracked the *flota's* movements with care, but lost his life after engaging Ducasse, now a Spanish protector, and the fleet got away. Although in 1702 the ships were captured and destroyed, at Vigo most of the treasure had been unloaded and ironically, as Henry Kamen noted, there was a bonanza for the Spanish Crown, which was able to appropriate illegal, as well as legal, bullion shipments.[39] Furthermore, the damage to Spain's naval capabilities meant that the fleet system was heavily undermined, and the threat of repetition meant that there was only one *galeones* and three *flotas* for the duration of the war. Wager attacked the *galeones* off Cartagena in 1708. One warship was captured, from which Wager himself reportedly made £50,000, and another was sunk; much of its treasure rests at the bottom of the sea today. The remainder of the fleet got home safely with 80 million pieces of eight (approaching £20 million).[40] Other prizes were taken, but overall the state's plunder proved disappointing. As John McNeill has stressed, war in the West Indies was costly in men and money, and was far from self-financing.[41]

Meanwhile war damaged commerce. English merchants were forced to withdraw from Cadiz in 1703, and that branch of trade collapsed, but figures for exports suggest that direct trade with Jamaica did not expand to fill the gap (Figure 4.1).[42] At the outset trade with the 'enemy' Spanish colonies was forbidden, which provoked protests from leading Jamaica merchants, led by Heathcote, who complained that they were losing out to the Dutch, who continued trading. The fall of Barcelona and the declaration of the Habsburg pretender as king in 1704 provided a pretext for lifting the ban. The Jamaican governor was instructed to make friendly overtures to his Spanish counterparts, now branded as 'allies', and to allow trade to resume. Renewed commercial activity on the Spanish colonial coast ensued, before the arrival of the *galeones* in 1706.[43] However, Jamaica's contraband trade was undermined both by vigorous French competition and the activities of its own privateers.

French merchants used the alliance with Spain to expand trade in both the Caribbean and the Pacific. As indicated above, after the damage at Vigo, and with the continued threat of enemy attack, the fleet

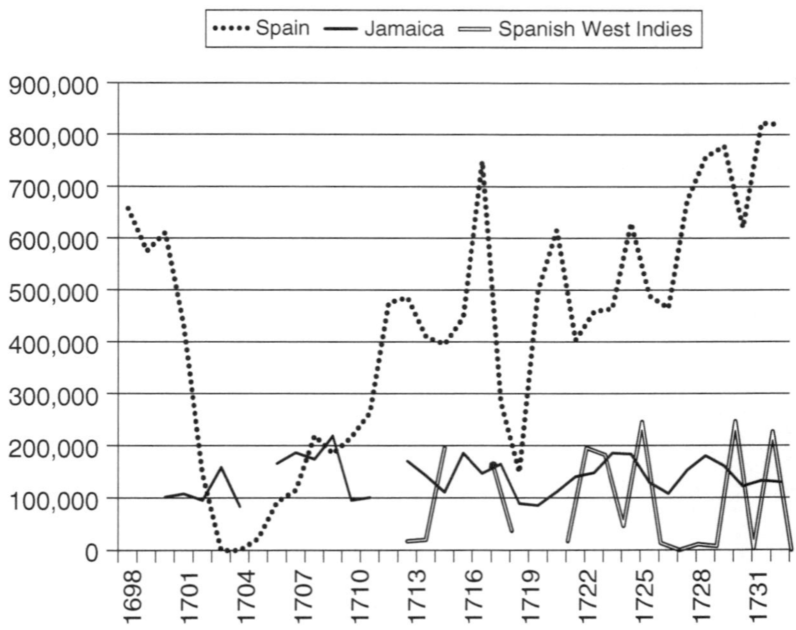

Figure 4.1 Exports of manufactures from England and Wales 1698–1733
Source: TNA CUST 3/1-32, 68, 82; Figures for 1705 and 1711 are missing but can be found for Spain in Cambridge University Library, CH (h), Paper 89/11.

system was curtailed. Spanish-American markets were starved of both European goods and transport home for their own produce. Spain's French allies took ample advantage of the opportunity. French merchants were given unprecedented concessions, and allowed to trade direct with the Spanish Indies. In May 1707 the Spanish merchants at Seville estimated that since the beginning of the war 30 French ships had traded to the ports of Campeche and Vera Cruz, and over 86 to the ports of Tierra Firme. As Kamen showed, these figures were not exaggerated, and the regular supplies led to complaints of overstocked markets and low prices.[44] To the further disquiet of Caribbean traders, the competing Pacific trade also expanded. Between 1704 and 1706, 17 ships left Saint Malo for the Pacific.[45] This trade received additional encouragement after 1707 with the arrival of the Francophile Marquis de Castelldosrius as Viceroy in Lima. He saw the contraband trade as a way to enrich himself, and allowed it to expand while taking a 25% cut for himself. Even with this burden the prices were a third, or even half, those charged at the Caribbean fairs, and further undermined the weak fleet system.[46] In 1709 the French were reported to have handed 30 million pesos of South Seas bullion (approaching £7.5 million) to their mint.[47] The English were bitterly aware that such treasure replenished France's war chest, and that if similar funds could be channelled in their own direction they would make a substantial dent in their own enormous war debt of £9 million.[48]

England's inability to expand its share of Spanish-American wealth by direct trade was compounded by the metropole's inability to control its agents on a distant periphery. On the outbreak of hostilities the governor of Jamaica followed the usual practice of issuing commissions of war to any 'with a reputation as honest men and gave security'.[49] The privateers claimed more prizes than the navy and brought in 30 of the 43 ships condemned in 1703, including eight with Spanish colours.[50] After 1704, when the initial ban on trade with the Spanish colonies was lifted, the governor followed orders from home and issued instructions that no privateers should 'meddle' with the Spaniards unless they were carrying French goods or contraband, but this had little impact on the ground (or the sea).[51] Jamaican privateers routinely attacked Jamaican or British ships which carried Spanish goods, and metropolitan merchants such as Heathcote accused the governor (like many of his predecessors) of colluding in this business to the detriment of peaceful commerce.[52] Furthermore, although the men-of-war stationed at Jamaica had instructions to protect the island's trade, they demanded prohibitive fees for providing convoys to the Spanish coast, and showed little interest in chasing privateers of any nationality who, as a rule, offered only 'dry business'.[53] Hovenden Walker, who took command of the naval station at the end of the war, was accused of treating Jamaican trading ships in the same way as privateers, as he wanted to limit competition for his own private trade in slaves and provisions.[54] The navy may well have been a 'branch of business', but the business it did was not always conducted on the state's behalf.

By 1710, with mounting expense and massive debt, most Englishmen were exhausted by war. Whig magnates with an interest in Jamaica, including Heathcote, continued to insist that there could be 'no peace without Spain', by which they meant open access to Spanish-American markets. However, with the downfall of the Whig ministry and the rise of Harley and the Tories, a resolute move was made to curb further expenditure, and negotiations were begun which led to the Peace of Utrecht in 1713.[55] Of course, government supporters presented the Treaty as a great victory, but apart from having incurred a huge debt, the British fell far short of realising their war aims, and secured little advantage in the American trade, which had played a central role in precipitating the conflict. Philip retained the Spanish Crown, and as a result of family bonds France continued to have privileged access to the American empire without the heavy administration and defence costs of direct ownership. British merchants were readmitted to trade at Cadiz on similar terms to the French, but took some years to recover pre-war levels of trade, and certainly did not 'crowd out' competition in the Americas, as suggested by Nettels. In fact the French increased their share, and did not lose their dominance until the late eighteenth century, with the damage and disruption caused by revolution.[56] As argued by Allen in his contemporary analysis, it was competitive pricing, or 'better pennyworths', which opened markets, not forced contracts or diplomatic agreements.[57]

Despite Britain's declared war aims, France stood firm on the retention of Saint-Domingue, which continued to be used for backdoor trade with Spanish markets, where French merchants had established solid networks. Furthermore, France developed a new route into Mexico via Louisiana. The French also retained, and even strengthened, their commerce in the Pacific until the mid-1720s, and its demise is usually attributed to shifting business practices in France and the rise of smuggling via Buenos Aires, rather than the Succession War.[58] Meanwhile British smuggling, as reflected in exports to Jamaica, showed sluggish growth: they remained well behind those to Cadiz (Figure 4.1).[59]

After 1707 the English had harboured hopes that they would be able to imitate French success in the Pacific.[60] Discussions with Archduke

Charles produced promises that England could hold 'security ports' in Spanish America, including two on the west coast. This possibility of a fortified trade, as conducted in the East Indies, underpinned much of the initial enthusiasm for promoting a South Sea Company (SSC).[61] The new venture was designed to provide a Tory counterweight to the powerful Whig institutions (the Bank of England and the United East India Company) which had dominated public credit for almost two decades, and had allowed the shareholders to engross massive profits from war. The new Tory South Sea Company, chartered in 1711, aimed to secure its share of the public-credit pie by exchanging the floating debt of over £9 million for company stock, on promise of five per cent interest (to be paid by the Treasury), and had the additional sweetener of privileged access to the 'infinite treasures' of Spanish America via the promised Pacific 'security ports'.[62] By the time Philip, rather than the Archduke, made peace he had no need to make these concessions, which would have alienated the French and the Dutch, as well as his own Spanish citizens, and instead offered a lesser prize – the slave *asiento*.[63] This contract had long held a grip on the popular imagination, and needless to say the Tories presented their disappointment as a triumph, but when the contract was offered to the South Sea Company in substitution for privileged access to the Pacific, it was accepted only after some heart-searching.[64]

The Treaty required the Company to provide Spanish America with an annual supply of 4,800 *piezas de indias* (around 6,000 slaves) on much the same terms as had been given to the French company, although for a longer period of 30 years.[65] Around 70% of the Company's trade went via Jamaica, where SSC agents organised the sorting and refreshment of slaves before sending them to Spanish factories.[66] This was not forcing open a new trade, as suggested by Nettels, as the English at Jamaica had been openly supplying the *asientistas* (including the French) with slaves since the 1660s, and in 1707 the Spanish slave trade was estimated to absorb half the island's annual supplies.[67] What was new was the heavy costs of administering the contract and dealing directly with the Spanish crown, which were reflected in the difference between the wholesale price of a slave in Jamaica (100 pesos) and the average price in one of the Spanish factories (204 pesos), shown in Table 4.1. Transport and maintenance costs from Jamaica were minimal, ranging from seven to thirty pesos per head. The massive mark-up between Jamaica and the Spanish coast was largely associated with dead-weight costs (rents) associated with the monopoly, including a per capita tax of 33.5 pesos per head paid to the Spanish Crown, the maintenance of staff at the

Table 4.1 First trading period of SSC slave introductions and returns, 1714–18

	Number of slaves	*Number sold*	*Gross*	*Average price, pesos*
Cartagena	1,588	1,295	274,778	190
Havana	1,479	1,246	307,437	207
Portobello	3,767	3,671	925,216	242
Vera Cruz	523	389	93,052	199

Source: Sorsby, Victoria: 'British trade with Spanish America under the Asiento', PhD thesis, University of London, 1975, pp. 279–81.

factories in America, tax in Jamaica of £1 (four pesos) per slave, presents, fees, and bribes to Spanish officials, and the Jamaican agents' commission of ten per cent.

Agents' profits were further enhanced by the easy opportunities for illicit trade in dry goods. Only SSC ships carrying slaves, and no others, had a right to be in Spanish ports and coastal waters, but a handful of slaves could be used as cover for a ship to enter a Spanish port with a full cargo of dry goods or flour. For example, the *Benjamin* left Jamaica with 30 slaves and £12,000 in illicit wares. However, while Company directors and employees had opportunities for personal enrichment, the shareholders saw little profit.[68] The disappointing returns no doubt help to explain why it fell far short of the contracted number, despite supplying, according to Palmer, 74,760 slaves to Spanish America between 1714 and 1739, and accounting for a third of Britain's recorded deliveries to America.[69]

It was allowed at Utrecht that the slave business was a 'losing trade', which made it tempting to use it as cover for smuggling activity. Thus, a preventative sweetener was conceded with permission to send an annual ship laden with 500 tons of manufactured goods (later increased to 650 tons to compensate for a delayed start) which could be sold tax-free at the fairs held on arrival of the *galleons* at Portobello or the *flota* at Vera Cruz. The celebrated annual ships proved almost as disappointing for the shareholders as did the slave trade. The first permission ships (the *Bedford* and the *Elizabeth*) were fitted out in 1715, and a further seven sailed before 1739.[70] Over the entire period they carried goods valued at £2,101,487, or an average annual value of £50,000, and probably as much again in contraband. Individual ships recorded massive profits, but overall, with delays, confiscations, administrative problems, and disputed payments, the company admitted paltry returns amounting to £6,000 per annum, and its trade remained well behind that to both Old Spain and the West Indies (Table 4.2).[71] The SSC's privileges did not

Table 4.2 Exports from England and Wales, 1715–1729, five-year average (£)

To:	*Spain*	*West Indies*	*South Sea Co.*
1715–19	407,660	430,000	120,851
1720–24	534,333	471,000	158,326
1725–29	633,610	473,000	58,070

Source: TNA Cust 3/17–30.

translate into a greatly increased share of the Spanish-American trade, and crowd out others. It may even have constrained British growth as the dead-weight costs of trading within the monopoly reduced flexibility and competitiveness.

Although the South Sea Company's trade did not generate great profits for any but its directors and employees, it did damage Jamaican planters and contraband traders.[72] Company agents had large purchasing power, and bid up island prices. In 1715, after the *asiento* was settled, one observer reported that slave prices soared to 140–160 pesos (£35–40) per head.[73] This peak was not maintained, but the available statistics show that between 1715 and 1719 slaves fetched 25% more in Jamaica (£22.7) than in Barbados (£18.3), although transport costs were only ten per cent higher.[74] As the governor proclaimed in 1717, 'planting is the mother of trade and negroes the support of planting', and islanders complained that the SSC was draining labour away from the island and raising prices.[75] The relatively high price of slaves may help to explain the sluggish growth of Jamaica's sugar exports in the 20 years after the Treaty of Utrecht, which fell far short of the rapid growth in the 20 years before the Glorious Revolution.[76] It has also been suggested that it fell behind the performance of French Saint-Domingue, where, after 1713, planting proceeded apace.

High prices for both dry goods and slaves also raised costs in Jamaica's long-established contraband commerce, as well as increasing risks, as governors became less inclined to 'wink at their proceedings'.[77] Many who had pursued careers as privateers during the long period of war, but might have made a living through contraband after the peace, in fact resorted to government employment as coastguards, which as smuggling escalated were in high demand on both sides of the imperial line.[78] Violence flourished, and almost continuous informal war earned the period its reputation as the 'golden age of pirates', albeit broken by formal warfare in 1718–20, 1727–9, and 1739–48.[79] Assessing the impact of the Peace of Utrecht and the working of the *asiento* treaty,

Lilian Penson concluded that 'the phantom of a legal trade deflected colonial enterprise and in the end brought it to ruin'.[80]

The settlement at Utrecht allowed the SSC's directors and employees to make substantial profits, but as Davenant forcefully remarked in his critique of Whig corruption, private interests should not be conflated with public gain.[81] Recent literature has emphasised the importance of the British navy in extending empire and defending commerce, but during the War of the Spanish Succession the navy proved unable to deliver a mortal blow to French competitors in either the Caribbean or the Pacific.[82] If anything it helped it to flourish by discourage the traditional fleets. Distance made it difficult for the state to control the behaviour of its agents at the periphery; governors, privateers, smugglers, and naval captains pursued their own interests and undermined imperial strategies, as they continued to do in peacetime. It is difficult to support Nettels's view that the Peace 'brought the advantages for which the English had gone to war'. The war did not achieve its principal aims of removing the Bourbon king and neutralising French influence, nor did it achieve any of the more particular commercial aims designed to increase England's share of Spanish-American trade. It did not 'crowd out' the French from the trade in Cadiz, nor constrain French competition in American markets, and it did not 'force open' new trades. Nor did it promote the best use of Jamaica as either entrepôt or plantation. Although the English failed to learn the lesson, and continued for a hundred years to interrupt their constructive commercial expansion with debilitating conflicts, the outcome of the War of the Spanish Succession in Jamaica demonstrated clearly the merit of Cobden's belief that more than military might was needed to increase the size of the commercial cake.

Notes

1. Pares, Richard: *War and trade in the West Indies, 1739–1763*, Oxford: Oxford University Press, 1936, p. viii.
2. O'Brien, P. K.: 'Inseparable connections: trade, economy, fiscal state and the expansion of empire, 1688–1815, in Marshall, P.: *Oxford history of the British Empire, vol. 2: the eighteenth century*, Oxford: Oxford University Press, 1998, pp. 53–77.
3. Nettels, Curtis: 'England and the Spanish-American trade, 1660–1715', *The Journal of Modern History*, Vol. 3, No. 1 (1931), pp. 1–32. For dissenting voices see Penson, L.: 'The West Indies and the Spanish American trade, 1713–1748' in Holland Rose, J, Newton, A.P, and Benians, E.A.

(eds): *Cambridge History of the British Empire, vol. 1*, Cambridge: Cambridge University Press, 1929, p. 30.

4. National Library of Jamaica, MST 390, Letter to Nottingham, Mar. 1689.
5. Estimates of bullion production are tabled in Stein, Stanley J. and Stein, Barbara H.: *Silver, trade and war: Spain and America in the making of early modern Europe*, Baltimore: Johns Hopkins Press, 2000, p. 24. Estimates of sugar production are provided by Deerr, Noel: *History of sugar*, 2 vols, London: Chapman and Hall, 1949–50.
6. For a contemporary account of Jamaica's role in Spanish American trade see Allen, Robert: *An essay on the nature and methods of carrying on a trade to the South-Sea* (1712), pp. 18–22.
7. On the business of privateering see Zahedieh, Nuala: '"A frugal, prudential and hopeful trade": privateering in Jamaica, 1655–89', *Journal of Imperial and Colonial History*, Vol. 18 (1990), pp. 145–68.
8. Ibid. The contribution to knowledge is reflected in the popular pirate literature of the late seventeenth century, including the diaries of Bartholomew Sharpe and William Dampier.
9. Osorio calculated that the fleets supplied enough goods for one million people, whereas there was effective demand from six million people of distinction. Stein and Stein, *Silver, trade and war*, pp. 97–9.
10. Walker, Tamara: '"He outfitted his family in notable decency": slavery, honour and dress in eighteenth century Lima, Peru', *Slavery and Abolition*, Vol. 30 (2009), pp. 383–402.
11. Haring, C.H.: *Trade and navigation between Spain and the Indies in the time of the Habsburgs* (1947), pp. 201–30. In 1702 Joseph de Veita Linage's *Norte de la Contratacon de las Indias Occidentales* (1672), the key book for understanding the operation of Spanish commerce, was translated into English as *The Spanish rule of trade in the West Indies.*
12. Cadiz's exports to the colonies were estimated to be worth £1.5 million per annum in the late seventeenth century. McLachlan, Jean O.: *Trade and peace with Old Spain, 1667–1750: A study of the influence of commerce on Anglo-Spanish diplomacy in the first half of the eighteenth century*, Cambridge: Cambridge University Press, 1940.
13. Critics of the fleet administration claimed that this was a deliberate strategy designed to starve colonial markets and raise prices, but in fact it was counterproductive in serving to encourage contraband trading, which had the effect of stocking markets. Stein and Stein, *Silver, trade and war*, pp. 97–9.
14. Lynch, John: *Spain under the Habsburgs*, 2 vols, Oxford: Oxford University Press, second ed., 1981, II, p. 208.
15. Stein and Stein: *Silver, trade and war*, pp. 71, 72, 78.
16. Cary, John: *Essay on the state of trade*, Bristol, 1695; Nettels, *England and the Spanish American trade*; Zahedieh, Nuala: 'The merchants of Port Royal, Jamaica, and the Spanish contraband trade, 1655–1692', *William and Mary Quarterly*, Vol. 63 (1986), pp. 570–93.
17. TNA CO 138/3, fo. 407, Letter from Beckford to Williamson, 6 Dec. 1675.
18. Walker, Geoffrey J.: *Spanish politics and imperial trade 1700–1789*, London: Macmillan Press, 1979, p. 12. The Treaty of Tordesillas divided the world into Spanish and Portuguese spheres, with the line drawn down the middle of the Atlantic. Africa was in the Portuguese sphere, and Spain did

not attempt settlement, as this would have undermined its own claims to monopoly.

19. A detailed account of the slave *asiento* is provided in Scelle, George: *La traite Negriere aux Indes de Castille,* two vols, Paris, 1906. Portuguese contractors provided slaves until the system collapsed when war broke out between the two nations in the 1640s. The *asiento* was renewed in 1662 and held by a series of merchant partnerships.
20. Osborne J.F.: 'James Castillo – asiento agent', *Jamaican Historical Review,* vol. 8 (1971), pp. 9–18; Zahedieh, *Merchants of Port Royal,* p. 390.
21. Somerset Record Office, Heneage MSS DD/WHh 1089, John to William Helyar, 16 Sept. 1688.
22. Inchiquin to Lords of Trade, 12 Aug., TNA CO 138/7, fo. 19; 'An estimate of what value is shipt every year from Jamaica to England', TNA CO 138/10, fo. 76.
23. Defoe, who had himself been interested in Spanish trade, described those who interfered with the established channel of commerce to America as 'a Crew of Peace-Breakers, Thieves and Pyratical Traders (for there are pyracies in trade) and deserve the gallows', *Mercator,* No. 174, 3 July 1714; *Mercator,* No. 171, 26 June 1714. Defoe argued that the traditional routing of goods to America via Cadiz was far better than the clandestine trade via Jamaica, which did not increase commerce, but merely changed its channels. Defoe, Daniel: *The interests of the several princes and states of Europe consider'd* , London, 1698. Defoe, Daniel: *A plan of the English commerce,* 1927, p. 244. For commentary on Defoe's views, see Novak, Maximilian E: 'Colonel Jack's 'Thieving Roguing' trade to Mexico and Defoe's attack on economic individualism', *Huntingdon Library Quarterly,* Vol. 24, No. 4 (1961), pp. 349–53.
24. National Library of Jamaica, MST 390, Letter to Nottingham, Mar. 1689.
25. Pritchard, *In search of empire: the French in the Americas, 1670–1730,* Cambridge: Cambridge University Press, 2004, pp. 362–3.
26. Walker, *Spanish politics,* p. 19.
27. *Calendar of State Papers, Colonial, 1689–91.*, No. 835, 839. In 1695, Cary noted that the *asiento* 'hath for some time stood on Tiptoe, ready to waft itself to another island'. Cary, *Essay on Trade,* p. 77. In 1699 an English observer at Portobello noted the arrival of three ships belonging to the French Guinea Company. TNA CO 137/5, No. 8, Extract of a letter from Portobello of 8 Oct. 1699 about the French trading with the Spaniards.
28. Jean Baptiste de Gennes, a naval officer, had already attempted a similar trading voyage to the Pacific in 1695–7. Although this voyage ended in total failure, it provided valuable information about the difficulties of navigating in that region. Dunmore, John: *French explorers in the Pacific: I, the eighteenth century,* Oxford: Oxford University Press, 1965, pp. 10–11. For a full account of de Beauschesne's expedition see Dahlgren, E.W: *Les relations commerciales et maritimes entre la France et les côtes de l'océan Pacifique: commencement du XVIIIe siècle,* Paris: Librairie Ancienne, 1909, pp. 123–46.
29. Allen, *Essay on trade in the South Sea,* p. 23.
30. TNA CO138/10, fo. 148, Letter from Sir William Beeston, 13 April 1700.
31. A copy of the agreement as signed by Ducasse in 1701 appears in the *Daily Courant,* 10–13 Dec., 1712, TNA CO 388/15, No. 95.
32. Allen, *Essay on trade in the South Sea,* p. 32. 'The French have quite altered the channels of that trade – that they already have totally discouraged the

trade by way of Portobello and Panama and consequently from Old Spain and Jamaica.'

33. Ibid, p. 22.
34. Nettels, *England and Spanish American trade*, p. 17; TNA CO 324/30, p. 60 'An account of France in Spanish America', Hedges to Handasyd, 17 Jan. 1706.
35. James Drake quoted in Bourne, Ruth: *Queen Anne's navy in the West Indies*, New Haven: Yale University Press, 1939, p. 20.
36. *Proposals for carrying on an Effectual War in America against the French and Spaniards*: London, 1702, p. 8.
37. A fleet of four to seven ships was stationed at the island throughout the war. Nettels, *England and Spanish American trade*, p. 27. In 1708–11 the Crown spent £1,934,000 on warships assigned to the colonies. Approximately 30% of this amount went to Jamaica alone. TNA CO 390/5, No. 46, 'Estimate...'
38. TNA CO 137/7, No. 22, Capt. Gardner to Lords of Trade, 10 Apr. 1706. Sperling, J: 'The international payments mechanism in the seventeenth and eighteenth centuries', *Economic History Review*, Vol. 14 (1962), pp. 446–68; Nettels, Curtis: *The money supply of the American colonies before 1720*, Maddison: University of Wisconsin Press, 1934, p. 38; Paul, Helen: *The South Sea Bubble: an economic history of its origins and consequences*, New York: Routledge, 2011.
39. Kamen, Henry: *The War of Succession in Spain 1700–1715*, London: Weidenfeld and Nicolson, 1969, p. 179.
40. Phillips, Carla Rahn: *The treasure of San Jose: death at sea in the War of the Spanish Succession*, Baltimore: Johns Hopkins Press, 2007.
41. McNeill, J.R.: *Mosquito empires: ecology and war in the Greater Caribbean, 1620–1914*, Cambridge: Cambridge University Press, 2010, pp. 144–9.
42. TNA Cust 3/6–8.
43. In August Gilbert Heathcote, as island agent in London, petitioned the Lords of Trade for permission to trade on the Spanish coasts, as otherwise the trade would be engrossed by the Dutch. The fall of Barcelona in September gave a pretext for lifting the ban. Carlos III was declared king and the Jamaican governor was instructed to make friendly overtures to his Spanish counterparts, or allies, and allow trade to resume. TNA CO 324/8, fos. 265, 268–70, 271, 340, 370–1, 374. TNA CO 137/45, No. 76, Handasyd to Governor of Cartagena, 31 Mar. 1706.
44. Kamen, *War of Succession*, pp. 144–8.
45. 'If there can be a method found out to prevent the French trading to Lima and the South Seas, trade here will soon be in a flourishing condition'. TNA CO 138/12, fo. 336, Handasyd to Board of Trade, 20 July 1708.
46. Legitimate goods from Portobello would sell at two or three times the price of those brought into Peru illegally by French ships. Walker, *Spanish politics*, pp. 34–5, 42. Allen reported that the merchants of Peru expressed unwillingness to go to the fair at Portobello unless relieved of tax, Allen, *Essay upon Trade*, p. 25.
47. Stein and Stein, *Silver, trade, and war*, p. 113. Between 1703 and 1715 Malouin merchants may have brought to French ports as much as 99 million pesos, and in the process doubled France's monetary stock. Lespagnol, Andre: *Messieurs de Saint-Malo: une elite négociante au temps de Louis XIV*, Saint-Malo: Presses Universitaires de Rennes, 1991.

48. 'This is so well known that in the news papers we have from time to time had it published how many millions they have brought and are daily bringing from the South Seas'. Allen, *Essay on trade in the South Sea,* p. 23.
49. TNA CO 137/6, No. 54, Handasyd to Lords of Trade, 19 June 1704.
50. TNA CO 137/6, No. 24 (vii), Handasyd to Lords of Trade, 19 June 1704.
51. TNA CO 324/8, fos. 437–9, Instructions to privateers, 4 May 1704.
52. TNA CO 137/7, No. 58, 'Account of the ill practice of Jamaican privateers.' Received from Gilbert Heathcote by Board of Trade, 31 July 1707. In 1708 Parliament took new measures to prevent attacks on Spanish ships with the *Act to Encourage Trade in America,* which created a zone along the Spanish coast from the River Chagre to Rio de la Hacha and reaching five leagues into the sea. In this area English privateers were not to molest Spanish vessels, even though the latter carried materials of war. The Act was not easily enforced, as the governor complained: 'to the prejudice of a trade much more beneficial to this nation than whatever can be gained by privateering in those parts'. Nettels, *England and Spanish American trade,* p. 25.
53. Kerr demanded more than the merchants would pay, and allowed ships to set out alone, resulting in losses above 40,000 pesos. The case came before the House of Commons. TNA CO 389/19, p. 221, Handasyd to Board of Trade, 29 Aug. 1707; CO 137/7, No. 70, Memo from Mr Wood, 13 Jan. 1708. Nettels, *England and Spanish American trade,* p. 27.
54. TNA CO 137/9, fo. 79, Jamaica merchants to Board of Trade, 3 Jan. 1713; CO 137/10, No. 13, 14, 19, Hamilton to Board of Trade, 3 Jan. 5 Mar. 1713, 26 Apr. 1715.
55. Heathcote, who was also an MP and twice governor of the Bank of England, joined three other Bank Whigs in lobbying the Queen against peace, and threatened to withdraw Bank funds if it went ahead. The threat backfired and the Tory ministry concluded a peace at Utrecht.
56. A census of 1751 highlighted continued French strength, with them dominating the top tax bracket among merchants and accounting for around 50% of the fleet cargoes. England trailed behind with around 20% of the trade and income. McLachlan, *Trade and peace,* pp. 24–5; Pearce, Adrian: *British trade with Spanish America, 1763–1808,* Liverpool: Liverpool University Press, 2007, pp. 7–9.
57. Allen, *Essay on Trade,* p. 34.
58. Dahlgren lists names of 62 French ships which departed for the South Seas between 1713 and 1724, Dahlgren, M.E.W.: *Voyages Francais a destination de la Mer du Sud, 1695–1749,* Paris: Imprimerie Nationale, 1907.
59. TNA CO 388/17, No. 87, Richard Harris to Board of Trade, received 3 Jan. 1715.
60. Against this background several proposals promoted Spanish–American trade – including on the isthmus of Darien. Stein and Stein, *Silver, trade, and war,* pp. 131–2.
61. A pamphlet of 1707 reflects that 'the benefits the French have found from it already are but shadows of what we may from this undertaking reasonably expect', *A Letter to Sir William Robinson in Relation to a Proposal for Trade in the Spanish West Indies,* London, 1707. In July 1711, as negotiations began for peace, the British demanded real securities in the form of four fortified ports in North and South America. To Whig disgust, this was abandoned in

September in return for other privileges, *The Assiento Contract Considered*, London, 1714.

62. Defoe, Daniel: *An Essay on the South Sea Trade with an Enquiry into the Ground and Reasons for the Present Dislike and Complaint against the Settlement of a South Sea Company*, London, 1712; Sperling, J.G.: *The South Sea Company: An Historical Essay and Bibliographical Finding List*, Boston: Harvard Graduate School of Business Administration, 1962.
63. In 1712, unable to secure trade bases in the Pacific through negotiations, the South Sea Company began to plan an expedition to the South Seas to establish a settlement with government support. BL Add MS 25,562, fos. 1–3.
64. Davies, K.G.: *The Royal African Company*, London: Longmans, Green & Co, 1957, pp. 133–52; 133–52; BL Add MS 25,562, fos. 4–6, 19.
65. The term was introduced in the mid-seventeenth century and referred to a prime male slave in good physical condition between 14 and 30 years of age and above 4 feet 8 inches tall. By the time the British *asiento* was signed this measurement was the standard used to measure a cargo of slaves. It was rare for the number of *piezas* to equal the number of slaves on the ship's manifest. Deductions were made for physical defects as well as for women, children and old people. King, James: 'Descriptive data on negro slaves in Spanish importation records and bills of sale', *The Journal of Negro History*, Vol. 28 (1943).
66. TNA CO 137/12, Pt. 1, No. 5, Hamilton to Stanhope, 12 June 1716; Ibid., No. 40, Letter from Governor of Havana, 8 Oct. 1716.
67. Report of Board of Trade to Her Majesty in consequence of a petition from the Royal African Company setting forth many inconveniences they labour under in that trade from the private traders. BL Add MS 14,034, fo. 108b.
68. According to a financial summary of the trade in slaves for the period between 1 May 1714 and 31 March 1721 the Company made a loss of £212,515 on the trade in negroes. Carmona, Salvator; Donosa, Rafael; and Walker, Stephen: 'Accounting, international relations and treaty verification: Britain, Spain and the *asiento*', *Accounting, Organizations and Society*, 2010, vol. 35, issue 2, p. 264. Palmer suggests that the Company was dissembling. On analysis of surviving records he suggests that the trade was very profitable although the money could not always be collected. Palmer, Colin: *Human cargoes: the British slave trade to Spanish America, 1700–1739*, Urbana: University of Illinois Press, 1981, p. 155.
69. Palmer, *Human cargoes*, p. 110; *Transatlantic Slave Trade Database*.
70. Walker, *Spanish politics*, p. 81.
71. Brown, V.L.: 'The South Sea Company and contraband trade', *American Historical Review*, Vol. 31 (1926), pp. 662–78. According to Nelson, between 1730 and 1739 goods shipped in the Company's slave ships and the annual ships had a total value of £5.5 million sterling (£550,000 a year), but not all were European-manufactured goods. They included large quantities of flour, meat, and naval stores from North America. Nelson, George: 'Contraband trade under the Asiento, 1730–1739', *American Historical Review*, Vol. 51 (1945), pp. 55–67.
72. In 1713 the Assembly complained that 'the island is now likely to become the meanest of all your majestie's colonies in America, since we have lost the benefit of sending dry goods, the manufactures of Great Britain, and

the produce of Your Majestie's northern colonies, as well as negroes to the subjects of Spain in America by the asiento lately settled'. McLachlan, *Trade and peace*, p. 62; Palmer, *Human cargoes*, pp. 65–8.

73. The importance of the Spanish trade was illustrated when it came to a sudden halt in 1713, and the price fell from £20 per head to £12 in three months. On resettlement of the *asiento*, prices soared. TNA CO 137/12, No. 12, Extract of letter from Mr Onslow, Aug. 1716.
74. Eltis, David and Richardson, David: 'Prices of African slaves newly arrived in the Americas, 1673–1865: new evidence on long-run trends and regional differentials', in Eltis, David; Lewis, Frank D; Sokoloff, Kenneth L. (eds), *Slavery in the development of the Americas,* Cambridge: Cambridge University Press, 2004, pp. 181–218.
75. TNA, CO 137/12, Part 2, fo. 311, Sir Nicholas Lawes to Lords of Trade, Nov. 1717.
76. Sheridan, Richard: *Sugar and slavery: an economic history of the British West Indies 1623–1775*, Barbados: Caribbean Universities Press, 1974, pp. 216–22.
77. *Some observations on the asiento trade*, p. 20.
78. 'It is very melancholy to think that the peace is re-established yet the effect and intent thereof is lost when lawful navigation is rendered so precarious'. Letter from Governor of Havana, 8 October, 1716, TNA CO 137/12, Part 1, 40 (i). Hamilton admits issuing commissions, Hamilton to Stanhope, 12 June 1716, TNA CO 137/12, Part 1, fos. 5–10.
79. Earle, Peter: *The pirate wars*, London: Methuen, 2003, ch. 9.
80. Penson, *West Indies and the Spanish American trade*, p. 331.
81. Whitworth, Charles: *The political and commercial works of that celebrated writer Charles D'Avenant*, 5 vols, London, 1771, vol. IV, p. 125.
82. O'Brien, *Inseparable connections*, pp. 73–4.

5

Baltimore and the French Atlantic: Empires, Commerce, and Identity in a Revolutionary Age, 1783–1798

Manuel Covo

In the aftermath of American independence, what kind of commercial relationships could exist between the French colonial empire and its new ally in the Americas – the United States? This question breaks away from traditional diplomatic history, which tends to regard the US and France as the nation-states they would become in the nineteenth century. Indeed, analysis of the connections between these revolutionary countries has almost always been presented within national frameworks. The US, however, was a vulnerable state entangled with European rivalries in the Caribbean, the economic engine of the time, in which French stakes were particularly high. After 1763 the French Empire mattered mostly thanks to the economic power of Saint-Domingue (now Haiti), whose sugar and coffee production led the world. But Saint-Domingue was no isolated colony. It was part and parcel of a truly French imperial project involving different areas, including the North American continent. In this perspective, what could be the interaction between the US, which was emerging as a postcolonial independent state, and a colonial empire centred around the Caribbean?[1]

To answer this question, we need not analyse the 1783 Paris conference again; it is also useless to examine Jefferson or Madison's correspondence, or to recount Toussaint Louverture's rise to power in Saint-Domingue. Instead, this problem can be understood through a different lens, an Atlantic lens – which underscores transnational circulations that were not determined by the so-called centres, but which happened at the so-called peripheries. To do so, this chapter focuses on a place which, at first sight, could seem counter-intuitive: Baltimore. On the one hand, this Maryland port city had been very much part of the North American colonial experience since its early days, but on the other, at the end of the eighteenth century it became the premier US port-of-entry for Caribbean commodities, making it a northern confine of a circum-Caribbean region.

Traditional historiography has tended to obscure this last point. Indeed, most of the local history of Baltimore connects it to the national destiny of the US. What drove most of those publications was a desire to understand how and to what extent Baltimore participated in the growth of the US as a nation. In that respect, Richard Smith Chew's recent dissertation defined the history of Baltimore in the Early Republic as one of transition from postcolonial dependence to national independence; from an Atlantic economy dominated by British commerce and credit to a domestic economy turned to the West and manufacturing. Chew complicated the usual trajectory (from colony to nation) by introducing the idea of an American postcoloniality. Even so, however, the nation still looms large.[2]

By breaking away from usual notions of territorial sovereignty we can change the point of view. What if Baltimore had not been a crucial location of the French Atlantic and the French Caribbean at the end of the eighteenth century? The hypothesis invites reflection on non-territorial polities which could transcend national boundaries, like diaspora. This also questions the specificity of a port-city that was not always the mirror of a 'national fate'. As Mark Peterson convincingly argues, 'individual British North American cities often had more features in common with their competitors and counterparts in the greater Atlantic world than they shared with their fellow cities in the "thirteen colonies".' That this statement could be extended to post-Independence US is a useful starting point to understand the complexity of diverse urban paths.[3]

During the revolutionary decades, flows of goods, the migration of people, and the circulation of ideas connected Baltimore to the French Atlantic, and integrated the North American port to a network of other nodal locations in the Caribbean. Although Baltimore was a less obvious candidate than Charleston, Philadelphia, or even Boston, the city found itself at the juncture of French imperial endeavours, merchant networks, and developing nation-states. First, this chapter will demonstrate that Versailles had its own colonial project towards the US during and after the revolutionary war. It will then explain why most of those global schemes had a completely unexpected local impact on Baltimore. Indeed, the French-Haitian Revolution and maritime war between France and Britain created tensions that tore Baltimore between imperial, national, and Caribbean forces.

The US within the French imperial formation

After the loss of Canada and Louisiana in 1763 France faced a huge challenge to provision and protect its booming colonies in the Antilles, especially Saint-Domingue. Just like most Caribbean colonies, they were intended to import commodities of primary necessity that were not produced on their soil, primarily timber, salted beef and fish, vegetables, wheat and flour, and so forth. The colonies were to be mostly devoted to capitalist agriculture and lucrative crops (sugar, coffee, indigo, cotton) which relied upon slavery. Since the time of Colbert, the system of the colonial *Exclusif* had granted a monopoly over colonial trade to the metropolis: European France was to feed its colonies exclusively, and to import the totality of sugar, coffee, and indigo they produced. Locally, however, smuggling had been tolerated, and provisional admission of foreign trade was always implemented in times of natural disasters such as hurricanes or earthquakes, and in times of war.[4]

After the Seven Years' War the French government did not think that this dual system of rigid law on the one hand, and broad permissiveness on the other, could stand any longer. After a fierce debate that involved colonists, chambers of commerce, and physiocrats (Quesnay and Mirabeau in particular), the Minister of the Navy accepted some relaxation of the system, by allowing the trade of non-strategic commodities in a few Caribbean ports. However, the American Revolutionary War, and, most importantly, the 1778 Treaty of Amity and Commerce between France and the US, changed the nature of a dialogue which previously had been conducted exclusively between the metropolis and her colonies. A great debate of political economy took place involving members of the State Council, journalists, philosophers including Denis Diderot, and American envoys to France, including Benjamin Franklin. One of the most important controversies pitted Anne Robert Jacques Turgot, the economist and *Contrôleur général des finances*, against Charles Gravier de Vergennes. They disagreed about the commercial impact on the French colonies of the creation of an independent US. Would it be the best move to counter British hegemonic ambitions in the Americas, and her predatory views on Saint-Domingue? Or would it create a new kind of metropolis on the continent which could become the new commercial master of all European colonies? Vergennes thought that dismantling the British empire would contain Albion, preventing her from conquering new territories in the Antilles. On the opposite side, Turgot feared that helping the US would spur independence movements in all the Americas, since most colonists would be willing to trade with this new power, and would eventually resort to violence if necessary. Both Vergennes and Turgot, however, shared a common belief in the centrality of trade, which was regarded as the core of a powerful navy, and of a powerful state in general. Everyone had in mind the expansion of French commercial power, without territorial gains.[5]

This purpose entailed two consequences: first, France was to snatch the American market from the British; second, French colonial trade was to be preserved without humiliating vital new allies. France could not ignore that North Americans would be rivals in its Caribbean trade, but the government imagined ways to achieve a balance between political, military, and economic interests. Through commerce, the US was to regard France as its new 'metropolis by adoption', as one memoir addressed to the *Bureau des colonies du ministère de la Marine et des Colonies* put it. In the case of a new war with Britain, France would need to have bases on the continent, as transatlantic connections were very likely to be sketchy at best. With the compromise reached by the 30 August 1784 decree which allowed foreign trade in major colonial ports in the Antilles, but enforced a strict monopoly on trade in sugar, coffee, and flour, the government intended to mark a strict boundary between legal and illegal exchange. Eventually this policy was meant to make the US a commercial satellite, and a defensive hinterland of Saint-Domingue.[6]

In the eyes of the US government, the trade restrictions were all the more disappointing because American vessels had been banned from the British West Indies by strict orders-in-council. But merchants were used to circumventing such legal barriers. Indeed, escaping European regulations in the Caribbean was the most common feature of American trade. That is why, at first, most shipowners and captains thought they would get back to business as usual – that is to say, carrying on the contraband trade. They did not mind employing fake sea letters, declaring false destinations, and circuits involving Dutch, Danish, or Swedish islands.[7]

Still, 1784 was not to be 1763, for two reasons: one, the French government was determined to enforce laws; second, France now had consuls in most North American ports. These French officials were first established during the American Revolution, and had to fulfil different maritime and economic goals. Their mission encompassed strengthening commercial links between France and the US and controlling the circulation of ships navigating to and from the French colonies. In particular, they wanted to make sure that brigs and schooners did not load contraband commodities in Martinique, Guadeloupe, and Saint-Domingue. That is why Versailles decided that no American vessel could enter a French port in the Antilles without a passport from the consul. Much information was to be written on the passport: the name, the size and height of the ship, the name of the captain, the number of sailors on board, and the nature of the cargo. This step was more rigid than all previous regulations since Louis XIV and Colbert's 1681 *Ordonnance maritime*. This sophistication of the regulation was another symptom of the 'identification revolution' of the eighteenth century. Moreover, metropolitan merchants were pushing for ever-more repression, as they had lost much in bad speculations on the market.[8]

In June 1786 the French consuls went public in American gazettes. They launched a wide communications campaign targeting French and American captains who were trading with the islands. The following text was published multiple times in Boston, New York, Philadelphia, Baltimore, and Charleston:

> **Notice is hereby given,**
>
> THAT Orders have been transmitted to the Administrators of the French West-India Islands, to refuse admittance to all foreign Vessels, which will not present a Passport from the Consul, Vice Consul, or Agent of his Most Christian Majesty, residing in the Port of their departure. In consequence thereof, all Owners or Masters of such Vessels, in the States of New-Jersey, New-York, and Connecticut, as are intended for the French West-India Islands, are requested to apply for Passports, to the Vice-Consul-General of France, residing in New-York.
>
> DE LA FOREST.
>
> ☞ All the Printers in the three above mentioned States are desired to publish this Advertisement.
>
> *New-York, June 12, 1786*

The registers of different consulates show that officers made sure that captains scrupulously complied with the regulations. Furthermore, some consuls showed zeal and imagination. Such was the case of Gaspard-Joseph-Amand Ducher, vice-consul in Portsmouth, who found that the ministerial initiative was 'an excellent idea', but that it entailed setting up more sophisticated procedures. In order to thwart the trafficking of falsified documents, he suggested designing a system of encrypted certifications. The consul in Charleston suggested that the French consuls send copies of American gazettes to the French officers in the colonies, so that the latter would be able to double-check what was declared at the customs against what was reported as declared entering US ports, which would shed light on discrepancies arising from smuggling. Requiring guarantee deposits to ensure the veracity of the information was also envisaged, and the French officers established new administrative procedures in Saint-Domingue. Thus the quartermaster in charge explained that to control the traffic

> he would send back to his office the movements of all ports. They would mention the dates of arrival and departure of each ship [and] of the quality and nature of its freight. Likewise the names of the captains and owners would be recorded. All those pieces of information would be notated and recorded into specific registers. Each vessel would have, so to speak, an account open so as to know scrupulously each and every movement. Each and every change of captain or owner was also mentioned.

The Saint-Dominguan journal *Affiches américaines* completed this package, stating precisely the movements of all foreign ships, and listing the passengers leaving the colony. The effort of the administration in this security procedure was considerable, and unheard of in its practice, more than in its form: from then on state officers enforced those rules. According to a merchant in Port-au-Prince, the trade had gone from 'bad to worse' because of 'the most obnoxious system formed in the colony'.[9]

The main consequence was that they outraged the American public over the commercial policy of France, provoking a real debate on the relevance of this legislation and its enforcement. The policy brought about a general outcry against what was deemed a betrayal of the American alliance, and an offence to the honour of the new nation. Some even called for 'remonstrating against this new procedure with that manly dignity and spirit which actuated [the Americans] while struggling for and finally hath given us, a rank among the nations of the earth'. This gendered comment suggested that the masculinity of the US was at stake. While the revolutionaries had fought against Britain to defend their economic independence, French regulation was regarded as an insult to the very identity of the new country. For many, liberty of commerce was at the core of what it meant to be American, and the French were not to imagine that they could abolish the natural law that had placed the Caribbean closer to the North American continent than to Europe.[10]

Confronting French regulations

Protest against the regulation was formulated in national terms. As Max Edling demonstrated, the need for a stronger commercial power fuelled the 1787 federal revolution. French policy, however, had very different effects at a local level. In Baltimore the policy was more cruelly felt than elsewhere in the US. The revolutionary war had played a significant role in the expansion of the port, which had remained, thus far, rather second-rank compared to Philadelphia, Boston, and Charleston. Even locally, Baltimore was not clearly ahead of Annapolis before the late 1770s. But new French connections contributed to enhancing the economic importance of the port. Three factors had already linked Baltimore to France and French trade: the tobacco trade with General Farmers in France; the provisioning of the French Navy which was stationed there in 1780; and, most important, the legalised West Indian trade during the maritime war, since Maryland was a great producer of flour. True, the Confederation suffered from an economic crisis in the post-war period, but Baltimore was even more badly hurt, because foreign flour was specifically forbidden entry into the French Antilles, whereas rice from Carolinas and salted fish from New England were authorised commodities. Baltimore had more obstacles to overcome than its northern or southern neighbours, and with only 13,000 inhabitants in 1790 it remained much smaller than Philadelphia and New York.[11]

To know exactly how important the trade with the French Antilles was for Baltimore is a thorny issue. One needs to reconstruct fragments of figures through records of customs, which leaves many holes.

Table 5.1 Vessels entered in Baltimore from the Caribbean, 1784–1785

Rank	*Colony*	*Empire*	*Ships*	*Burden (tons)*
1	St. Eustatius	Dutch	25	1,641
2	St. Domingue	French	24	1,552
3	Barbados	British	16	1,342
4	Jamaica	British	15	1,181
5	Curacao	Dutch	8	547
6	Dominique	British	8	502
7	St Croix	Danish	7	480
8	Antigua	British	6	339
9	Cayenne	French	1	230
10	St Thomas	Danish	3	193
11	Martinique	French	4	129
12	Guadeloupe	French	5	105
13	St Kitts	British	3	101
14	Bermuda	British	2	95
15	Tobago	French	1	70

Source: Data reconstructed through: National Archives and Records Administration, Mid-Atlantic Regional Branch, Philadelphia, RG 36, 1149, vol. 2.

Between September 1784 and October 1785, 33 small vessels officially went to French Saint-Domingue, roughly more than 20% of the Caribbean trade of the city. 34 vessels entered from the French Antilles in the same period.

One must not take these figures naively: most ships heading for neutral islands (Danish St Thomas, Dutch St Eustatius and Curacao) were in reality smuggling either with British or with French colonies or traders. French consuls kept warning colonial officers about this illegal trade, but they also sometimes preferred to turn a blind eye, so as not to compromise good relationships with the merchant community. The contradictory mission of improving French trade and repressing smuggling made their position extremely uncomfortable, as they were to gain the trust of the very people they had to control.[12]

The business community involved in the trade was very diverse. Beyond a few well-established merchants, such as Colonel Samuel Smith of John Smith & Sons, a large number of the smugglers were fresh immigrants arrived from France. The consul registered 28 'French people', either merchants or sea captains, and many of them from French ports or the Antilles. As the Saint-Dominguan market seemed glutted, some merchants thought that American trade would become a lucrative option. Several had arrived during the Revolutionary War, and had participated in the campaign. Such was the case of a Jewish merchant from Bayonne, Paul Bentalou, first established in Baltimore in 1780. Bentalou kept kinship connections with merchants in French Aquitaine, who in turn had major interests in the Antilles. Indeed, most merchants hoping to thrive on the periphery of the French empire established multi-centred networks. Zachary Coopman & Company, for instance, was established in both Cap-Français and Baltimore. Etienne Zacharie arrived from Lyon in the mid-1780s, and took up business on the mainland while his partner from Flanders, François Coopman, headed the firm in Saint-Domingue. It was a vulnerable and very risky trade, one which often led to bankruptcy, especially as the *intendant* in Saint-Domingue was repressing smuggling more than ever in the history of French colonies. Since the judicial tools of contract enforcement were not available to illicit traders, many merchants lost much to the hands of crooks. Bentalou's 1784 venture in Saint-Domingue, for example, was catastrophic, because his correspondent in Jamaica, Alexander Lindo, refused to recognise his debt to Bentalou for smuggling slaves into Saint-Domingue. As the metropolitan trade with the US had proved to be disappointing, the Caribbean trade was also regarded as a delusion.[13]

Table 5.2 American vessels admitted in Saint-Domingue, 1788

Rank	***Port***	***Vessels***
1	Philadelphia	59
2	Charleston	35
3	Newbury	34
4	New York	33
5	Boston	32
6	New London	31
7	Norfolk	27
8	Salem	25
9	Savannah	23
10	Baltimore	22
11	Middletown	21

Source: *Affiches américaines,* 1788.

Baltimore was not a leading port, and its contribution to US–Saint-Domingue trade remained rather small. By compiling data from a Saint-Dominguan newspaper, it is possible to identify the official origins of American vessels that entered the three major ports in the colony, Cap-Français, Port-au-Prince, and Aux Cayes.

Among American Atlantic ports Baltimore ranked only tenth. Philadelphia, the capital of Saint-Dominguan trade as Alec Dunn has demonstrated, New England, and the South were clearly ahead. The modest community of French smugglers, however, was burgeoning, and took advantage of the new environment of the 1790s. The Revolution in both France and Saint-Domingue, and the maritime war with Britain from 1793, transformed the scene completely, making Baltimore a crucial place in the French Atlantic, yet not in the way that had been expected by Versailles.[14]

Baltimore, revolution, and world war

The former French imperial designs crumbled piecemeal after 1789. Meanwhile another system was taking shape in the 'periphery', one which redefined the relationship between Baltimore and the French Atlantic on the one side, and between Baltimore and the US on the other. Different unforeseeable parameters elevated the port of Maryland to a centrality that it had not achieved before, and contributed to its rising economic power.

It has often been argued that the 1788 subsistence crisis in France started the French Revolution, but that it caused the breakdown of

French colonial regulation has almost never been mentioned. The crisis interrupted flour imports from the metropolis; the Supreme judicial Court in Bordeaux, the *Parlement de Guyenne*, even prevented ships loaded with flour from leaving the harbour, a decision which spurred massive turmoil in Saint-Domingue. To prevent famine, or at least the inflation of prices, colonial officers were forced to suspend prohibitive legislation on flour and exports of colonial produce. While French governors in the islands deemed this relaxation temporary, they had to extend the opening of the ports to foreign ships even further in August 1791, after the slave insurrection in Saint-Domingue commenced. These decisions were a boon for Baltimore, which was now permitted legally to export flour from its hinterland. The number of barrels of flour exported from Baltimore to the Caribbean jumped from 47,500 in 1786 to 88,540 in 1789, with the French Antilles representing more or less 50% of all the city's Caribbean trade. Thanks to revolutions in both France and Saint-Domingue, commerce now flourished in Baltimore.[15]

The port city also benefitted greatly from the naval war between France and Britain. The declaration of war in February 1793 de facto abolished French mercantilist regulations. At the same time British privateers cut transatlantic French trade from the Antilles – and henceforth competition. In spite of the harsh policies of both imperial powers towards neutral trade, this new context proved to be instrumental in the growth of American commerce. American merchants shipped Caribbean produce to Europe under neutral cover, and obtained a monopoly over the carrying trade in coffee and sugar. In this global framework an unpredictable event projected Baltimore to the role of a key port. The burning and destruction of Cap-Français led to a massive influx of refugees from Saint-Domingue into the city. As Bentalou explained, 'shipping, people, and wealth of that great mart, had poured into Baltimore'.

One of the reasons for this advance was that the French consuls made required commercial vessels arriving from Saint-Domingue to station in Baltimore for almost a year. French ships had been blocked in the colony by war, and had gathered in Cap-Français before the destruction of the port forced them to depart. The vessels which navigated between France and the colonies were five times larger than the American brigs that circulated in the Caribbean, and their crews much more numerous. Most importantly, this led to a great arrival of major traders, of captains, and of capital in the city. Some settled and founded new houses of commerce with English-speaking partners. They brought with them their financial and social capital, and their knowledge of Saint-Domingue, which were great assets for economic success. The most important commercial firm in Cap-Français, Foäche Morange & Co., which was funded by Stanislas Foäche, a major merchant in Le Havre, relocated to Baltimore in 1793. The firm was to re-export to Europe what it could garner from Foäche's plantations in Saint-Domingue. Many others sold coffee and sugar, and were willing to work for American firms as super-cargoes in the Caribbean.[16]

Baltimore's war economy flourished. According to a French merchant from Saint-Domingue, the market was a 'constant game in stocks and merchandise'. New semi-legal activities took place. Privateers sold their prize-vessels in Baltimore. Between 1793 and 1795, 24 vessels were sold, most of which came from the Antilles. Through the French consul in Baltimore, the French government signed huge contracts to provision both the metropolis and colonies. This new context took root in established networks, but transformed the status of the francophone merchant community in Baltimore, and its relationship with both the metropolis and the public authorities. Great merchants of Bordeaux and Nantes looked for business partners that would cover the trade to France with neutral flags, and were eager to invest capital in such houses. For example, former smuggler Bentalou became a partner in the Bordeaux firm Léon Changeur et fils. At the same time, thanks to his intimacy with François Moissonnier, the new republican consul, Bentalou even became an 'agent of the Republic', and had charge of the convoy which left Cap-Français for Baltimore in 1793. Many others contracted with the French legation to provision Saint-Domingue. Men deemed criminal under the monarchy in 1783 became models of republicanism in 1793, as Baltimore became a key location of the French presence in the Americas.[17]

Table 5.3 Contracts made by the French legation in the US with Baltimore firms, April–May 1794

Trading firm	*Cargo of flour (bls)*
Barney and Hollins	9,650
Oliver & Thompson	7,440
Samuel & John Smith	2,900
Paul Bentalou	1,500
George Stiles	1,250

Source: *CADN, Philadelphie, Consulat général*, 107.

Table 5.4 Ships entering US ports from Saint-Domingue (1792–1798)[19]

Years	Baltimore		New York		Philadelphia	
	Ships	Burden	Ships	Burden	Ships	Burden
1792	74	8,580	52	4,921	155	14,653
1793	153	26,248	84	10,518	169	n/a
1794	94	9,272	90	6,522	106	n/a
1795	156	13,523	104	n/a	228	n/a
1796	179	17,503	116	n/a	282	n/a
1797	199	19,076	117	n/a	203	n/a
1798	166	17,466	94	n/a	119	n/a

All this led to a boom in Baltimore's Caribbean trade. The city overtook Philadelphia in the course of the revolutionary wars to become the primary trading partner of Saint-Domingue. The trade of 1797 was nine times what it was in 1788 in terms of number vessel numbers alone. The growth in terms of value must have been much higher, due to enormous imports of coffee.[18]

The politicisation of commerce

The commercial growth of Baltimore had political stakes. On the one hand the interaction of institutional and social actors redefined the French empire beyond the national boundary. On the other, on a national scale, networks of trade collided with the polarisation of Federalists and Republican democrats, who defined themselves by fighting about the war between the superpowers of the day. The very independence of the US seemed endangered, and caused a series of controversies within the Chamber of Representatives and the Senate, but also on the street, through popular societies, and in print. In this respect the issue of American trade, and the redefining of relationships with Britain and France, were completely intertwined, and posed many questions: What was to be the attitude of the government towards the new French plenipotentiary, Edmond-Charles Genet, who disregarded American neutrality by fitting out privateers in US ports in the spring of 1793? What should be the reaction to offensive British orders-in-council which resulted in the plundering of more than 300 American vessels in the Caribbean? How should the US respond to the 19 November 1794 Jay Treaty, and rapprochement with the former metropolis? Was it legitimate to turn the nation's back on revolutionary France, because the French government restricted American trade after the ratification of its treaty with Britain? With the development of the revolution in Saint-Domingue, race and slavery complicated the controversy even more. All these contentious issues drew attention to merchants and commercial interests, but while the historiography tends to analyse the debate at the national level, the issue was framed differently according to the specificities of each city.[20]

The influx of refugees from Saint-Domingue was significant in all American ports from Boston to Savannah, but, as we have seen, its economic impact was nowhere more important than in Baltimore. The merchant community there firmly divided over the adequate response to the challenge posed by imperial rivalries in the Caribbean. The city's major shipowner, Samuel Smith, played a leading role as a Representative in Congress, but he did not lean towards Federalism, and even emerged as one of the leaders of the Democratic-Republicans in the mid-Atlantic states. He was instrumental in philanthropic federal help to the refugees from Saint-Domingue; he was one of the major suppliers of the French Antilles, and a close friend of Genet's consul in Baltimore. Smith organised a real political Francophile machinery around him, while most merchants of Baltimore supported the Federalist agenda. For this he was attacked in Baltimore journals by other traders who thought he did not defend the interest of the port properly.

In Congress, at different stages of the rising tensions with France, Smith tried his utmost to save the French alliance, as he had been outraged by 'that piratical nation... that King of Sea Robbers – that Leviathan, which aims at swallowing all that floats on the ocean – that monster, whose only law is power, and who neither respects the rights of nations nor the property of individuals!' His understanding of private, local, and national interests did not have Hamiltonian accents, although Smith viewed himself as a quintessential American merchant, one who thought trade, republicanism, and Francophilia were intimately interconnected. Still, when the 1798 Quasi-War started, he chose America against France, and Toussaint Louverture against the French Directory.[21]

This was not the case with another so-called national hero of Baltimore: Joshua Barney. The 'tensions of scales' which pulled the city in multiple directions defined Barney's trajectory – which somehow embodied all the contradictions that Baltimore was going through in those times of crisis. According to the Maryland Historical Society Museum, Barney was a hero of the American Revolution and of the War of 1812. His achievements as a privateer captain in the 1770s, and his abnegation to defend Washington at Bladensburg, against all odds, made him a hero in 1812. In public history Barney is the main character connecting Baltimore to the glorious cause of the nation. Yet there is a serious gap in this biography: the 1780s and the 1790s.

The truth is that his life does not fit this teleological trajectory. During this period Barney first established a house of commerce in Baltimore. After American independence he organised a few adventures in the Caribbean from Baltimore, smuggling cargoes and slaves. While doing business in the Antilles he found himself in the middle of the revolution in Saint-Domingue and in Cap-Français in 1793. It was he who almost rescued Leger-Felicité Sonthonax, the French envoy who abolished slavery on the island in August 1793. Barney became an intimate friend of the French Jacobin, and struck contracts with colonial administration to provision Saint-Domingue from Baltimore. Back in the city, he even expanded his trade by obtaining new contracts to supply metropolitan France with flour. Barney was both a slave-trader and a supplier of Jacobin France, which abolished slavery on 4 February 1794.[22]

But his economic mission to France entailed a political one. In September 1794 he went to Paris with the new American envoy, James Monroe, and he himself presented the Union of two flags to the National Convention. From then on he became involved in the French Revolution, was named *officier premier* in the French navy, and went back to Saint-Domingue in 1796 as head of the colonial station there, all the while lending money to the French legation through his Baltimore network. He even became a naturalised citizen of the French Republic. In the Caribbean he was to provision the island again, and went back to the US in September, but was attacked by Federalists who accused him of national betrayal. He answered to this criticism in a newspaper by asserting his loyalty to the American revolutionary cause.

> however faithfully I may execute the orders of the government of France, whose ships I have the honour to command ; yet my private interest has not the least weight in my conduct [...] Much has been said lately, in certain N. York and Philadelphia papers of insults, piracies, robberies &c. The whole of which I regard as the venom of a party in this country, who never dared to show their faces in the glorious revolution of 1776, the greatest part of whom were actually fighting against America.

In 1798 and 1799, during the Quasi-War, the national hero from Baltimore joined the French forces and preyed on American ships, thinking he remained true to his interest and convictions by choosing France, which according to him had remained loyal to the spirit of 1776, whereas the Federalists had abandoned it. Barney never perceived a contradiction between his commercial activities, his staunch Anglophobia, his privateering in favour of the French Navy, and his patriotism. His understanding of the national identity was clearly embedded in an Atlantic republicanism that was now challenged by a more exclusive conception of the nation. Barney was an offspring of a port-city that was a receptacle of different yet connected revolutions in the US, France, and Saint-Domingue, a scene of imperial rivalry and a commercial confine of the northern Caribbean.[23]

The identity of an Atlantic port in the Americas at the end of the eighteenth century was complex. It would, of course, be absurd to make Baltimore a French city; French influence faded as quickly as it had appeared, and New Orleans became the great US repository of Frenchness. But this chapter has demonstrated that Baltimore's history cannot be reduced to a teleological narrative surging towards national integration. In the 1810s, Baltimore was to be involved in the Spanish Atlantic as its numerous privateers played an important role in the South American revolutions. At the same time, links with foreign people and foreign sovereign states were not always signs of neocolonial dependence. True, the French imperial government had designed a reorganisation of French colonial commerce in the US, a project which faced many obstacles. The course of war and revolution went far beyond everything the French government had envisioned beforehand. Somehow unexpectedly, Baltimore became a crucial nodal point in this respect, and came to be a place of local, national, and international tensions produced by the circulation of commodities, merchants, administrators, privateers, refugees, and soldiers from the Caribbean and Europe.[24]

The case of Baltimore illuminates shifting processes that were in many ways unpredictable. Its location within the national framework of the US changed in this period. Whereas Charleston is usually regarded as the ideal-type of a Caribbean port on the Southern Atlantic coast, Baltimore was 'Caribbeanised' during the revolutionary era. The big bang created by the revolution and war in Saint-Domingue, by far the wealthiest colony in the Caribbean, redistributed the cards of each and every port in the region. We may wonder to what extent this commercial reorientation transformed the urban culture of Baltimore, but we can already safely say that it shaped the political climate in which the merchant community tried to make sense of its national trajectory. This example calls for a greater attention to the geographic imagination of the historic actors – a sometime volatile and evanescent geography that questions the obviousness of rigid and anachronistic frameworks.

Notes

1. On the importance of Saint-Domingue at the end of the eighteenth century, see: Geggus, D.P. and Fiering, N., (eds): *The world of the Haitian Revolution*, Bloomington: Indiana University Press, 2009. On the diplomacy of the American Revolution in the classical historiography: Albert, P.J. and Hoffman, R. (eds): *Peace and the peacemakers: the treaty of 1783*, Charlottesville: University Press of Virginia, 1986. On the transformation of the French empire in the aftermath of the Seven Years' War, see: Reinert, S. and Røge, P.: *The political economy of empire in the early modern world*, Basingstoke: Palgrave Macmillan, 2013. On Toussaint Louverture, see, for instance: James, C.L.R.: *The black jacobins*, New York: the Dial Press, 1938; Pluchon, P.: *Toussaint Louverture: un révolutionnaire noir d'Ancien régime*, Paris: Fayard, 1989. For a reflection on Frenchness in a colonial context, see: Vidal, C., (ed.): *Français? La nation en débat entre colonies et métropole (XVIe–XIXe siècle)*, Paris: EHESS éditions, 2014.
2. Baltimore was a recent city that started to grow in the 1750s only. Browne, G.L.: *Baltimore in the nation, 1789–1861*, Chapel Hill: University of North Carolina Press, 1980; Chew, R.S., 'The measure of independence: from the American Revolution to the market revolution in the Mid-Atlantic', PhD diss., The College of William and Mary, 2002. For a discussion on the relevance of the postcolonial to understand the history of the early US, see: Armitage, David, 'From colonial history to postcolonial history: a turn too far?', *The William and Mary Quarterly*, Vol. 64, No. 2 (2007), pp. 251–4.
3. On non-territorial forms of polities, see: Mancke, Elizabeth, 'Polity formation and Atlantic political narratives', in Canny, N. and Morgan, P. D. (eds), *Oxford handbook of the Atlantic World, c. 1450–1820*, Oxford: Oxford University Press, 2011, pp. 382–99. On the diasporic nature of merchant networks in the Atlantic, see for instance, Hancock, David, 'The trouble with networks: managing the Scots' early-modern Madeira trade', *Business History Review*, Vol. 79, No. 3 (2005), pp. 467–91; Kagan, R.L. and Morgan, P.D., (eds): *Atlantic diasporas: Jews, Conversos, and crypto-Jews in the age of mercantilism, 1500–1800*, Baltimore: Johns Hopkins University Press, 2009. On the American Revolution from the point of view of major cities and merchants, see, in particular: Tyler, J.W.: *Smugglers and patriots: Boston merchants and the advent of the American Revolution*, Boston: Northeastern University Press, 1985; Doerflinger, T.: *A vigorous spirit of enterprise: merchants and economic development in revolutionary Philadelphia*, Chapel Hill: University of North Carolina Press, 1986; Matson, C.: *Merchants and empire: trading in colonial New York*, Baltimore: Johns Hopkins University Press, 1998. The diplomatic resolution of the American Revolution is viewed by Jack Rakove as a tale of three cities: Rakove, J.: *Revolutionaries: a new history of the invention of America*, Boston: Hougton Mifflin Harcourt, 2010. For an overview in recent scholarship: Carp, B.: *Rebels rising: cities and the American Revolution*, Oxford: Oxford University Press, 2007. The quote is from Peterson, M.A., 'The war in the cities', in Kamensky, J., (ed.), *The Oxford handbook of the American Revolution*, Oxford: Oxford University Press, 2013, pp. 194–215, p. 195. To show this, Peterson analyses the history of Boston by emphasising the ancient and strong connections between the city, France, and the French West Indies: Peterson, Mark: 'Boston à l'heure française: religion, culture et commerce à l'époque des révolutions atlantiques', *Annales historiques de la Révolution française*, No. 363 (2011), pp. 7–31.

4. The 1717 and 1727 *lettres patentes* were the foundational legal bases of the French *Exclusif*. On the commercial legislation and its discontents, Mims, S.L.: *Colbert's West India policy*, London: H. Frowde, 1912; Frostin, C.: *Les révoltes blanches à Saint-Domingue aux XVIIe et XVIIIe siècles: Haïti avant 1789*, Rennes: Presses universitaires de Rennes, 2008. On smuggling between New England and French Antilles, Goebel, D.B.: 'The "New England trade" and the French West Indies, 1763–1774: a study in trade policies', *The William and Mary Quarterly*, Vol. 20, No. 3 (1963), pp. 332–72; Truxes, T. M.: *Defying empire: trading with the enemy in colonial New York*, New Haven: Yale University Press, 2008.
5. On the debates about colonial trade at the end of the eighteenth century in France, Tarrade J.: *Le commerce colonial de la France à la fin de l'Ancien Régime: l'évolution du régime de l'exclusif de 1763 à 1789*, Paris: Presses universitaires de France, 1972; Cheney, P.: *Revolutionary commerce: globalization and the French Monarchy*, Cambridge: Harvard University Press, 2010; Fraser Terjanian, A.: *Commerce and its discontents in eighteenth-century French political thought*, New York: Cambridge University Press, 2013. On the Americans' intervention in the debate, Nussbaum, F.L.: 'The French colonial arrêt of 1784', *The South Atlantic Quarterly*, No. 27 (1928), pp. 62–78. On Vergennes' political economy, see Cheney, Paul: 'A false dawn for Enlightenment cosmopolitanism? Franco-American trade during the American war of Independence', *The William and Mary Quarterly*, 2006, Vol. 63, No. 3, pp. 463–88; Turgot, A.R.J., *Mémoire sur les colonies américaines, sur leurs relations politiques avec leurs métropoles et sur la manière dont la France et l'Espagne ont dû envisager les suites de l'indépendance des États-Unis de l'Amérique, par feu M. Turgot*, Paris: impr. de Du Pont, député de Nemours, 1791.
6. The quote is from Archives nationales d'outre-mer (ANOM), col. F2b8, 'Mémoire sur le commerce entre les isles françoises et les Etats-Unis de l'Amérique', fo. 222. At least 54 memoirs debated the issue in the wake of American independence.
7. On the division of the British empire, the exclusion of the US from the West Indies, and the debate that followed, Bell, Herbert C.: 'British commercial policy in the West Indies, 1783–93', *The English Historical Review*, Vol. 31, No. 123 (1916), pp. 429–41; O'Shaughnessy, Andrew Jackson: *An Empire divided: the American Revolution and the British Caribbean*, Philadelphia: University of Pennsylvania Press, 2000; Sadosky, Leonard: 'Reimagining the British Empire and America in an Age of Revolution: The Case of William Eden', in Sadosky, L., O'Shaughnessy, A. J., and Onuf, P. (eds): *Old World, New World: America and Europe in the age of Jefferson*, Charlottesville: University of Virginia Press, 2010, pp. 83–104. Stephen Girard, who smuggled sugar, coffee, and flour between Philadelphia and Saint-Domingue, was rather confident that his activities would remain unchallenged. See McMaster, John Bach: *The life and times of Stephen Girard*, two vols., London: J.B. Lippincott Company, 1918, I, pp. 122–65; Albert J. Gares: 'Stephen Girard's West Indian Trade 1789–1812', *The Pennsylvania Magazine of History and Biography*, Vol. 72, No. 4 (1942), pp. 311–42. On smuggling in the Caribbean, Pérotin-Dumon, Anne: 'Cabotage, contraband and corsairs: the port cities of Guadeloupe and their inhabitants (1650–1800)', in Knight, F.W., and Liss, P. K. (eds), *Atlantic port cities: economy, culture, and society in the Atlantic world, 1650–1850*, Knoxville: University of Tennessee Press, 1991, pp. 58–86; Klooster, Wim (ed.), *Illicit riches: Dutch trade in the Caribbean, 1648–1795*, Leiden: KITLV Press, 1998; Salvucci, Linda K: 'Atlantic

intersections: early American commerce and the rise of the Spanish West Indies (Cuba)', *The Business History Review*, Vol. 79, No. 4 (2005), pp. 781–809; Klooster, W.: 'Inter-imperial smuggling in the Americas, 1600–1800', in Bailyn B., and Denault, P. L. (eds), *Soundings in Atlantic history: latent structures and intellectual currents, 1500–1830*, Cambridge: Harvard University Press, 2009, pp. 141–80.

8. On French consuls in the United States, Mézin, A.: *Les consuls de France au siècle des Lumières: 1715–1792*, Paris: Ministère des affaires étrangères, 1997; Bégaud, S., Belissa, M., and Visser, J.: *Aux origines d'une alliance improbable: le réseau consulaire français aux États-Unis, 1776–1815*, Paris: Direction des archives, Ministère des affaires étrangères (2005). ANOM, col., B 192, fos 59–61; Mathieu, J.: *Le commerce entre la Nouvelle-France et les Antilles au XVIIIe siècle*, Montréal: Fides, 1981, pp. 42–5; Torpey, J.: *The invention of the passport: surveillance, citizenship and the state*, Cambridge, New York: Cambridge University Press, 2000; Denis, V.: *Une histoire de l'identité: France, 1715–1815*, Seyssel: Champ Vallon, 2008. On merchants' disappointment with the American trade: Marzagalli, Silvia: 'The failure of a transatlantic alliance? Franco-American trade, 1783–1815', *History of European Ideas*, Vol. 34, No. 4 (2008), pp. 456–64.
9. The consuls' article was published for instance in: *Columbian Herald*, No. 184 (12 June 1786) p. 3; *New Haven Chronicle*, Vol. I, No. 10 (20 June 1786). For Ducher's project and the suggestions of the consul in Charleston: Ministère des Affaires étrangères (MAE), correspondance consulaire et commerciale, Portsmouth, 1786, fo. 25; Archives nationales de France (AN), AE BI 372, 1786, fo. 216. On official bookkeeping in Saint-Domingue, ANOM, C9a 157, 1787; Historical Society of Pennsylvania, Claude Unger Collection, Dutilh & Wachsmuth, Box 9, to Etienne Dutilh, 28 June 1787.
10. *Cumberland Gazette*, 24 August 1786, p. 2. On the culture of honour in the early American Republic: Freeman, J.B.: *Affairs of honor: national politics in the new republic*, New Haven: Yale University Press, 2002. The natural proximity of the continent and the islands was clearly expressed by John Adams in a famous letter he wrote about the disappointing attitude of European powers towards the United States in 1783. See Taylor, R.J.and Lint, G. L. (eds): *Papers of John Adams, June 1783–January 1784*, Cambridge: Harvard University Press, 2010, p. 49.
11. On the role of commercial regulation in the 1787 revolution, Edling, M.: *A revolution in favor of government: origins of the U.S. constitution and the making of the American state*, Oxford, New York: Oxford University Press, 2008. On the economic importance of Annapolis in Maryland, Papenfuse, E.C.: *In pursuit of profit: the Annapolis merchants in the era of the American Revolution, 1763–1805*, Baltimore: Johns Hopkins University Press, 1975. On the commercial impact of the war in the US, especially in Maryland and on the economic relationship to France, Sullivan, K.: *Maryland and France, 1774–1789*, Philadelphia: University of Pennsylvania, 1936, pp. 65–77; Crowl, P.A., *Maryland during and after the Revolution, a political and economic Study*, Baltimore: Johns Hopkins Press, 1943; Price, J.M.: *France and the Chesapeake, a history of the French tobacco monopoly, 1674–1791, and of its relationship to the British and American tobacco trades*, Ann Arbor: University of Michigan Press, 1973; Gilbert, G.: *Baltimore's flour trade to the Caribbean, 1750–1815*, New York: Garland, 1986; Hunter, B.: 'Wheat, war, and the American economy during the Age of Revolution', *The William and Mary Quarterly*, Vol. 62, No. 3 (2005), pp. 505–26. On the economic depression that occurred during

the Confederation period, Bjork, G. C.: *Stagnation and growth in the American economy, 1784–1792*, New York, London: Garland, 1985.

12. AN, F2b8, 'Mémoire sur la contrebande', 1787. In this perspective, the consul's attitude can be compared with the various compromises taken by colonial officials at the 'periphery of empire'. See: Koot C.: *Empire at the periphery: British colonists, Anglo-Dutch trade, and the development of the British Atlantic, 1621–1713*, New York: New York University Press, 2011.
13. Bentalou's diary contains a wealth of information about an itinerary full of ups-and-downs: Maryland Historical Society, Ms 125. See also: MHS, Ms 1152 'John Smith & Sons letterbooks'. On the French community in Baltimore, Centre des archives diplomatiques de Nantes (CADN), Baltimore, 111. For a theoretical approach of multi-centred and self-organized networks, Hancock, D.: 'L'émergence d'une économie de réseau (1640–1815)', *Annales. Histoire, Sciences Sociales*, Vol. 58, No. 3 (2003), pp. 649–72. On the state of the job market in Saint-Domingue at the time, Covo, M.: 'I, François B.: merchant, Protestant and refugee – a tale of failure in the Atlantic world', *French History*, Vol. 25, No. 1 (2011), pp. 69–88.
14. Sources: *Affiches américaines*, 1788. On the Saint-Domingue trade in Philadelphia, see: Dun, J.A.: '"What avenues of commerce, will you, Americans, not explore!": commercial Philadelphia's vantage onto the early Haitian revolution', *The William and Mary Quarterly*, Vol. 62, No. 3 (2005), pp. 473–504.
15. Tarrade, *Le commerce colonial de la France*, II, pp. 706–11; Gilbert, *Baltimore's flour trade*, p. 86.
16. On trade regulations in the course of the war, Clauder, A.C.: *American commerce as affected by the wars of the French revolution and Napoleon, 1793–1812*, Clifton: A. M. Kelley, 1932. On the expansion of the American carrying trade, Marzagalli, Silvia.: 'Guerre et commerce: le rôle des neutres dans les relations entre l'espace antillais et l'Europe occidentale au tournant du XVIIIe siècle', in Butel, P. and Lavallé, B. (eds): *L'espace Caraïbe: théâtre et enjeu des luttes impériales, XVIe–XIXe siècle*, Bordeaux: Maison des Pays ibériques, 1996, pp. 219–37; Marzagalli, Silvia: 'Establishing transatlantic trade networks in time of war: Bordeaux and the United States, 1793–1815', *Business History Review*, Vol. 79, No. 4 (2005), pp. 811–44. Bentalou's quote, MHS, Ms 125, August 1793, fo. 16. On the influx of Saint-Dominguan refugees and the political stakes of the battle in Cap-Français, White A.: *Encountering revolution: Haiti and the making of the early republic*, Baltimore: John Hopkins University Press, 2010; Popkin, J.D.: *You are all free: the Haitian revolution and the abolition of slavery*, Cambridge: Cambridge University Press, 2010; on the major traders in Saint-Domingue: Begouën Demeaux, M.: *Mémorial d'une famille du Havre: 1737–1806, II: Stanislas Foäche*, Paris: Larose, 1951.
17. AN, 505 Mi 82, Morange to Foäche, 6 October 1795; CADN, Philadelphie, consulat général, p. 107.
18. Craig McDonald, Michelle: 'The chance of the moment: coffee and the new West Indies commodities trade', *The William and Mary Quarterly*, Vol. 62, No. 3 (2005), pp. 441–72.
19. Sources for Baltimore, NARA, Mid-Atlantic Regional Branch, RG 36, 1149, Vols 6–7; for New York, NARA, New York regional Branch, RG 36, 903–4; New York Historical Society (NYHS), United Insurance Company Papers; for Philadelphia, Dun, *What avenues will you Americans....*

20. DeConde, A.: *Entangling alliance: politics & diplomacy under George Washington*, Durham: Duke University Press, 1958; Ammon, H.: *The Genet mission*, New York: Norton, 1973; Elkins, S. and McKitrick, E.L.: *The age of federalism*, New York: Oxford University Press, 1993, esp. pp. 381–96; Waldstreicher D.: *In the midst of perpetual fetes: the making of American nationalism, 1776–1820*, Chapel Hill: University of North Carolina Press, 1997; Newman, S. P.: *Parades and the politics of the street: festive culture in the early American republic*, Philadelphia: University of Pennsylvania Press, 1999; Estes, T.: *The Jay treaty debate, public opinion, and the evolution of early American political culture*, Amherst: University of Massachusetts Press, 2006; White, *Encountering revolution*, 142–3, 154–65; Young, Christopher J.: 'Connecting the president and the people: Washington's neutrality, Genet's challenge, and Hamilton's fight for public support', Vol. 31, No. 3 *Journal of the Early Republic* (2011), pp. 435–66; Cotlar, S.: *Tom Paine's America: the rise and fall of transatlantic radicalism in the early republic*, Charlottesville: University of Virginia Press, 2011.
21. *Annals of Congress*, 3d. Cong., p. 536, 27 March 1794. Smith had not been a supporter of Madison's pro-French regulation, however. He changed his mind only in 1794. On Smith's political leverage, see Cassell, F.: *Merchant congressman in the young republic: Samuel Smith of Maryland, 1752–1839*, Madison: University of Wisconsin Press, 1971, pp. 64–72, 85–6.
22. These biographical elements are compiled from a hagiographic historiography that tends to downplay the French part of Barney's life in order to glorify him as a national hero. Weller, M. I.: *Commodore Joshua Barney: the hero of the battle of Bladensburg*, Washington, 1911; Adams, W.F.: *Commodore Joshua Barney: many interesting facts connected with the life of Commodore Joshua Barney, hero of the United States Navy, 1776–1812*, Springfield: Priv. print, 1912; Delahaye Paine, R.: *Joshua Barney, a forgotten hero of blue water*, New York, London: The Century Co., 1924; Norton, L.A.: *Joshua Barney: hero of the Revolution and 1812*, Annapolis: Naval Institute Press, 2000. More specific documents on the relationship between Sonthonax and Barney are found in archival materials: AN, DXXV 11, file 102; 64, file 640.
23. ANOM, CC9a 14, the civil commission to the ministry of the Navy, 18 vendémiaire year V (9 October 1796); CADN, Philadelphie, consulat général, 107, Létombe to Barney, 1 pluviôse year V (20 January 1797); *Daily Telegraph*, 26 January 1797; *Federal gazette*, 22 December 1797. Barney is a quintessential Atlantic go-between who can be compared with other brokers of the kind: Racine K.: *Francisco de Miranda: a transatlantic life in the age of revolution*, Wilmington: Scholarly Resources, 2003; Safier, N.: 'A courier between empires. Hipolito da Costa and the Atlantic world', in Bailyn and Denault, *Soundings in Atlantic history*, pp. 265–93. Barney was also one of these 'republicans vagabonds', as Vanessa Mongey phrased it to describe privateer captains involved in South-American revolutions in the 1810s–1830s. Mongey, V.: 'Les vagabonds de la république: les révolutionnaires européens aux Amériques, 1780–1820', in Morelli, F., Thibaud, C. and Verdo, G. (eds), *Les empires atlantiques des Lumières au libéralisme, 1763–1865*, Rennes: Presses universitaires de Rennes, Réseau des universités de l'Ouest Atlantique, 2009, pp. 67–82.
24. On French Louisiana in an Atlantic context, Vidal, C. (ed.): *Louisiana: crossroads of the Atlantic World*, Philadelphia: University of Pennsylvania Press,

2013. On privateers from Baltimore in South America: Griffin, Charles C.: 'Privateering from Baltimore during the Spanish American wars of independence', *Maryland Historical Magazine*, Vol. 35, No. 1 (1940), pp. 1–25; Head, David: 'Baltimore seafarers, privateering, and the South American revolutions, 1816–1820', *Maryland Historical Magazine*, Vol. 105 (2008), pp. 269–93; Arlyck, Kevin: 'Plaintiffs v. privateers: litigation and foreign affairs in the federal courts, 1816–1822', *Law and History Review*, Vol. 30, No. 1 (2012), pp. 245–78; Head, David, 'New nations, new connections: Spanish American privateering from the United States and the development of Atlantic relations', *Early American Studies*, No. 11 (2013), pp. 161–75.

6
Modernity and the Demise of the Dutch Atlantic, 1650–1914

Gert Oostindie

It has become somewhat of a cliché to emphasise that the emergence of an integrated Atlantic space in the early modern period can be properly understood only by taking trans-imperial connections seriously. The emergence of an Atlantic World is seen as a crucial formative phase in the wider process of globalisation that would accelerate in the nineteenth and twentieth centuries. Yet words should be chosen carefully. Surely the early modern process of 'globalisation' did not result in anything like a perfectly integrated market for capital and labour, let alone a commonwealth of cultures and values. Much of the emergence of an Atlantic World took place in the context of state policies intended to hold a mercantilist grip on as large a share as possible of the newly explored overseas territories and trade routes, and minimise the impact of outsiders. But, as Jan de Vries suggested, if one European player presented characteristics presaging later processes of globalisation, it was the Dutch Republic. 'Perhaps the Dutch experience of the seventeenth and eighteenth centuries', he wrote, 'offers a test of the viability of the concept of an "Atlantic World". More than other European states, the Dutch Republic was prepared to participate in an international economy with colonies populated by a polyglot population of settlers. "Prematurely" modern, the Dutch launched multinational enterprises before the age of mercantilism had ended.' In de Vries' view, the Dutch were far more cosmopolitan than the rest of the Atlantic World in which they operated, and their state was increasingly weak – and yet they 'found and exploited weak spots in the armour of mercantilism whereby their economy might prosper'.[1] But, as this chapter will show, much in the early modern Dutch Atlantic World was neither particularly progressive, nor flexible, nor successful, despite a general academic understatement of its importance.

Much of this chapter supports de Vries' hypothesis, underlining the flexibility of Dutch state institutions, as well as private business and individuals operating in the Dutch orbit. The expansion of Dutch colonies and networks was indeed facilitated by relatively open migratory, commercial, financial, and information networks. These networks in turn undermined mercantilist ideas and practices, and enabled actors in the Dutch orbit to play a more important role in the development of an increasingly integrated Atlantic World than one would expect on the basis of the small size of their homeland or the number and dimensions of their colonies. An analysis of the structure of the Dutch state and its colonial institutions likewise supports the thesis that the Dutch model had several distinctively modern features, with a state relegating governance and trade in the tropics, to semi-public institutions dependent on private shareholders and allowing for various administrative models.

The various Dutch companies clearly preferred mercantilist policies in their own domains, and gave them up only reluctantly when such policies proved untenable. Moreover, while the Dutch did play a logistical and financial role of some importance in the early expansion of sugar plantations from Brazil to the Caribbean, their innovative contribution to plantation production was limited to the introduction of the polder system, an advance relevant only to their own colonies in the Guianas. And whereas around 1750 the Dutch capital market did develop a new and ambitious system for extending plantation loans, and offered them to non-Dutch planters and colonies, this highly speculative system of subprime mortgages resulted in a financial meltdown, marking the beginning of the demise of the Dutch plantation complex, and facilitating the British takeover of what was to become British Guiana. By then, the frailness of the Dutch decentralised state and its maritime and military apparatus was evident, as was the demise of the Dutch intermediary role in the Atlantic World.

Most of this chapter is dedicated to the period up to 1800, when the Dutch Atlantic really mattered, and not only to its metropolis: prior to the Napoleonic Wars the 'Dutch' Atlantic functioned as a network economy linked to the more extensive Atlantic imperia of her various European competitors. Demographically, the Dutch colonies in the Atlantic were dependent not only on the supply of enslaved Africans, but equally on migration from elsewhere in Europe. In the long nineteenth century the relevance of the Dutch Atlantic withered to near insignificance not only in the global economy, but even within the Dutch orbit, as its centre now moved entirely to the Dutch East Indies. The closing sections of this chapter discuss the demise of the Dutch Atlantic in the context of the simultaneous boom of new Caribbean sugar frontiers and also of Java, and recapitulate the relevance of the Dutch case for Atlantic history and the study of early globalisation at large.[2]

Contours, organisation, and populations of the Dutch Atlantic

In contrast to all other European colonial powers, Dutch territorial and commercial expansion in the early modern period was not primarily directed to the Atlantic, but rather to Asia. A rough periodisation would posit the heyday of Dutch Atlantic colonialism in the period between 1600 and 1680, followed by a period of consolidation and expansion until the late eighteenth century, and demise thereafter.

The Dutch Republic emerged between the mid-sixteenth and the early seventeenth centuries in a revolt against Habsburg Spain, simultaneously rising to 'primacy in world trade' (Jonathan Israel) and attaining the status of 'world-hegemonic power' (Emmanuel Wallerstein).[3] It is not really surprising considering its relative small size that the Republic could not hold on to this position for long. Even though the young state remained comparatively prosperous, its international status declined in the later seventeenth century. At the 1713 Peace of Utrecht, the Dutch state was no longer an equal to the European monarchies defining a new European order, and by extension, an Atlantic one.

The Republic's economic strength was initially built on both its local agrarian potential and its function as a nodal point in European trade networks. Expansion to the continents beyond Europe started around 1600. The chartered Dutch East India Company (VOC) was established in 1602, financed through shareholders' investments, and granted a monopoly over both trade and governance in all of the Dutch orbit east of the southern African Cape. The VOC has often been described as the world's first multinational company, and a major innovation in the history of early-modern globalisation. The company proved to be a major success until the late eighteenth century, even if its contribution to the metropolitan economy did not match the impact of the domestic economy or the European trades. Dutch expansion in the Atlantic started around the same time, and was intimately linked to the Dutch Revolt against Spain. Through the influx of Flemish Protestant and Portuguese Jewish refugees from the Iberian world, the nascent Dutch Republic became aware of the economic potential of Africa and particularly the Americas.[4] Warfare and colonisation in the Atlantic would have the dual function of challenging the enemy and advancing Dutch economic and geopolitical status.[5]

The Dutch West India Company (WIC) was established as a joint-stock company in 1621, based on the innovative model of the VOC, again with a monopoly on both trade and governance. Acting in a highly assertive manner, the WIC developed Manhattan, initially discovered by VOC captain Henry Hudson (1609), and its surrounding areas as New Netherland; captured the Spanish armada in the Cuban Bay of Matanzas (1628), the proceeds of which served to finance the takeover of northeast Brazil from Portugal (1630); conquered the Portuguese fortress of Elmina on the African Gold Coast as a foothold to engage in the Atlantic slave trade (1637); and, in between, explored the Guianas and conquered several Caribbean islands in the 1630s. These, then, were the 'heroic' decades of Dutch Atlantic expansion, corresponding to the Republic's zenith as a European power. By 1680 the Dutch heyday in Atlantic history was over, and, hence, much of this chapter focuses on what became a minor player in the burgeoning Atlantic World. Portugal reconquered Pernambuco (1654) and the British took over New Netherland (1667), leaving the Dutch only Elmina in West Africa, six tiny Antillean islands unsuitable for profitable export agriculture and disqualified as *islas inútiles* by Spain, and a few adjacent territories in the Guianas with some promise as potential plantation economies.

This geographical contraction, coupled with the poor financial results of the WIC and the stark contrast with the remarkable history of the VOC, have helped to cement the idea of Dutch insignificance in the Atlantic. This is a notion in need of revision, also for the period after 1680. There is no doubt that Dutch expansion and commerce were more important in Asia than in the Atlantic, nor is there any question that the Dutch were minor European players in the Atlantic. Even so, the commercial significance of the Atlantic was more important to the Republic than has generally been realised, and, likewise, Dutch colonies and actors were more intensely embedded in larger Atlantic networks than has been acknowledged. Transcending national divides, these Dutch actors made vital contributions to the emerging Atlantic World as intermediaries by offering openings in the mercantilist system which were highly useful for non-Dutch actors. Paradoxically, by providing such outlets they helped to prolong the period of mercantilism.

The modernity of the Dutch Republic was not brought about by a highly centralised state run by a visionary absolute monarch. The Dutch Republic emerged from its revolt against Spain as a decentralised state made up of seven provinces, of which Holland was by far the most important. Within Holland, Amsterdam was the dominant city. The Republic was ruled as an oligarchic pseudo-democracy with a constant, mostly tacit, struggle for dominance between urban patricians and the semi-monarchic Stadtholders, between the various provinces and cities, and between free trade and mercantilist interests. The States-General were the sovereign power of the state, but much power was delegated to institutions at lower levels, whether provincial or municipal.

The Republic experimented with a mix of free trade and mercantilist policies. The fairly flexible, decentralised organisation of the metropolitan state translated into several ad hoc arrangements for international trade, and particularly for colonial expansion. The delegation of monopoly powers to the VOC is a case in point. The system adopted for the Atlantic was far less centralised, and again reflected a high level of flexibility. The WIC was organised along the lines of the VOC, and initially the idea was for the company to wield a monopoly as well. This policy soon floundered. It proved impossible to enforce a monopoly on and around the easily accessible Atlantic shores. Moreover, metropolitan interest in investing in the WIC withered because much of the company's effort was directed at costly combat against Spain, rather than profitable trade. In the end the States-General consented to the parcelling out of territories in the Americas to enterprising individuals and consortia. The WIC ended up governing only Elmina and the Antilles directly, while a patchwork of institutional arrangements was devised for the Dutch Guianas.

Whereas in other Atlantic empires the metropolitan state was directly responsible for governance, this was not the case for the Dutch Atlantic. From the start trade with the Guianas was delegated to individual companies such as the *Sociëteit van Suriname,* while the WIC's trade monopoly to and from its own territories in West Africa and the Caribbean was abolished in the 1730s. Long before that date interloping in the Dutch transatlantic slave trade had been endemic, while the WIC itself had realised that free trade rather than mercantilism could be the way forward for the Antilles, particularly Curaçao and St. Eustatius. Theoretically, the Guianas did function in a mercantilist framework, but it soon turned out that direct connections with the British West Indies, and increasingly the North American mainland colonies, were indispensable for the functioning and growth of these nascent plantation colonies.[6]

None of the various European powers in the Atlantic operated closed mercantilist economies, and smuggling and other forms of illicit border-crossings were essential for the development of the entire Atlantic World.[7] Much modern scholarship therefore rightly emphasises how deeply British, French, and Iberian colonies were entangled with other powers' colonies and networks. After all, it takes two to tango, and even more to play ball, so smuggling is never a one-sided affair. Its widespread occurrence throughout the colonial Americas only underscores that no European state could hope or even aspire to attain perfect mercantilism. The difference between the Dutch and the bigger Atlantic powers is, therefore, one of degree, but of a significant degree indeed, as the Dutch policy emphatically endorsed contra-mercantilist practice as a means to enrich the Republic. This policy did not necessarily reflect an aversion to mercantilism, as the initially mercantilist prerogatives given to the West India companies indicate, but rather it responded to an awareness that given the modest size of their empire and maritime strength, the Netherlands were not well-positioned to act as mercantilists. Hence, the choice to benefit from other nations' mercantilist policies by undermining them. As this implied the establishment of free trade zones, the attraction of non-Dutch settlers, and, it follows, the establishment of relatively cosmopolitan port cities, not to mention the extension of credit beyond the narrow Dutch colonial world, this indeed made the Dutch harbingers of a modern interconnected world.

A short description of the way the Dutch Atlantic colonies were populated illustrates this 'non-national' modernity. Up to Emancipation in 1863 the overwhelming majority of the inhabitants of the Dutch Atlantic colonies were enslaved Africans and their descendants, with Amerindian and European populations forming minorities. The African slave trade is of singular importance to the demographic history of the Atlantic, and this includes the Dutch. From the late sixteenth up to the early nineteenth centuries Dutch ships transported some 605,000 enslaved Africans across the Atlantic, some five per cent of the total Atlantic slave trade.[8] The Dutch share in the overall slave trade peaked early, in the decades after 1650, when merchants in the Republic directly or indirectly benefitted from the Spanish slave *asiento*. The Dutch slave trade is exceptional because of its dual character. On the one hand was the standard trade geared towards providing the Dutch plantation colonies in Brazil and later the Guianas with the necessary labour force. On the other was considerable transhipment of enslaved Africans to the surrounding Spanish and French colonies via Curaçao and St Eustatius. Nearly half of the total number of Africans shipped in Dutch vessels ended up in non-Dutch colonies, a proportion higher than for any other slave-trading nation except Denmark and Sweden. This is a crucial illustration of the Dutch role as a lubricant of the Atlantic system.[9]

As for European migration, in the entire Dutch colonial orbit only two colonies developed into real settlement colonies: New Netherland, where the number of settlers on the eve of the British takeover in 1667 was roughly 7,500, and the Cape Colony, in which the European population numbered some 20,000 by 1800. The number of Europeans living in Asia exceeded the figure in the Dutch Atlantic, but by what margin is impossible to establish.[10] After the loss of Dutch Brazil and New Netherland, the Dutch Atlantic colonies were located in the Guianas and two groups of islands in the Caribbean, and hence only in the tropical Americas, in contrast to the Republic's European competitors.[11] The share of the Dutch Caribbean in the total Caribbean population was low, some eight per cent around 1750. The share of Europeans in the Dutch Caribbean was nine per cent, compared with a Caribbean average of 24%. One more striking contrast is that among the rival European powers, only the Dutch Republic sent its colonial migrants mainly to Asia. The whites of the Iberian, British, and French colonies were all heavily concentrated in the Atlantic, where the centres of gravity of their empires lay.

A second distinction concerns the 'national' composition of the European population in the Dutch Atlantic colonies and settlements. The Dutch Republic itself, of course, had a much smaller population than its major European competitors in the Atlantic World. Moreover, after the swift increase in the 'Golden Age' between 1600 (1.4 million) and 1650 (1.85 million), its population stagnated for an entire century (1.9 million in 1750) and resumed growth only slowly thereafter (to 2.1 million in 1800). Immigration accounted for much of this modest growth.[12] Immigrants were indispensable to the labour market and a vital part of society, and the same applies to the European populations of the Dutch colonies. The Dutch colonial companies welcomed settlers from other European countries, and offered economic as well as religious liberties. The Dutch Atlantic colonies hosted a remarkable heterogeneity of national origins, with large numbers of German and Scandinavian, but also British, French, and Swiss migrants, as well as Iberians, nearly all Portuguese Jews. The latter group, but also British settlers, contributed to, and, indeed, embodied the diversification of trade networks beyond the strict confines of bilateral relations between the metropolis and the colony. These migratory patterns, and the toleration of religious minorities (always restricted to whites) that was integral to them, made the Dutch Atlantic distinctively modern, a prelude to the age of globalisation.[13]

The demographics of the Dutch Atlantic colonies up to the Napoleonic Wars are easily summarised.[14] Over the course of the eighteenth century the European population of Elmina declined from a few hundred

Europeans to a couple of dozens; reliable numbers are not available. In the Caribbean, Surinam had by far the most extensive population, peaking at some 63,000 (excluding Amerindians and Maroons) in 1774. The lesser Guianas had some 30,000 inhabitants around 1780, and exploded thereafter as the British took over. Of the six Antillean islands, only Curaçao, and later St Eustatius (21,000 and 8,000 by 1790), had significant populations. Given their small size, their population densities were well above regional standards. The composition of these various populations was diverse, and reflected the divergence of their economic structures. Enslaved Africans and their descendants made up the great majority of the colonial populations in the plantation colonies of the Guianas. In the commercial centres of Curaçao and St Eustatius, the proportion of enslaved Africans was lower, while the proportion not only of whites but also of free non-whites was considerably higher. This added to a relatively high intra-Caribbean mobility of the non-white population of these islands, particularly Curaçao, where sailors were mainly of (part) African origins, some free, some enslaved.

Plantation production

Atlantic and particularly Caribbean economic growth was intimately linked to the expansion of plantation agriculture, and with it African slavery. All of this is early globalisation writ large: the transfer of tropical crops and the necessary technologies from the Old World to the New, massive (and mainly forced) migrations of labour, the cross-imperial circulation of capital, commodities, and expertise. Within the field of plantation production itself, for the early modern Dutch Atlantic after the demise of Dutch Brazil, this is predominantly the story of Surinam. Plantation production on the Antilles was commercially insignificant, while the plantation boom of the lesser Guianas took off only in the second half of the eighteenth century. As for Surinam, this was not a history only of sugar (Figure 6.1). Sugar was the dominant export crop before 1750 and again after 1820, but in between coffee dominated the colony's export values. In addition Surinam exported cotton between the 1760s and the mid-nineteenth century, and small quantities of cacao.[15]

The role of individuals and institutions based in the Dutch Atlantic World in the development of plantation technology has sparked some debate in the historiography. The basic organisation of the 'plantation complex' was developed in the Old World and transported to the Americas without significant Dutch involvement. The short intermezzo of Dutch Brazil would mark the beginnings of direct Dutch

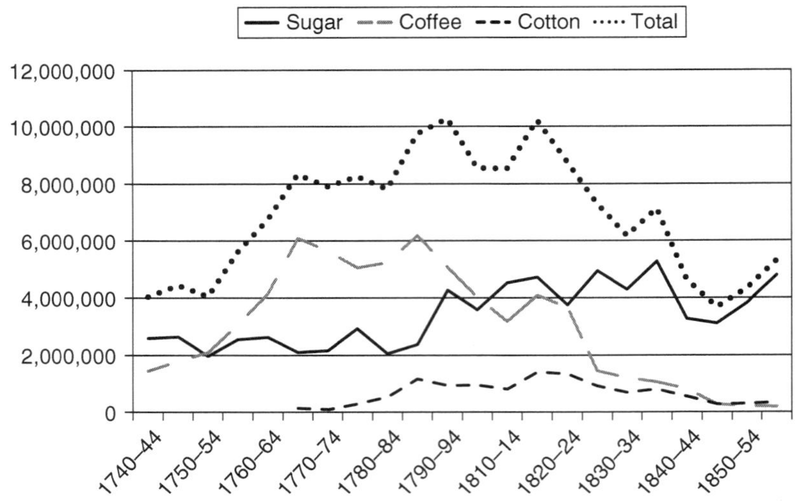

Figure 6.1 Surinam export values of sugar, coffee, and cotton 1740–1864
Source: van Stipriaan, *Surinaams contrast.*

involvement. In Brazil, the Portuguese had adapted the technology of Mediterranean and Atlantic sugar to local conditions, and had introduced the important innovation of the vertical three-roller mill. During Dutch reign sugar output fell short of the region's capacity due to warfare and the strained relations between the Portuguese planters and the Dutch, who themselves did not manage sugar *engenhos* and therefore did not play an innovative role.[16]

It has long been suggested that Dutch actors were pivotal in bringing the sugar revolution from Brazil to the Caribbean and particularly Barbados, transmitting the technological know-how, adequate finance, and enslaved Africans to the British and French Caribbean isles, as well as to the Guianas, and also taking care of shipment and marketing across the Atlantic. This transfer began in the 1640s during the Dutch reign in Brazil, since Dutch Brazil alone could not satisfy the needs of the Amsterdam sugar market. After the fall of Dutch Brazil in 1654 the process gained further momentum as Portuguese Jews who fled Brazil for fear of renewed religious prosecution settled either in the Republic or in Caribbean colonies, thereby transporting an intricate network of expertise and finance to the north, and giving additional momentum to the emerging Caribbean sugar revolution. A 1663 English memorandum claimed 'that the Dutch loosing Brasille, many dutch and Jews repairing to Barbadoes began the planting and making of sugar, which caused the Dutch with shipping to relieve them and [supply] Credit', while the Dutch also delivered the necessary enslaved Africans: 'having lost brasille not knowing where to vent them they trusted them to Barbadoes'.[17]

This would all contribute to a powerful argument depicting the Dutch as harbingers of early globalisation in an industry fundamental to the rise of the Atlantic World, but the Dutch role has been convincingly nuanced by McCusker and Menard, who emphasise the proactive role of British financers and planters on Barbados, in addition to the Dutch. It is still commonly accepted that during the early phase of the development of the plantation Americas, Dutch (including Portuguese Jewish) actors helped to spread the technologically and financially innovative – and morally degrading – sugar industry to the northern Americas, but it should be underlined that, at this stage, the Dutch had not contributed anything to the innovation of the process of sugar production itself, but only, and critically, to transatlantic logistics and finance.[18]

The one unique Dutch contribution to the technology of sugar production was developed in their own colonies on the 'Wild Coast' of the Guianas. Establishing their plantations along the swampy lands on the Atlantic coasts and the inland rivers of the Guianas, Dutch planters were responsible for an innovation adapted from metropolitan agriculture – the polder system. The best soils for agriculture were the lands adjacent to the Atlantic and the various rivers of the Amazon basin, but these lands were continuously swamped through the workings of the ebb and flow of the tides. The first sugar plantations, established during the short British rule of the colony, were, therefore, started on drier but less fertile grounds further inland. An adaption of the metropolitan institution of the polder proved to be the way forward. Plantations were laid out in quadrangles with the short side alongside the rivers, stretching a mile or more inwards. Enslaved Africans were put to the back-breaking work of building dikes to control the level of water within the plantation, as well as the delving of an internal system of dikes and trenches to regulate water levels, allowing for both irrigation and drainage. This multifunctional system of canals and trenches also facilitated easy bulk transport in flat barges. On sugar plantations there was an additional innovation. Sugar technicians learned to make use of the considerable tidal variation of the rivers as a source of hydraulic energy for the sugar mills, by taking in water as the tide was rising, and during ebb tide letting the water recede from the canals to the river through a water wheel powering the sugar mill.

The innovation of the polder system spread fast in the Dutch Guianas for all types of plantations, whether producing sugar, coffee, or cotton. The adaption of a water-powered sugar mill was costly but crucial for a plantation's chances for survival. By the mid-eighteenth century animal traction was still more prevalent than hydraulic energy, but by the 1770s water mills were dominant, and would remain so until the adoption of the steam mill in the nineteenth century.[19] The human cost to the enslaved population was high, as the labour input was extremely demanding, but for much of the eighteenth century the productivity of the Surinamese plantations was spectacular indeed, well above regional averages into the nineteenth century. In his *Histoire philosophique et politique,* Abbé Raynal praised the Dutch for 'having domesticated the ocean in the New World just as they did in the Old World'. In the later eighteenth century the prospects of this unique technological complex were promising enough to attract increasing numbers of planters from the 'depleted' British West Indies to Essequibo and Demerara, thus setting the stage for the eventual British takeover of these colonies.[20]

By then the Dutch planters were no longer at the vanguard of Caribbean plantation production. Apart from their use of polder technology, they had been keen followers rather than initiators of innovations in sugar and coffee production throughout the eighteenth century.[21] As was the case elsewhere, the introduction of the steam engine in the next century caused a substantial increase in the average scale, capital input, and productivity of Surinamese plantations, but by then the colony could not keep up with the new Caribbean frontiers, particularly Cuba and, ironically, British Guiana, once a Dutch colony.[22]

Trade and finance

Whereas the planters of the Dutch Guianas may have benefitted from the high productivity of their polder plantations, they suffered from the absence of a protected home market, as their competitors sheltered by mercantilism enjoyed. The Dutch Republic's substantial sugar-processing industry was not only supplied with sugar from the Guianas, but bolstered by metropolitan re-exports from the French West Indies. Without a protected market for their sugar and other tropical commodities, there was no question of a guaranteed price level for Dutch producers. As a consequence price levels on the Dutch market were generally below those in competing European markets.

This lack of metropolitan protection was inherent in the idiosyncratic organisation of trade in the Dutch Atlantic. Just as the metropolitan staple market, Curaçao, and St Eustatius were already operating as free trade zones in the seventeenth century, in the 1730s the WIC had officially relinquished its untenable trade monopoly. The Guianas, in contrast, were operated by ad hoc institutions with mercantilist ambitions. Such ambitions, however, were not honoured in the metropolis, nor feasible in the colonies. The States-General had approved the delegation of governance, and as a rule did not interfere in regulations reserving bilateral transatlantic trade to Dutch ships enfranchised by the various companies, the *Sociëteit van Suriname*, the *Sociëteit van Berbice*, and the Essequibo branch of the WIC, dominated by the Chamber of Zeeland. But this did not imply that these same companies had any influence on price-setting, or, rather, the lack of it, in the Republic's staple market. No matter how often metropolitan West India interest groups lobbied for such protection, the principle of an open staple market was never given up. In this sense the Dutch Republic remained a champion of early modern globalisation.[23]

Even within their own jurisdiction, the Dutch Guiana companies could not enforce a mercantilist framework, and soon acknowledged that attempting to do so would ruin the colony. Trade with Barbados, the mother colony of the short-lived British colony of Surinam until it was conquered by the Dutch in 1667, continued to be indispensable for the further development of the colony. In 1704 the *Sociëteit van Suriname* grudgingly approved of regional trade with the British West Indies and North American colonies on non-Dutch ships. As the century progressed, the British North American colonies, and later the United States, became vital trade partners for Surinam, exchanging essential foodstuffs, livestock and commodities for local produce, particularly molasses.[24] The remaining restrictive regulations imposed by the company were neglected or circumvented. Throughout the eighteenth century the number and tonnage of non-Dutch ships exceeded bilateral Dutch shipping between Surinam and the Netherlands by a clear margin.[25] The spectacular development of Berbice, and particularly of Essequibo and Demerara, in the second half of the eighteenth century likewise depended on trade from and to the British colonies – but in this case also British investment and migration.[26] It is therefore safe to conclude that immersion in non-Dutch Atlantic commercial networks was absolutely vital for the development of the Dutch Guianas. Bare necessity rather than ideology dictated their participation in an empire-crossing economical and demographic Atlantic space.

The total volume and value of trade from the Dutch Guianas to the metropolitan market grew for most of the period from their establishment to the late eighteenth century, but the significance of Dutch Atlantic trade to the metropolis is only fully appreciated if the value of the trade with Curaçao and, particularly for the later eighteenth century, St Eustatius, are added.

Neither island had a significant local production, but both emerged as vital hubs in a wider commercial network. Curaçao linked particularly the Spanish Main and the French Caribbean to Europe through the Dutch staple market, while Statia was a booming centre of re-exports connecting the French and British West Indies both to North American and Dutch markets. Between 1700 and 1780, and with some fluctuations over time, the average value of the two islands' trade with the Dutch Republic slightly exceeded metropolitan trade with Surinam.[27] In *Wealth of Nations* Adam Smith praised the islands' subversion of the reigning mercantilism, pointing at them as 'free ports open to the ships of all nations; and this freedom, in the midst of better colonies whose ports are open to those of one nation only, has been the great cause of the prosperity of those two barren islands'.[28]

Undermining mercantilism

Inclusion of the trade via these islands, as well as trade to and from Elmina and the lesser Guianas, significantly increases projections of the trade between the Dutch Atlantic and Europe. Most estimates point to strong growth in the value of Dutch Atlantic trade in the eighteenth century, surpassing overall Dutch economic or trade growth. According to de Vries' figures the annual value of these imports had been around 4.5 million guilders in the 1640s, collapsed to 2.5 million in the 1680s, then recovered to 4.3 million in the 1700s, rising to 10.3 million in the 1740s, and peaking at 22.4 million in the 1770s. The figures would be higher if direct trade between the North American colonies and the Republic was included, but even without taking that trade into account, the annual value of Atlantic imports into the Republic increased steadily throughout the eighteenth century, reaching a peak just prior to the devastating Fourth Anglo-Dutch War (1780–84). In the 1770s, for the first time ever, this trade surpassed Asian imports into the Republic, which had shown almost uninterrupted growth from the 1640s (7.9 million guilders) through the 1760s (21.4 million), but decreased to 20 million in the 1770s. This unprecedented peak of Atlantic imports, then, was to a large extent due to Dutch transhipping of tropical staples produced in the colonies of competing European states. In this respect the Dutch surely lived up to their image of middlemen linking various parts of an integrated Atlantic, and undermining other nations' mercantilist pretentions.[29]

These revisions show that Dutch Atlantic trade was more important than has long been acknowledged both for the Dutch Republic and in the wider Atlantic system. However, the point should not be stretched too far, as a second look at de Vries' calculations bears out. Whereas in the 1780s the average annual value of imports into Britain (151 million guilders), France (171 million), and the Dutch Republic (147 million) were more or less in the same league, a major difference is that the share of European trade was far higher for the Netherlands (71%) than for France (51%) and Britain (46%). Imports from Asia were more or less similar for Britain (16%) and the Netherlands (14%), but far less so for France (5%). In contrast the share of the Atlantic trade of Great Britain (38%) and France (42%) was far more important than for the Dutch (15%). This leaves us with the conclusion that even the upwards revisions do not make the Dutch a major Atlantic trading nation. Klooster estimates that the Dutch share in total Atlantic trade was about 10 per cent around 1750. According to de Vries' calculations, the Dutch share in the Atlantic imports of Britain, France and the Netherlands combined was barely 15%, but of course this proportion in overall European Atlantic trade drops when figures for Spain and Portugal are included.[30]

In the end, then, the clear increase in importance of Dutch Atlantic trade throughout much of the eighteenth century was only partly related to production growth in the Republic's own plantation colonies, and does not fundamentally alter the image of the Dutch as a player of limited importance in the Atlantic. After the passing of its Golden Age, the Dutch Republic was simply a nation with fewer resources and other priorities in Asian and, particularly, in European trade, than the ones nurtured by its European competitors. At the same time, the figures do support the idea that the Dutch were disproportionally engaged in border-crossing, thus facilitating the integration of the Atlantic.

What did all of this mean for the Dutch economy? Of course, even if constituting only a modest part in overall trade, the Atlantic endeavours did contribute to the national economy. Why then the claim that 'the 200 year history of Dutch Atlantic economy is one of repeated cycles of hope, frustration, and failure', as de Vries and van der Woude had it?[31] The answer lies not as much in the failure of plantation production in the Guianas, but rather in all-too-frequent poor returns to investment in governance, trade, as well as plantation production. Part of the poor reputation of the Dutch Atlantic economy derives from the results of the WIC. Much in contrast to the VOC, its Atlantic counterpart was bankrupted as early as 1674, while the scaled-back second WIC again produced poor returns on investment before its collapse in the late eighteenth century.[32] The Surinam and Berbice companies did not fare much better, even if they managed to survive until they too were dissolved at the end of the eighteenth century. The results of the most prominent private company engaged in the Atlantic slave trade, the *Middelburgsche Commercie Compagnie* (MCC), fluctuated greatly, but certainly did not produce spectacular profits.[33] This then leaves the unresolved issue of the profitability of Atlantic trade and investments in plantations for individuals and private businesses. Against anecdotal evidence of enormous profits achieved, particularly in the early period of Dutch Atlantic engagement, the few quantitative studies available suggest an increasing indebtedness of the Guianas' plantation sector, and a widespread inability to finance amortisation from plantation income, particularly in the later eighteenth century.[34]

Evidence for the financial failure of the Dutch Guianas' plantation sector is provided by the extremely poor results of the *negotiatiestelsel*, a unique and highly speculative system of providing substantial loans for the operation of plantations. Under this system, mortgages were extended to (increasingly absentee) owners with the plantation itself as security. The loan was supposed to be amortised in two decades, and during this period all plantation provisions and produce were to be handled and shipped by metropolitan merchant houses who themselves had secured the funding by selling shares in the mortgage system. This *negotiatiestelsel* was developed in the early 1750s, in the paradoxical context of rising indebtedness of planters against high expectations regarding further expansion of the Surinamese plantation sector and high demand for loans to finance the increasingly large and capital-intensive modern plantations. The end result was dramatic. By the 1770s, the frailty of the system was evident, as planters could not realise the necessary profits to pay off their huge debts. Within a short period the whole system collapsed. Between 1753 and 1795, around 80 million Dutch guilders were extended as credit to start and operate plantations, just over half of these for Surinam, a quarter to the other Dutch Guianas, and the rest to non-Dutch colonies. By 1800 not even a quarter of the initial debt was paid off, and a significant proportion – possibly even a majority – of these plantations was bankrupted and sold, or reluctantly taken over by the financing parties.[35]

Several answers have been advanced for the puzzle of why credit continued to be made available in spite of early signs of recurring losses. Obviously considerable amounts of good money were thrown after bad in the vain hope of turning the tide. Intermediaries, dependent not on net results, but rather on their provisions for services – merchant banks, administrators, overseers – had a vested interest in continuing plantation operations irrespective of the benefits to the owners (something disturbingly familiar today).[36] One thing stands out, though. With the *negotiatiestelsel*, the Dutch had developed an innovative and highly risky financial product that would become typical of today's global economy, and again, we may characterise the Dutch as early exponents of modern capitalism. At the same time their predominantly national investors were victims of this financially daring experiment in international finance.

Care is required in drawing conclusions about the profitability of both the slave trade and the plantation sector itself, but it may be safely concluded that investments in these sectors were insecure at best, particularly in the later eighteenth century. On the other hand, benefits arose from spin-off effects ranging from shipping (including shipbuilding and insurance), the supply of commodities in exchange for captive Africans and of provisions and other commodities for the plantations, and of course the processing, consumption, and re-export of imported tropical staples. No economic historian so far has ventured to calculate such multiplier effects. Yet, as in most of Western Europe, domestic consumption of Caribbean (but not necessarily Dutch Caribbean) commodities such as sugar, coffee, and cocoa increased swiftly during the eighteenth century. The growing share of Atlantic commerce in overall Dutch trade meant business opportunities for the elites, and employment for the lower classes. So surely life in the Republic itself was affected by Dutch expansion in the Atlantic too.

Networks of trade and knowledge

The logistics of transatlantic and regional trade included intensive networks of communication. Ships carried not only products, but equally letters, and later newspapers. The people aboard carried news from one place to another. Businessmen communicated with one another on a permanent, even if often interrupted, basis. Through these same channels, correspondents in the colonies learned of major political developments in Europe and in the Dutch Republic – wars, financial crises, regime changes, and so on – and vice versa. They informed their counterparts on the latest from the colonies. Indeed, to historians of the Surinamese plantation sector, the phenomena of both absenteeism and the emergence of the *negotiatiestelsel* have proved to be of crucial importance, as meant paperwork crossing the ocean. This correspondence focused on the economic aspects of plantation routines, trade, and finance, but included broader information on the state of affairs in the Guianas, news about other colonies, and the like. Far less paper seems to have been preserved on regional trade, and certainly this correspondence has been less used in studies of Atlantic trade.

News also travelled by mouth, and while correspondence was mainly the domain of educated white men, and only occasionally women, this involved a much broader group of people. Within the Dutch Atlantic and particularly the insular Caribbean, white, free coloured, and even enslaved artisans working in the port cities of Curaçao and St Eustatius facilitated the transfer of information from one colony to another. More evidence has become available on the increasingly complex labour market of the islands, and specifically on the economic role of non-white actors in the regional trade, most of these participating as petty traders, but some acquiring considerable property, particularly in Curaçao.[37] In St Eustatius, in contrast, the most remarkable feature of the merchant class was its predominantly white expatriate character, with Dutch nationals forming only a minority in a cosmopolitan circle of merchants entertaining intensive commercial relations within the Caribbean region and beyond.[38]

Among the commercial networks constitutive of the Dutch Atlantic, those forged by Portuguese Jews have attracted most attention. These networks were of seminal importance in the first phase of Dutch settlement and colonisation, both during and in the aftermath of Dutch Brazil. Portuguese Jews remained major players in the Curaçao trade, and to a lesser extent in St Eustatius, expanding their networks based on kin and faith to other parts of the Western hemisphere, while consolidating links to the Jewish merchant community in the Republic. In Surinam, the only colony ever where Jewish settlers were widely engaged in plantation agriculture, Portuguese Jewish and soon also Ashkenazi Jews were deeply involved in transatlantic as well as regional trade, and surely not exclusively with coreligionists.[39]

Other networks were relevant. Merchants of Dutch origin in British North America and Barbados were particularly important in the first phases of the making of an increasingly integrated Atlantic World, as were economic relations between Lisbon and Amsterdam.[40] Early on, Dutch and Flemish merchant communities settled in cities such as Cadiz and Nantes, thus securing direct Dutch involvement in the Spanish and French Atlantic trade circuits.[41] Conversely, British West Indian and North American ship captains and merchants were prominently active in the port cities of Paramaribo (Surinam), Willemstad (Curaçao), and Philipsburg (St. Eustatius), and throughout, a vibrant Anglo-Dutch merchant community was established on both sides of the North Sea.[42]

Commercial networks also facilitated the circulation of people and ideas potentially subversive to the economy and social order in the Atlantic. This circulation was integral to, and at times undermined, the emerging integrated Atlantic World. The participation of non-whites, whether enslaved or not, in intraregional shipping could have this impact, as in the continuous maritime marronage from Curaçao to the Spanish Main. More spectacular was the impact of the Haitian Revolution. Just as news brought in from France had sparked this momentous episode in Atlantic history, so did news conveyed by sailors working on the trade route between the two islands about the events in the French Caribbean inspire a major, eventually brutally suppressed, slave revolt in Curaçao in 1795.[43]

Political concerns were rarely debated in the correspondence linking Europeans within the Dutch Atlantic. There was no serious metropolitan debate about the justification of the Atlantic slave trade or slavery itself, much less a genuine abolitionist movement. The two issues that continued to be debated concerning the Guianas were the degree to which mercantilism was feasible and acceptable, and the question to what extent the private companies could rightfully demand military assistance of the States-General in times of war or domestic turmoil. By 1780, as the Fourth Anglo-Dutch War began, widespread discontent in the colonies emerged over the inability of the Dutch state to protect its own territories. This added to very critical debates about the ailing WIC and the semi-mercantilist character of the Guiana companies, all of which eventually contributed to the demise of the latter and the assumption of direct control over them by the Dutch state. Then again, just as the abolition of the slave trade was eventually imposed by the British during the Napoleonic Wars, so too the transition from company to state control of the Dutch Atlantic formed part of a wider regime change. In a bloodless revolution in 1795, the *ancien régime* Dutch Republic was replaced, with the support of revolutionary France, by the so-called Batavian Republic, which launched a policy of centralisation which has been continued ever since, both under French rule and in the subsequent Kingdom of the Netherlands.

A reminder of the languages of communication is appropriate, as linguistics also show how the early modern Dutch Atlantic prefigured the cultural diversity that would become characteristic of the modern world. Historically, the emergence of large commercial networks has implied a lot of linguistic border-crossing, and mercantilist regulations were far more important in carving up the Atlantic in several sub-imperia than language divisions. Even so, the dominance of Spanish, Portuguese, English, and French in the various national components of the Atlantic World facilitated commercial and financial transactions within their own sphere. In the Dutch Atlantic, in contrast, Dutch was dominant in only the Republic and in the tiny European enclave of the Elmina fortress, while in the colonies the major actors must have been at least bilingual. The language spoken on the plantations of Surinam was the English-based Creole language Sranantongo. Most European locals in Paramaribo must have spoken both Dutch and Sranantongo, but bearing in mind the ubiquity of British and American merchants and sailors, many must have been able to speak some English as well. In the Southern Antilles the vernacular for all classes and ethnicities became the Iberian-based Creole language Papiamentu early in the eighteenth century. Probably the local free population communicated easily in Spanish, but many must have had working knowledge of French and English too, bearing in mind the multifarious trade networks. In the Northern Antilles English was the dominant language. Even if the language of correspondence with the Republic remained Dutch, trade transactions in cosmopolitan St Eustatius were conducted mainly in English. In short, linguistic heterogeneity sustained the boundary-crossing character of the Dutch Atlantic, while adding to its internal fragmentation.

The demise of the Dutch Atlantic

The Fourth Anglo-Dutch War heralded the demise of the Dutch Atlantic. The immediate effect seemed limited, as the colonies briefly occupied by the British (St Eustatius, Berbice, Demerara, and Essequibo) were restored to the Republic at the closing of the war, but the long-term consequences were serious. The war had exposed the ineptitude of both the WIC and the Dutch Republic as a whole to protect Dutch shipping and colonies against Great Britain, by now the unrivalled maritime power in the Atlantic. Dissatisfaction with the decline of the once-glorious Republic caused strong political dissent within the country between 'Patriots' opposing the ancient regime state and its institutions, including the colonial companies, and 'Orangists' convinced that the Stadtholder would be able to revive the nation's fortunes. A patriot coup d'état failed in 1787, but in 1795 the intervention of the French revolutionary state led to the establishment of the Batavian Republic and the beginnings of a period which lasted until *c.*1815, in which the Netherlands was supervised and later occupied by France, while the colonial companies were dissolved and the bulk of the Dutch colonies were held in 'protective custody' by Great Britain, with the consent of the exiled Stadtholder Willem V.

In the years 1813 to 1815 the Netherlands was reconstituted as a kingdom, and the country's first king – the same Stadtholder now styled Willem I – was determined to capitalise on his colonial possessions, to usher the nation into a new era of prosperity. For geopolitical reasons the dominant European powers had decided to enlarge the Dutch state in Europe by adding the Southern Lowlands, a situation that would be aborted in 1830 by the Belgian Revolt. Meanwhile the extent of the Dutch colonial empire was seriously reduced by Great Britain. The various peace treaties concluded at the end of the Napoleonic Wars restored the Dutch East Indies, Surinam, the Antilles, and Elmina to the Dutch, but the British consolidated their takeover of the Cape Colony, Ceylon/Sri Lanka, Berbice, Demerara, and Essequibo, henceforth known as British Guiana. Great Britain also saw to it that the Netherlands accepted and implemented the abolition of the Atlantic slave trade. The Dutch slave trade had already come to a virtual standstill in the late eighteenth century anyway.

Willem I, now exclusively responsible for colonial affairs, embarked upon an ambitious programme to revive his empire. Initially, his ambitions included both the East and the West Indies, and to some degree even Elmina. Within one or perhaps two decades, however, it became clear that the world had changed, and that only the Dutch East Indies held enormous promise. Elmina lost whatever value it once might have had with the abolition of the Atlantic slave trade, and the Dutch eventually ceded the port city to Great Britain in 1871. More importantly, the days of mercantilism were over. This meant that there was no longer any use for free trade zones such as Curaçao and St. Eustatius as escape valves in a mercantilist world. The Dutch Antilles lost their significance both in the region and to the metropolis, where their cost to the national treasury led some parliamentarians to ponder the option of selling the islands to the highest bidder. Only between the 1920s and the 1950s did the islands briefly regain strategic value because of the establishment of oil refineries on Aruba and Curaçao.

This left only Surinam as a potential asset, but again, the high expectations of the king and his entourage were not met. Throughout the nineteenth century and up to the 1930s colonial policy was directed towards the development of plantation agriculture, facilitating this sector in all possible ways except for protectionism in the home market. The most conspicuous support for the plantation sector was labour market policy. The Dutch state did not abolish slavery until 1863, adding a period of state supervision (*Staatstoezicht*) to secure a continued supply of Afro-Surinamese labourers to the plantations. This period was inspired by the British post-Emancipation experiment with 'apprenticeship', but whereas this was a four-year period in the British West Indies, the Dutch opted for no less than ten years. The system failed; a considerable part of the freed population preferred either peasant agriculture or urbanisation to poorly paid plantation work reminiscent of slavery.

At the behest of the local planter class, and following British practice in Guiana and Trinidad, Dutch colonial policy then embarked on the recruitment of indentured labour from British India, totalling some 34,000 from 1873 to 1917. Next some 32,000 Javanese were recruited between 1891 and 1938. Over time the composition of the plantation work force changed dramatically, with the share of Afro-Surinamese falling from 100% in 1863 to roughly half in the mid-1880s and to below ten per cent after 1900. In the next decades British Indian and Javanese recruits became the dominant plantation workforce, but, at the same time, the share of the sugar sector in overall employment went down. The share of sugar in total exports fell from just over 70% in the 1860s, to 66% in the 1870s, then to around or even below 30% in the entire period from the 1880s until the First World War.[44] The demographic and social consequences of indentured labour lasted longer. Today roughly half of Surinam's population is of Asian descent, making the country the most multi-ethnic in the Caribbean.

Perhaps because the traditional – as opposed to the new rice-producing – Surinamese plantation sector collapsed in the twentieth century, a stubborn idea lingers that the credit crisis of the *negotiatiestelsel* marked the beginnings of an uninterrupted downward slope. This is not correct. The nineteenth-century history of the plantation sector up to the abolition of slavery in 1863 discloses concerted specialisation, with a move away from coffee to concentrate solely on sugar production; a tendency towards converting small-scale plantations into larger, more capital-intensive units utilising state-of-the-art technology; and, indeed, a considerable growth in sugar production up to the 1860s, followed by a slow decrease in the next decades and a return to the mid-1810s levels at the end of the century. A major break with the past was the reorientation of marketing: since the 1860s most of the sugar has been exported to the US, rather than to the Netherlands.[45]

Clearly the Surinamese sugar sector could not keep up with the most successful Caribbean competitors, Cuba, British Guiana, and Trinidad. The economic success story of these three colonies is intimately linked to the opening of new production frontiers, massive investments both in innovation and in the recruitment of both enslaved (Cuba) and indentured labour, and preferential access to huge markets. The comparison with British Guiana is particularly intriguing. The decisive period in the divergence between the two colonies seems to have been the period between 1780 and 1815, and may be explained primarily by massive imports of enslaved Africans into British Guiana. Surinam's population decreased from over 60,000 to 50,000 in these years, while the population of Berbice, Demerara, and Essequibo skyrocketed from some 30,000 to 100,000. In 1815, with both colonies no longer able to draw on the Atlantic slave trade to boost the labour supply, British Guiana had considerably more inhabitants. The proportion of the total colonial population which was enslaved was higher (90% to 95% against just over 80% in Surinam), and the economically active segment of the slave population of British Guiana was structurally greater as a result of recent massive imports. Moreover British Guiana still had a moving production frontier. In Surinam, only the western part of the colony offered a new frontier – and precisely there, much capital was invested by British planters, just as British capital had been decisive in the Guianas' plantation boom during even the last decades of Dutch rule.[46]

But why then were there no similar large Dutch investments in Surinam? Part of the explanation may indeed be found in the financial crash of the 1770s and the collapse of the *negotiatiestelsel*. This crisis left many plantations bankrupt, with investors losing their capital. In many other cases investors were convinced by bankers and merchant houses to re-invest in these plantations, in the unrealistic hope that the products of the plantations would suffice at least to pay off the debts. The profitability of the Surinamese plantation sector in the remaining period of slavery remained meagre at best, and hence it was difficult to attract new capital.

An altogether different side to the equation should also be considered. In 1830 the Dutch colonial state in the East Indies introduced the Cultivation System, forcing Javanese peasants to dedicate part of their time to the production of export crops under the supervision of local rulers. In terms of exports and benefits this system was tremendously successful, and taught the Dutch that on Java native labour could be forcibly recruited and exploited at far lower cost and in far larger numbers than in Surinam – by 1850 the total population of Java was over

9.5 million. In 1830 the Netherlands imported roughly the same quantity of cane sugar from Surinam and Java; in 1850 Java produced five times as much; in 1860, fourteen times.[47] A translation into Dutch of Alvaro Reynoso's then-leading Cuban study on the cultivation of sugar cane was published only three years later in Dutch, clearly aiming to boost the sugar production in Java, not Surinam.[48]

Java's production of sugar cane continued its spectacular growth after the substitution of a liberal policy for the Cultivation System by 1870, giving way to private Dutch enterprise. Between 1900 and 1940 Java had replaced Cuba as the world's largest producer of cane sugar. In 1800 Java's cane sugar production had been a mere 3,700 metric tons; in 1850, 102,000; and in 1900 a staggering 744,000, more than double the Cuban figure. By then, as beet sugar became increasingly important on the European market, Javanese cane sugar exporters had found new Asian vents in the globalising world market, exporting to China, Japan, and India. Incidentally, the Cultivation System had also produced a more modest boom in Javanese coffee production, which increased almost tenfold between 1800 (7,316 tons) and 1850 (69,144), decreasing thereafter (to 42,752 tons in 1900), as sugar was given priority.[49]

Clearly the introduction of the Cultivation System was a watershed in colonial history, and sealed the demise of Surinam as an exporter of tropical produce. Irony is inherent in the multiple ways which Surinam and Java were linked during this transitory phase. First, the later governor-general of the Dutch East Indies, Johannes van den Bosch, masterminded the Cultivation System, partly on the basis of a visit to the Dutch West Indies in 1827–8. In Surinam, he introduced legislation for amelioration policies, and stimulated Christianisation of the enslaved population, but he apparently also concluded that successful production of tropical export crops could only be accomplished in Java on the basis of some sort of bonded labour, as was the case in the West Indies. Appointed governor-general of the Dutch East Indies in 1828, van den Bosch was able to implement this idea himself. The Cultivation System resulted in a spectacular rise in income for the Dutch treasury, making up between 32% and 54% of total tax income from the 1830s through the 1860s.[50] This bonanza not only allowed for the redemption of the huge state debts and the financing of infrastructure such as the first railroads, but also for the financing of emancipation in 1863. Thus, exploited Javanese peasants paid the bill for the indemnification granted to the owners of the freed slaves in the Dutch Caribbean.

Small wonder that by then Dutch capital flowed between the metropolis and the East Indies in ever larger quantities, while colonial officials in the West Indies pleaded mostly in vain for private and public investments. The two major investments in which the Dutch state acted as intermediary were, first, the indemnification of former slave owners, and secondly the establishment of the system of indentured labour. Both responded to the claims of planters and their metropolitan business partners for support, but neither helped to make the plantation sector competitive and rewarding in the long run – even if the major Surinamese plantation, Mariënburg, was technologically state-of-the-art.[51] Whereas in the East Indies both Dutch and foreign private initiatives eagerly took over after the Cultivation System gave way in an era of liberal economic policies, in Surinam the state tried in vain to attract private investment. In parliament the Dutch Minister of Colonial Affairs, Hendrik Colijn, stated desperately in 1935 that 'Everything that has been tried in Surinam has all simply failed.'[52] Bauxite would soon alleviate such concerns, albeit temporarily, just as the establishment of oil refineries would make the Antilles into useful parts of the Kingdom for some decades, but in the end only the East Indies mattered. The Dutch public and private sector alike remained wary of the prospect of giving up the archipelago. Dutch interest in the Caribbean, in contrast, had mainly become a governmental concern, driven more by geopolitical than by strictly economic motives.

A new frontier

As the process of globalisation intensified in the long nineteenth century Dutch interest in the Atlantic colonies waned, while the Dutch East Indies provided a new economic and geopolitical frontier. Gradually, the awareness faded that the Dutch Atlantic had ever mattered, and that the Dutch had played a role of some significance in the emergence of the early modern Atlantic World.

Without falling into the trap of overestimating the modernity of the early Atlantic World per se, and construing what Emmer referred to as the 'myth of early globalisation', the early modern history of the Atlantic may be characterised as a watershed, the beginnings of a long-term process of integration, starting in, but not limited to, the economic sphere.[53] Between 1600 and 1800 the Dutch were actively involved in this process. They were not primarily conspicuous in the field of plantation production, but they certainly were in commerce and finance, which also stimulated demographic and cultural border-crossing. Dutch economic policies were a mixture of mercantilist and free trade principles and practices, inspired more by pragmatism than by ideology. Dutch free trade ports lubricated the predominantly mercantile Atlantic economy and were, therefore, tolerated by their more powerful competitors. In the later eighteenth century Dutch decline set in, first because of failing credit arrangements, and hence a crisis in the plantation colonies, and next due to a lack of naval power to protect Dutch trade colonies and networks: by then it was all too obvious that, all along, the Dutch had been at the mercy of the French, and particularly the British, fiscal-military states.[54] After the Napoleonic Wars, as the days of mercantilism came to an end, the free trade zones of Curaçao and St Eustatius became obsolete. The Dutch had also lost the most promising frontier colonies in the Guianas, so the prospects of further plantation growth in the West Indies withered. Ironically, the establishment of British Guiana was the finishing touch to a process of informal British takeover initiated in the previous century.

In the preceding two centuries the Dutch Atlantic had been characterised by remarkable institutional heterogeneity, free and often unregulated flows of capital, strong cross-imperial commercial and demographic linkages, and hence high proportions of non-Dutch settlers among its white populations, which added to its ethnic and cultural diversity. Much in this period suggests a remarkably modern and flexible capitalist spirit, and as such the Dutch Atlantic passes the test of being a champion of early-modern 'soft' globalisation, as suggested by de Vries.[55] But in the end neither the Dutch Republic nor its colonies – and much less the enslaved populations labouring in these same colonies – seem to have benefitted particularly from this precocious globalisation.

Notes

1. de Vries, Jan: 'The Dutch Atlantic economies', in Coclanis, Peter A. (ed.): *The Atlantic economy during the seventeenth and eighteenth centuries: organization, operation, practice, and personnel*, Columbia: University of South Carolina Press, 2005, p. 21.
2. The chapter leans heavily on my work in the project 'Dutch Atlantic Connections, 1680–1800' subsidised by the Dutch Organisation for Scientific Research NWO and the Dutch Institute for Advanced Study NIAS, and especially two articles co-authored with Jessica Vance Roitman, 'Repositioning the Dutch' and 'Introduction', in Oostindie, G. and Roitman, V. (eds.): *Dutch Atlantic connections, 1680–1800: linking empires, bridging borders*, Leiden: Brill, 2014. Many thanks to Jessica Roitman for commenting on a draft for this chapter.
3. Israel, Jonathan: *Dutch primacy in world trade, 1585–1740*, Oxford: Clarendon Press, 1989; Wallerstein, Immanuel: *The modern world-system, Vol. II: Mercantilism*

and the consolidation of the European world-economy, 1600–1750, New York: Academic Press, 1980, pp. 36–71.

4. Dutch whaling and other fisheries are generally not included in studies on Dutch Atlantic history. Fisheries were important to the Dutch economy, but territorial expansion was not involved, and the WIC did not intervene in this field.
5. Schmidt, Benjamin: *Innocence abroad: The Dutch imagination and the New World, 1570–1670,* Cambridge: Cambridge University Press, 2001; Usselinx, Willem: *Naerder bedenckingen, over de zee-vaerdt, coophandel ende neeringhe, als mede de versekeringhe vanden staet deser vereenichde Landen, inde teghenwoordighe vrede-handelinghe met den coninck van Spangnien ende de aerts-hertoghen,* s.l.: s.n., 1608.
6. The empire-crossing characteristics of the Dutch Atlantic are emphasised in Enthoven, Victor and Postma, Johannes (eds): *Riches from Atlantic commerce: Dutch Atlantic trade and shipping, 1585–1817,* Leiden: Brill, 2003, and particularly in Oostindie, Gert and Vance Roitman, Jessica (eds), *Dutch Atlantic connections, 1680–1800: Linking empires, bridging borders,* Leiden: Brill, 2014.
7. Klooster, Wim: 'Inter-imperial smuggling in the Americas, 1600–1800', in Bailyn, Bernard and Denault, Patricia L. (eds), *Soundings in Atlantic history: latent structures and intellectual currents, 1500–1830,* Cambridge: Harvard University Press, 2009, pp. 141–80. See also Zahedieh, this volume.
8. Roughly 517,000 captives were disembarked in the Americas by Dutch ships; the average mortality rate during the middle passage was 14.5 per cent. Trans-Atlantic Slave Trade Database, www.slavevoyages.org and Dutch Atlantic Connections database, www.dans.knaw.nl.
9. www.slavevoyages.org.
10. Kruijtzer, Gijs: 'European migration in the Dutch sphere', in Gert Oostindie (ed.), *Dutch colonialism, migration and cultural heritage,* Leiden: KITLV Press, 2008, pp. 97–154; Lommerse, Hanneke: 'Population figures', in Oostindie, Gert (ed.), *Dutch colonialism, migration and cultural heritage,* Leiden: KITLV Press, 2008, pp. 315–42.
11. Greene, Jack P.: *Pursuits of happiness. The social development of early modern British colonies and the formation of American culture,* Chapel Hill: University of North Carolina Press, 1998, pp. 178–9 (British), Pritchard, James: *In search of empire. The French in the Americas, 1670–1730,* Cambridge: Cambridge University Press, 2004, pp. 423–5 (French), Engerman, Stanley L., and Higman, Barry W.: 'The demographic structures of the Caribbean slave societies in the eighteenth and nineteenth centuries', in Franklin W. Knight (ed.), *General history of the Caribbean. Vol. III, The slave societies of the Caribbean,* London: Macmillan/UNESCO, 1997, pp. 48–9 (Spanish).
12. de Vries, Jan and van der Woude, Ad: *The first modern economy: success, failure, and perseverance of the Dutch economy, 1500–1815,* Cambridge: Cambridge University Press, 1997, p. 50.
13. Kruijtzer, Gijs: 'European Migration in the Dutch sphere', in Oostindie, *Dutch colonialism,* p. 108.
14. Lommerse, *Population figures.*
15. van Stipriaan, Alex: *Surinaams contrast: Roofbouw en overleven in een Caraïbische plantagekolonie 1750–1863,* Leiden: KITLV Uitgeverij, 1993, pp. 35, 429–33, 437.

16. Schwartz, Stuart B.: 'A Commonwealth within itself. The early Brazilian sugar industry, 1550–1670', in Schwartz, Stuart B. (ed.): *Tropical Babylons: sugar and the making of the Atlantic world, 1450–1680,* Chapel Hill: University of North Carolina Press, 2004, pp. 162–70.
17. Cited in McCusker, John J. and Menard, Russell R.: 'The sugar industry in the seventeenth century: a new perspective on the Barbadian "sugar revolution"', in Schwartz, *Tropical Babylons*, p. 295. See Emmer, P.C.: '"Jesus Christ was good, but trade was better": an overview of the transit trade of the Dutch Antilles, 1634–1795', in Paquette, Robert L. and Engerman, Stanley E. (eds), *The lesser Antilles in the age of European expansion,* Gainesville: University Press of Florida, 1996, pp. 211–12; Galloway, J.H.: *The sugar cane industry: an historical geography from its origins to 1914,* Cambridge: Cambridge University Press, 1989, pp. 76, 79–83; Schwartz, A Commonwealth within itself, p. 7.
18. McCusker and Menard, *Sugar industry*, pp. 295–7.
19. van Stipriaan, Alex: 'The Suriname rat race: labour and technology on sugar plantations, 1750–1900', *New West Indian Guide,* Vol. 63, Nos 1–2 (1989), p. 97.
20. Raynal, G. F.: *Tafreel van de bezittingen en den koophandel der Europeänen in de beide Indiën, getrokken uit de wysgerige en staatkundige geschiedenis van de bezittingen en den koophandel der Europeänen in de beide Indiën,* Amsterdam: M. Schalekamp, 1784, p. 336 (my translation); Oostindie, Gert: *Roosenburg en Mon Bijou: Twee Surinaamse plantages, 1720–1870,* Dordrecht: Foris, 1989, p. 128; van Stipriaan, *Surinaams contrast*, p. 139, Oostindie, Gert and van Stipriaan, Alex: 'Slavery and slave cultures in a hydraulic society: Suriname', in Palmié, Stephan (ed.), *Slavery and slave cultures in the Americas,* Knoxville: University of Tennessee Press, 1995; Oostindie, Gert: '"British capital, industry, and perseverance" versus Dutch "old school"? The Dutch Atlantic and the takeover of Berbice, Demerara and Essequibo, 1750–1815', *BMGN/Low Countries Historical Review,* Vol. 127, No. 4 (2012), pp. 28–55.
21. Davids, Karel: 'Sources of technological change in the Dutch Guiana, c. 1670–1860', in Lafuente, A., Elena, A., and Ortega, M.L. (eds), *Mundialización de la ciencia y cultura nacional,* Madrid: Universidad Autónoma de Madrid, 1993; van Stipriaan, *Suriname rat race.*
22. See Tomich, this volume.
23. den Heijer, Henk: 'A public and private West India interest', in Oostindie, Gert and Roitman, Jessica Vance (eds), *Dutch Atlantic connections 1680–1800,* Leiden: Brill, 2104.
24. See Leonard, this volume.
25. Fatah-Black, Karwan: 'Suriname and the Atlantic World, 1650–1800', unpublished PhD thesis, Leiden University, 2013, p. 247.
26. Oostindie, *British Capital.*
27. Klooster, Wim: *Illicit riches. Dutch trade in the Caribbean, 1648–1795,* Leiden: KITLV Press, 1998, p. 176, see also pp. 89–104.
28. Smith, Adam: *An inquiry Into the nature and causes of the wealth of nations,* Amsterdam: MetαLibri, 2007, IV, 7.34. See also Luzac, Elias: *Hollands rijkdom [...],* Leyden: Luzac en Van Damme, 1780–1783, II, p. 257 and III, pp. 362–3, Raynal, G.F.: *Tafreel van de bezittingen en den koophandel der Europeänen in de beide Indiën, getrokken uit de wysgerige en staatkundige geschiedenis van de bezittingen en den koophandel der Europeänen in de beide Indiën,* Amsterdam: M. Schalekamp, 1784, pp. 338–9.

29. de Vries, Jan: 'The limits of globalization in the early modern world', *Economic History Review,* Vol. 63, No. 3 (2010), p. 729; de Vries, *Dutch Atlantic economies*, p. 19.
30. de Vries, *Dutch Atlantic Economies*, p. 19; idem. *Limits*, p. 729; Enthoven, Victor: 'An assessment of Dutch transatlantic commerce, 1585–1817', in Postma and Enthoven (eds), *Riches from Atlantic commerce*, p. 441. Klooster's estimate of 8.5 per cent (*Illicit riches*, p. 177, *Overview of Dutch Trade*, p. 379) does not include Berbice, Demerara, or Essequibo.
31. de Vries and van der Woude, *First modern economy*, p. 402.
32. Heijer, Henk den: 'The Dutch West India Company, 1621–1791', Postma Enthoven, *Riches from Atlantic commerce*, pp. 30, 108.
33. Emmer, P.C.: *De Nederlandse slavenhandel 1500–1850*, Amsterdam: Arbeiderspers, 2000, pp. 169–70.
34. Van Stipriaan, *Surinaams contrast*, pp. 266–75; Oostindie, Gert: 'The economics of Suriname slavery', *Economic and Social History in the Netherlands*, Vol. 5 (1993), pp. 7–11.
35. van de Voort J.P.: 'Dutch capital in the West Indies during the eighteenth century', *Acta Historiae Neerlandicae: the Low Countries history yearbook*, Vol. 14 (1981), p. 101; Van Stipriaan, *Surinaams contrast*, pp. 215–31.
36. Oostindie, *Roosenburg*, pp. 371–87, van Stipriaan, *Surinaams contrast*, pp. 276–309.
37. Jordaan, Han: 'Slavernij en vrijheid op Curaçao: de dynamiek van een achttiende-eeuws Atlantisch handelsknooppunt', unpublished PhD thesis, Leiden University, 2012; Rupert, Linda M.: *Creolization and contraband: Curaçao in the early modern Atlantic world*, Athens: University of Georgia Press, 2012.
38. Enthoven, Victor, '"That Abominable Nest of Pirates". St. Eustatius and the North Americans, 1680–1780', *Early American Studies*, Spring 2012; Klooster, *Illicit riches*, pp. 89–104.
39. van Lier, R.A.J.: *Frontier society. A social analysis of the history of Surinam*, The Hague: Martinus Nijhoff, 1971, pp. 85–95; Fatah-Black, *Suriname*, passim; Fatah-Black, Karwan: 'Paramaribo as Dutch and Atlantic nodal point, 1650–1795', in Oostindie and Roitman, *Dutch Atlantic connections*; Israel, Jonathan: 'Jews and crypto-Jews in the Atlantic world systems, 1500–1800', in Kagan, Richard L. and Morgan, Philip D. (eds): *Atlantic diasporas: Jews, conversos and crypto-Jews in the age of mercantilism, 1500–1800*, Baltimore: Johns Hopkins Press, 2009, pp. 3–17; Klooster, *Illicit riches*, pp. 65–7; Rupert, *Creolization*.
40. Koot, *Empire*; Schnurmann, Claudia: 'Atlantic trade and American identities. The correlations of supranational commerce, political opposition, and colonial regionalism', in Coclanis, *Atlantic economy*; Morgan, Kenneth: 'Anglo-Dutch economic relations in the Atlantic world, 1688–1783', in Oostindie and Roitman, *Dutch Atlantic connections*; Antunes, Cátia: *Globalisation in the early modern period: the economic relationship between Amsterdam and Lisbon, 1640–1705*, Amsterdam: Aksant, 2004, pp. 185–6.
41. Marzagalli, Silvia: 'The French Atlantic and the Dutch, late seventeenth-late eighteenth centuries', in Oostindie and Roitman, *Dutch Atlantic connections*; Crespo Solana, Ana: 'A network-based merchant empire: Dutch Trade in the Hispanic Atlantic context (1680–1740)', in ibid.

42. Fatah-Black, *Paramaribo*; Klooster, *Illicit riches*, pp. 89–104, Ormrod, David: *The rise of commercial empires: England and the Netherlands in the age of mercantilism, 1650–1770*, Cambridge: Cambridge University Press, 2003, p. 336.
43. Klooster, Wim and Gert Oostindie (eds): *Curaçao in the Age of Revolutions, 1795–1800*, Leiden: KITLV Press, 2011.
44. van Lier, *Frontier society*, pp. 223, 230, 240; Adhin, J.H.: *Development planning in Suriname in historical perspective*, Leiden: Stenfort Kroese, 1961, p. 83.
45. Average exports stood at 8.8 million kg in 1816–1825, 12.2 million in the 1860s, 7.5 million in the 1880s (Reesse, Jan Jacob: *De suikerhandel van Amsterdam van 1813 tot 1894: Een bijdrage tot de handelsgeschiedenis des vaderlands hoofdzakelijk uit de archieven verzameld en samengesteld door J.J. Reesse*, 's-Gravenhage: Nijhoff, 1911, p. xli). The number of sugar plantations decreased from 108 in 1830 to 14 in 1890 (van Stipriaan, *Surinaams contrast*, p. 31). Thereafter, only one sugar plantation, Mariënburg, continued to matter, reporting 33 profitable and 18 negative years in the 1888–1940 period (Hoefte, Rosemarijn: *In place of slavery: a social history of British Indian and Javanese laborers in Suriname*, Gainesville: University Press of Florida, 1998, pp. 16, 18).
46. Lommerse, *Population figures*, pp. 323–4; Oostindie, *British capital*, p. 35. On British investments in Surinam, see Van Stipriaan, *Surinaams contrast*, pp. 73, 231, 237, 239–42, 256.
47. Reesse, *Suikerhandel*, pp. xxxiii, xli.
48. Reynoso, Alvaro: *Verhandeling [...] over de cultuur van suikerriet*, Rotterdam: Nijgh, 1855.
49. Bosma, Ulbe and Curry-Machado, Jonathan: 'Two islands, one commodity: Cuba, Java, and the global sugar trade (1790–1930)', *New West Indian Guide*, Vol. 86, Nos 3–4 (2012), pp. 238–9.
50. van Zanden, J.L. and van Riel, A.: *Nederland 1780–1914: staat, instituties en economische ontwikkeling*, Amsterdam: Balans, 2000, p. 223.
51. Hoefte, *In place of*, pp. 15–23.
52. Quoted in Ramsoedh, Hans: 'Suriname 1933–1940: koloniale politiek en beleid onder gouverneur Kielstra', unpublished PhD thesis, Utrecht University, 1990, p. 5.
53. Emmer, P.C.: 'The myth of early modern globalisation: the Atlantic economy, 1500–1800', *European Review*, Vol. 11, No. 1 (2003) , pp. 37–47; Cf. Elliot, J.H.: 'Afterword. Atlantic history: A circumnavigation', in Armitage, David and Braddick, Michael J. (eds), *The British Atlantic World, 1500–1800*, Basingstoke: Palgrave Macmillan, 2002, pp. 246–9.
54. Ormrod, *Rise*, p. 43. See also O'Brien, Patrick, 'Mercantilism and imperialism in the rise and decline of the Dutch and British economies 1585–1815', *De Economist*, Vol. 148, No. 4 (2000), pp. 469–501; Emmer, *Jesus Christ*, pp. 219–20.
55. de Vries, *Dutch Atlantic economies*, p. 21; de Vries, *Limits*, pp. 714–15 and 732 for the distinction between 'hard' and 'soft' globalisation.

7

From Local to Transatlantic: Insuring Trade in the Caribbean

A.B. Leonard

Marine insurance is an old and very flexible financial instrument. Most of the fundamental characteristics of the policies underwritten today in international insurance markets were established in the fourteenth century, in the Italian city-states which then dominated extra-European trade. It was created by merchants as a tool to be employed amongst themselves, and was intended to spread the risks of ocean-going commerce as widely as possible between them, for the lowest possible cost. Unlike most credit instruments, which make advances of capital, marine insurance provides contingent capital which is paid to the buyer only in the event that an actual insured loss has occurred. This allows individual merchants to trade with less capital than the specific perils of individual adventures prudently demand, permitting them to maximise their investment in trade goods. As trade expanded and merchants' trading patterns took them to increasingly distant ports, they brought their practices of marine insurance with them, transferring and expanding their system of risk-spreading at each new location.[1]

The techniques and customs of marine insurance underwriting arrived, fully formed, in the Atlantic World with merchants and their trade. It was a critical component of that commerce, providing a contingent-capital safety net which allowed merchants to focus their cash and credit resources on their adventures, thereby oiling the machinery of Atlantic World exchange. The patterns of its use and spread highlight Atlantic World relationships, both within and outside imperial territorial constructs. This chapter examines such relationships through the lens of marine insurance, and illustrates circuits of money and trust upon which Caribbean trade – between both imperial centres and their dependent peripheries, but also between peripheries themselves – was secured. It begins with a brief outline of the mechanics of marine insurance and its early practice in London, before moving to a specific area of Atlantic World trade: that between emerging mainland colonial economies and the plantation-based economies of the Caribbean. Insurance underwriting in British North America adopted the patterns and practices of European, and especially London insurers. This chapter then examines in detail the underwriting activities of Obadiah Brown, an eighteenth-century merchant of Providence, whose insurance business followed his trade. It then returns to London, which continued to play a critical role in Atlantic World marine insurance, although one slightly different to that of New England merchant-insurers. Finally, the chapter considers briefly the decline of private marine insurance underwriting in the United States.

Early marine insurance

The structure of an insurance contract, and of the market institutions that support underwriting, is relatively simple. Following Italian practices, the model adopted in most trading centres by the fifteenth and sixteenth centuries was little different from that employed more than a century earlier in the Mediterranean. Multiple underwriters participated in each contract by assuming a proportion of its face value (their *line*), measured in the currency of the policy and expressed as a part of the total value of the sum insured, which was the maximum indemnity under the policy. This approach spread the risk broadly. Each participating insurer signed his name below the boilerplate text of the policy, making him and 'under-writer' or 'sub-scriber'. Underwriters charged a fee (the *premium*) which was expressed as a percentage of the sum insured (the *rate*), and which varied based on the characteristics of both the vessel and the voyage to be insured (the *risk*). The rate charged to the buyer (the *insured*) by all the underwriter-participants under an individual policy varied only very rarely; each charged the same price for their share of a specific risk, usually set by an underwriter who was expert in the risk-type (the *leader*). Rates were adjusted according to the loss experience, and to various threats related to a specific voyage, such as the season, or the activity of corsairs. Underwriters sometimes specified the broad perils which were to be insured under the policy, usually according to standard policy language (the *wording*), and included in the contract the name of the insured vessel, the nature of the cargo, and the details of its voyage. Policies were often arranged by intermediaries (*brokers*), and in some locations sealed by notaries. Most of the underwriters were merchants themselves (contemporaneously, *merchant-insurers*), although by the fifteenth century wealthy investors were also taking lines on insurance contracts. Insurers' salvage rights, which conferred upon them ownership of any property for which a claim (an *indemnity*) had been paid under a policy, were established in principle, if not in law. Custom and trust were critical ingredients. As Muldrew states, 'the Atlantic World was held together through ties of interpersonal credit between merchants trading over long distances'. Nowhere is this trust in the personal credit of co-venturers, colleagues, and competitors more evident than in marine insurance, where the insured buys only the underwriter's promise to pay, reversing traditional credit relationships: it is the buyer who must be concerned about the creditworthiness of the seller.[2]

With these characteristics more or less intact, marine insurance arrived in northwestern Europe from Italy in the late middle ages. It was traded regularly amongst merchants in Cadiz, Aleppo, Antwerp, Amsterdam, Hamburg, and other major European ports by the sixteenth century. It had reached London, at latest, by the early fifteenth century. The oldest surviving record yet found of conventional insurance transactions in the city is an entry in the Plea Rolls of the City of London. In 1426 Alexander Ferrantyn, a Florentine merchant resident in London, took an insurance dispute to the Lord Mayor and Aldermen. The case, heard at Guildhall, involved his underwriters' refusal of a claim for his vessel the '*Seint Anne of London*', which was carrying a cargo of wine to England from Bordeaux. Both the vessel and the cargo had been insured for £250 by 17 Italian merchants also resident in London. The ship and goods had been seized by Spaniards, but Ferrantyn, through an agent, had managed to repurchase them from the Flemish merchants to whom the privateers had sold their prize. The policy specified that the 'order, manner, and custom of the Florentines' was to govern the contract. Ferrantyn asked his insurers to pay up, citing specifically 'the *lex mercatoria*' (the law merchant), and the clause in his policy about Florentine custom. The disputants claimed respectively that this local practice required the indemnity to be paid in this circumstance, and that it did not. Both parties promised to produce notarised testimony from the Italian city which would outline the prevailing local custom. So confident were the defending insurers that they paid into the court the disputed £250, plus £100 as surety.[3]

Ferrantyn's case is important not only for its primacy, but also because it shows how marine insurance practice spread with trade around the world. It arrived in England with the Italian merchant community, who used the instrument according to the customs of their homeland.

England's judicial system was ill-equipped to decide questions such as that upon which the case turned, and so the uncodified, customary Law Merchant was imported with marine insurance practice. In any case, merchants preferred faster and cheaper methods of dispute resolution, such as arbitration. Panels of disinterested merchant arbiters who were familiar with insurance custom and practice were convened to decide outcomes. Thus, England's marine insurance market remained largely outside of the interest of the common law until the later eighteenth century, by which time London's market was already the world leader.[4]

The report of a parliamentary enquiry into marine insurance in London, published in 1720, reveals concretely the nature of the market at the turn of the eighteenth century. Individual underwriters (sometimes acting within private commercial partnerships) remained the driving force, and perhaps the whole local source of cover. 'Insurance is now in the Hands of private Persons, called Office-Keepers, carried on by Brokers', wrote Nicholas Lechemere, the Attorney General, in a document preserved in the report of the inquest. Their 'Employment is to procure for the Merchant, as his Occasions require, Persons to subscribe the Policies of Insurance on such Terms as shall be agreed on'. Eight brokers submitted to Lechemere a list of 163 individuals who 'had subscribed Policies of Insurance on Ships and Merchandizes' which they had arranged. Overlap between brokers and underwriters continued. For example, John Fletcher, one of the eight submitting brokers, subscribed to at least one of nearly fifty extant policies issued to the Turkey merchant Ralph Radcliffe, and is named as the insured party in a policy issued in 1716 to insure goods to the value of £1,000 on a voyage from London to China and back. The courts remained a secondary dispute resolution option: John Barnard, a wine merchant, leading underwriter, and future City MP, stated that 'Insurers are generally desirous to have Disputes about Losses or Averages adjusted by Arbitration, it being in their Interest to do so, and that the Insurers very often pay unreasonable Demands, rather than suffer themselves to be Sued'.[5]

The report also reveals that by 1720 London was attracting considerable foreign insurance custom. Barnard stated that 'Foreigners from almost all parts of *Europe* have continual Recourse to the Insurers of *London*, to be insured for very large Sums of Money'. London underwriters had established an international agency network to obtain this business. Barnard's agent in Cadiz, for example, would accept risks in the Spanish port on his behalf. By this time the international dispersion of London's underwriting business was already long established, according to evidence surviving from 1693, which reveals that it was commonplace that 'Assurances are made in Trust for divers Persons in remote Parts of the Kingdom, and Places beyond the Seas'. The 1720 enquiry itself related in part to applications by two joint-stock companies (the future Royal Exchange Assurance and the London Assurance) for royal charters granting a duopoly over the underwriting of marine insurance in London. This was awarded in 1720, under legislation better known to historians as the *Bubble Act*. However, underwriting by individuals – the traditional method – was permitted to continue, which served to strengthen that sector of the market by restricting the growth of corporate underwriting. Private underwriting received a further boost when the two new companies contracted their risk appetite in the decade after their charters, in part due to concentrated claims arising from the loss of twelve Jamaica ships. In 1741 the provisions of the *Act* restricting corporate underwriting were extended to include Britain's North American colonies.[6]

Nash wrote of London's merchants that they 'progressed in the classic sequence followed by other premier business communities in Europe, such as those of Antwerp and Amsterdam – that is, from merchandising and ship owning, to commission trading and insurance, and ultimately to finance and merchant banking'. Atlantic World merchants of the American mainland colonies also followed this path. For the insurance link in Nash's sequence, emerging American merchants emulated their London-based merchant-insurer predecessors, principally to satisfy the risks associated with their rapidly expanding trade with the Caribbean and the broader West Indies. From 1651 until independence, they conducted this trade within (or working around) the constraints of the Navigation Acts, and under the restrictions of the *Bubble Act*. A further constraint for merchants in New England, Pennsylvania, and New York, hungry for metropolitan imports, was the absence of a major staple commodity to balance their trade with Britain. They turned instead to the provision of services such as the carrying trade, and provisioning of the colonial Caribbean possessions of Britain, Spain, France, and the lesser Atlantic powers. Insurance is often cited as one of these 'invisibles', although the nature and extent of New England merchants' marine insurance practice has not been outlined in detail by historians.[7]

The New England–Caribbean trade

By the middle of the eighteenth century north Atlantic trade between European powers and their colonies was enjoying a particular growth. According to Richardson, the increasing buoyancy of this trade in

New England – an area comprising the modern US states of Maine, Massachusetts, New Hampshire, Vermont, Rhode Island, and Connecticut – stimulated the growth of regional New England trade, particularly with the Caribbean. A contemporary, the London merchant Michael Atkins, wrote in correspondence in 1751 that 'Traders at the Northern Colonies have all the West India business to themselves, Europeans can have no encouragement for mixing with them in the comodities of provisions and lumber. You [New Englanders] time things better than we and go to markett cheaper.' McCusker and Menard described the 'sugar islands' as 'indispensible to the development of the mainland colonies', their 'economies so tightly intertwined that full understanding of developments in one is impossible without an appreciation of developments in the other.' New England's trade links within the Caribbean were already well established by the mid-eighteenth century, and included the import of tropical produce, primarily molasses and rum, and the export of goods including livestock, fish, lumber, candles, ironware, and building and packaging materials. This trade, Richardson argues, gave rise to powerful merchant communities in Boston and other New England ports, and to a shipbuilding industry supplying, among others, the British merchant fleet. It also saw many merchants in the northern colonies become ship owners, reflecting another step in Nash's merchant-development continuum.[8]

The value of New England's exports to the Caribbean approximately quadrupled between the 1750s and the early 1770s, to exceed £300,000 per year. The significant rise in these estimates is supported by a parallel increase in the number of clearances of vessels from Salem, Boston, and Rhode Island in the period leading up to the Revolutionary War. Work by Shepherd and Walton shows that New England's trade with the West Indies increased steadily in the brief period for which detailed data are available, from £510,000 in 1768 to £750,000 in 1772, of which roughly half each year comprised New England exports. The importance of this trade to the pre-independence economy of the six states is easy to overlook (the transatlantic business receives much more scholarly attention), but Shepherd and Walton's figures put the West Indies share of New England's exports at 63.6% over the five-year period, much more than any of Britain, Southern Europe (primarily a vent for New England fish), or Africa (which purchased primarily New England rum). Richardson concludes that 'trade with the Caribbean... was the most dynamic sector in New England overseas trade in the period after the War of Austrian Succession.'[9]

Insurance arrived at the mainland colonies in the same way it had reached London, with the merchants who traded there. Men such as the Quaker merchant James Claypoole were regular buyers of insurance in London before they emigrated to British North America. From London, Claypoole had traded in a variety of commodities to Europe and the West Indies during the Restoration era, before leaving England with William Penn as a founder of Philadelphia. Brought and understood by men such as Claypoole, marine insurance was quickly adopted in the mainland colonies. In the seventeenth century, for example, Maryland planters shipping tobacco to England on their own account would frequently arrange insurances. However, in that century marine insurance was underwritten primarily in the venturing merchants' European home cities, or those of their European correspondents, for transatlantic voyages. In the years 1664 to 1667 the English merchant Charles Marescoe underwrote marine insurance in London for voyages from Virginia and New England to English ports. (Since his own, Baltic trading interests did not extend to North America, he must have been underwriting as an investment and risk-diversification exercise.) Extant correspondence shows clearly and repeatedly that merchants resident in America purchased insurance from their London suppliers and agents. For example, Joseph Cruttenden, a London merchant supplying apothecaries' materials to North America, wrote in 1710 to his Boston-based client Habijah Savage that 'I have complyd with your desire and charged but 6^{lt}5 pCent advance on all the things now sent... allowing for insurance which may be high'. Later he wrote to another Boston client, John Nichols, 'You see by the invoice I have charged you with the Ensurance which was done with your ffriends consent, for it was noe way reasonable for me to run the risque.'[10]

Limited private underwriting was almost certainly occurring along in the commercial centres of the north-eastern colonies at the time, but evidence which confirms that it occurred during the seventeenth century is scant. It is clear, however, that marine insurance practice developed in line with American trade to other British colonies and elsewhere. In Philadelphia in 1721 local demand for local insurance was sufficient to encourage one John Copson to launch an insurance brokerage office. His advertisement of 25 May that year in the *American Weekly Mercury* announced the opening by him of an 'Office of Publick Insurance on Vessels, Goods and Merchandizes'. It has been widely regarded as marking the advent of local insurance in the British American colonies. The announcement stated that 'the merchants of this City of Philadelphia and other ports have been obliged to send to London for such insurance, which has not only been tedious and troublesome, but ever precarious, and for the remedying of which this office is opened'. Copson promised that the underwriters would 'be Persons of undoubted Worth and Reputation', but no other record of the venture has survived. A second broking office was established by Joseph Saunders in 1748 (by which time Copson's appears to have ceased operations). Boston's first dedicated insurance office was perhaps opened in 1724 by a notary, Joseph Marion, although the date of establishment is disputed; a second was opened in 1739 by Benjamin Pollard. New York did not have an insurance broking office until 1759.[11]

Others followed, and insurance offices garnered a role as mercantile places of association. John Rowe, a merchant of Boston, wrote in 1768 that he had 'Spent the evening at Mr Barber's Insurance-Office.' The community of brokers and underwriters grew relatively rapidly. Gillingham has identified 22 brokers active in Philadelphia alone in the years 1721 to 1805, who were completing policies with 164 discrete underwriters. Insurance facilities were also established in the southern colonies. In Virginia in 1739 several merchants of Charleston formed a loose association for the purpose of mutual insurance, and John Benfield opened an insurance office in the city thirty years later. Broking offices were also established in Norfolk and Baltimore after 1750. The proliferation of formalised broking operations indicates that a relatively deep community of underwriters must have been operating. Where local offices had not been established, insurance buyers would source coverage from nearby centres. Pares reports that merchants of Portsmouth, New Hampshire wrote regularly to John Reynell of Philadelphia (which Pares describes as 'probably the chief centre for insurance on the continent') requesting that he arrange insurance.[12]

While these facts about broking offices have been long established, little work has been completed about the dynamics of the market, or about the individual underwriters themselves. Pares has observed that 'underwriting had some aspects of a mutual insurance scheme'. He recognised five of 23 underwriters taking lines on the Caribbean voyage risks of John Reynell as merchants, including the three most frequent underwriters of his policies. This, of course, reflects the merchant-insurer practice established in London, and that had been invented by Italian merchants centuries before.

Obadiah Brown: underwriter

Among British North America's fledging merchant-insurer community was Obadiah Brown (1712–62), a prominent merchant, shipowner, and manufacturer active in Rhode Island. Brown, a native of Providence and third-generation descendant of one of the colony's Baptist founders, carried on a trade in goods including imported cocoa, rum, and molasses, and the export of mainland commodities including lumber, livestock, and foodstuffs. He was also a manufacturer of spermaceti candles. Brown has been identified as the first Providence merchant to trade directly with England. He and his brother James, and later James's four sons (who had been raised by their uncle), were the colony's leading mercantile figures.[13]

Obadiah, 14 years his brother's junior, was educated in trade by his sibling. He began his career on the seas with a trading voyage to Antigua in 1733, as Master of his brother's sloop the *Dolphin*. In 1734 his second and third voyages took him as far as Surinam, which was to figure prominently in his later trade, and in his insurances. Surinam had been opened by its Dutch colonial masters to British trade in 1704, yielded large quantities of molasses for export, and was an eager purchaser of the northern colony's two key products: tobacco and horses. Brown also travelled, in his early voyages, to Hispaniola, where the port known in his records as 'The Mount' (modern Monte Cristi in the Dominican Republic) lay just a few miles from the border of French Hispaniola. The Mount was thus a regular location for illicit trade between British and Spanish Atlantic merchants and their French counterparts, with whom commerce was prohibited during the Seven Years' War, 1754 to 1763. This otherwise undistinguished port was also to be of future importance to Obadiah Brown.[14]

Brown's earliest insurance activity leaving a trace in the record is a transaction which would more correctly be described as a risk swap. In 1747 he and his uncle Elisha Brown, who like Obadiah was a shipowner and merchant, each agreed to accept £100 of risk upon the other's vessel, the brigantines *Desire* and *Wainscot* respectively. No doubt from Elisha, from his brother James, and from fellow merchants and published sources, Obadiah Brown learned the customary London practices of marine insurance. Amongst his papers survives his copy of a small book imported from London and entitled *A guide to Book Keepers according to the Italian Manner*. Brown will have taken note of its instructions regarding insurance. In explaining the Italian system of double-entry bookkeeping (which, like marine insurance, is a product of the Italian commercial revolution), it advises, for example, 'When you pay the Premio for the Ensurance of any Sum upon a Voyage, you enter that Payment on the Creditor-side of your Cash-Book'. It then describes how to account for a business-partner's share of premiums in joint voyages. Another entry describes the more complex entries required when accounting for claims and abatements under insurance policies arranged by the bookkeeper for third parties: 'When you Receive Money of the Ensurer, on Account of Ensurance by you made in Commission, and you make him an Abatement, you enter the whole as follows: First, In your Journal you Enter...'[15]

It seems unlikely that this was the only trade manual which Obadiah Brown owned (it survived perhaps because the volume's blank pages are filled with Brown's manuscript journals of his first voyages in the West Indies trade as Master of ships owned by his brother). Many other merchant manuals were published, some of which covered in great detail the customs and conventions of marine insurance underwriting (rather than simply the accounting for insurance transactions). These include Edward Hatton's much-reprinted *Comes Commercii: or the trader's companion*, published in London in the late seventeenth century and republished in multiple editions until the early nineteenth, and Charles Molloy's *Treatise of affairs maritime, and of commerce*, first published in London in 1676. Both included detailed sections on the methodologies of transacting insurance, and contributed to the dissemination of London practice throughout the merchant world. William Leybourn's 1693 book *Panarithmologia* includes a section *On insurances* which begins 'Suppose you ship 300 *l.* of Goods for *Jamaica*, you being unwilling to run so great a hazard your self...'[16]

Brown extended his mercantile activities in the 1750s, when he began trading directly from Providence to London, and expanded his core West Indies trade. It was at this time that he began routinely to underwrite. His *Marine Insurance Book* contains a record of the risks he assumed during the periods March 1753 to the spring of 1758, then from March 1760 until 1762, the year of his death (the break represents a halt in Brown's underwriting, rather than a gap in the record). The document illustrates what may be a typical insurance-risk portfolio for a local merchant-insurer of his time. It shows that Brown dabbled in underwriting, insuring primarily vessels involved in his own arena of commerce, the Caribbean trade, and especially to Surinam, Jamaica, and Hispaniola (Table 7.1). Often his clients were his close business associates; sometimes he underwrote insurances related to his own vessels (presumably for third parties, perhaps upon his captains' privilege cargoes). Thus, his insurance activity mirrored his mercantile work. Brown's underwriting ignored imperial boundaries, and flourished with New England's Caribbean trade. Policies he supported covered voyages from various mainland ports to or from Caribbean ports, including those of the Leeward Islands and other important British and foreign

Table 7.1 Voyages insured by Obadiah Brown

Place	***Total***	***To***	***From***	***Return***	***Unstated***
Surinam	**32**	9	21	2	
Jamaica	**23**	12	7	4	
Hispaniola (incl. Monte Christi)	**22**	7	14		1
South Carolina	**5**	3	1	1	
Carolina	**4**	1	1	2	
Barbados	**4**	4			
Guadeloupe	**4**	2	1	1	
Ocoya	4		2		2
Barbados	**3**	3			
Holland/Amsterdam	**3**	2	1		
Newfoundland	**2**	1		1	
Leeward Islands	2			2	
Antigua	**2**	1	1		
St Eustacia	2		1		1
Martinique	**2**	1	1		
Others*	6	1	5		
Europe or Africa	**2**	1	1		
Cross risks, time risks, and unspecified	39				

Notes: *Including Bay of Honduras, Havana, Essequibo, Montreal, and Georgia.
Source: Compiled from Obadiah Brown Papers, 'Marine insurance book, 1753–1762', RIHS MSS 315, E445/44 Part 1, Reel 23.

entrepôts. An occasional slave voyage from west Africa was insured, as were fisheries cargoes from Newfoundland, and sometimes vessels travelling to or from European ports.[17]

Brown's Risk Book reveals that many of the practices of marine insurance underwriting which had been developed in Italy and refined in London were employed in the American colonies by the mid-eighteenth century. At the most basic level, premium was charged as a percentage of the sum insured. Brown's currency of account was Rhode Island pounds, although he occasionally offered cover in Spanish dollars, and sometimes accepted premium payments in 'Melasses'. He sometimes charged an additional premium for vessels which made extra calls. For example, in 1753 he made an 'adition of 2 p^{r}ct more for tuching at ye Moscata shore' on an insurance he granted on the sloop *Indian King*, sailing from Honduras.

Although Brown's Risk Book is primarily an accounting record, it is clear that the insurances noted in it were supported by traditional policy contracts, since Brown sometimes adds notes to entries such as 'Pollacy canceled'. Sometimes the language of London policies was employed and has been recorded, for example when Brown noted that a vessel was permitted under the policy to travel 'to at and from' its destination. As well as the details of the voyage, the Risk Book typically (but not always) states the other standard underwriting information that was usually included on the face of London policies: the vessel name and type, and the name of its master. When some details were not known, Brown omitted them, sometimes leaving a blank space for future insertion. A handful of entries note explicitly that cover is specifically for cargo or goods. Also in common with London risk books of the period, Brown noted the fate of each vessel insured, whether 'Ar[r]ived', 'Lost', or 'Taken'.

The outbreak of the Seven Years' War (1756–63) and the privateering it brought to the eastern seaboard of the North American continent had clear impacts on Brown's underwriting. Premium rates rose dramatically. For example, Brown's 1753 rate for the voyage from Surinam began at three per cent. It rose to 16% in 1758, and dropped to 14% in 1760. His peacetime rate of three per cent for vessels sailing to Barbados rose to 15%; sloops sailing to Jamaica saw the price of cover rise from a peacetime rate of five per cent to a wartime peak of 12%, although the rate charged by Brown to insure a snow, a type of vessel typically less manoeuvrable and therefore less defensible, on that route in wartime reached 20%. In wartime Brown often charged rates between 20% and 30%, hinting at the large profits which wartime trade voyages must have yielded. Two insurances on vessels designated a 'flag of truce' (which were granted immunity from enemy attack, and were limited in their role to prisoner exchange, but which inevitably carried massive quantities of contraband) attracted rates of just five per cent, including one travelling to the French colony of Martinique. This shows that the entire uplift in the rates charged during wartime was a result of the increased likelihood of losses arising from the peril group known as '*gentium*' in traditional insurance parlance, the risks of men. Further, war (or the losses arising from it) appears to have compelled Brown to cease, briefly, his underwriting activities in the spring of 1758. When risks are intensified, adequate premium levels uncertain, and losses eroding earlier profits, such a decision may make sense, especially when a merchant such as Brown had alternative, potentially highly profitable investment opportunities outside marine insurance underwriting.

With war, the notation 'Taken' begins to appear in Brown's Risk Book. Some vessels are noted as 'retaken' or 'ransomed'. In one instance in 1761 Brown had insured £500 on the sloop *Pawtucket*, which was sailing for the port of Monte Christi in Spanish Hispaniola, where covert trade with the French was rife. The Risk Book states that the vessel was 'Taken and retaken', and that an 'aviraj' (average) claim was calculated, presumably to cover the costs of damage, pilfering, or an entitlement arising from the capture. Average is a Mediterranean insurance practice which had been adopted in London, and clearly was inherent in North American underwriting practice from an early stage.[18] In the wartime period Brown also began to insure privateering ventures, such as his £50 line granted 30 January 1762 'On Richd Jacksons Private Adventure' to Guadeloupe. He sometimes required that the insured bear an *'abatement'*, a condition of cover which limited the payment of any claims under the policy to an agreed and specified percentage of the total valid claim. This retained self-insurance was another practice common in London, and served to discourage fraud. Such a charge is noted, for example, in Obadiah Brown's 1762 line on James Brown's voyage from Surinam, which carries the clause 'in case of loss 5 pC^{t} deduct'; this policy abated five per cent, and thus paid 95%. At this time too Brown sometimes underwrote policies on an 'all Risques' basis, as noted in several entries, although it is not clear what perils were excluded in the absence of the extension. On 23 June 1761 Brown added a common European wartime warranty: his coverage of the 'Sloope Britannia Thos Greene master from Antigo' was required to travel 'with Convoy'. Two final clues about colonial underwriting practice appear in the final page of the document. Brown recorded that he had paid a total of £280 for the inspection of vessels, and contributed to an 'average' claim which arose under one of the policies in which he had participated.[19]

Brown's Risk Book also reveals insurance accounting conventions. On 1 January 1757 he noted £570 as his underwriting 'Balance Due to Prem^{m} Acc^{t}', having paid losses of exactly £2,000, under 89 risks assumed, which yielded a recorded premium of £2,570 (although the actual total was £2,555; Brown or his clerk had made an error of addition). This represents a loss ratio – the measure of loss expenses against premium income, but excluding other costs such as commissions and overheads – of 77.8% (Brown's records show no brokerage fees, which may indicate he underwrote directly and informally, rather than through an intermediary). Brown continued to underwrite, but at his next settlement date, in early 1758, the fortunes of his underwriting had changed. He had made a loss on the period, and transferred the total that had been reserved back to his current premium account (illustrating another customary and prudent insurance practice, that of *reserving*, or setting aside profits from past underwriting to meet future claims under current, open policies, and under those to be underwritten in the future). Brown then ceased underwriting, perhaps deterred by the loss. He resumed again in 1760, when rates had increased sharply, and continued to underwrite profitably until his death. He recouped his losses, and made a pure underwriting profit of £1,045 over his recorded experience as a marine insurer. When the only recorded expenses, £280 for the inspection of vessels, are included, his underwriting activities still remained profitable overall. A completed record, divided according to the accounting periods Brown adopted, is presented in Table 7.2. His lifetime underwriting profit was about 8.5% of premiums. Additional expenses are likely to have reduced the profit, however, while the income, direct or indirect, achieved against the accrued premiums he held before the retained capital was paid out as claims may have made the venture somewhat more profitable.

Brown made two forays into the slaving business. The first venture, led by his brother, appears to have been a moderate success. The outcome of the second such voyage, in 1759, is less certain. Brown's vessel for the purpose, *Wheel of Fortune,* had been insured. Hedges reports that Tench Francis Jr., a Philadelphia merchant-lawyer and son of a London-trained merchant jurist, acted as an intermediary with underwriters who offered to insure the vessel's round-trip to Africa at the rate of 25%. (The underwriters are not identified, and may have been from Philadelphia, London, or elsewhere; Francis was acting as a broker, but his choice of markets is opaque.) Later, cargo on the voyage was insured by Francis and Thomas Willing & Co. for £400 at 28%. In both cases a portion of the premium (10% and 13% respectively) was to be returned

Table 7.2 Obadiah Brown's pure underwriting balances, 1753–1762, £*

Accounting Period	*Premium in the period*	*Claims in the Period*	*Pure profit/ loss*	*Loss Ratio*	*Total Premium*	*Total Claims*	*Pure Overall Result*	*Total Loss Ratio*
29 Feb. 1753–1 Jan. 1757	2,555	2,000	555	78.3%	2,555	2,000	555	78.3%
1 Jan. 1757–1 Mar. 1758	1,070	1,246	−175.50	116.4%	3,662.50	3,246	417	88.6%
1 Jan.1760–28 Feb. 1761	1,912	2,000	−88	104.6%	5,574.5	5,246	329	94.1%
1 Mar. 1761–31 May 1762	3,370.33	2,654.2	716.12	78.7%	8,944.83	7,900.21	1,045	88.3%

Note: *'Pure' balances represent only the balance of premium over claims, without regard for underwriting expenses and overheads. Money balances converted to decimal.
Source: Compiled from Obadiah Brown Papers, 'Marine insurance book, 1753–1762', RIHS MSS 315, E445/44 Part 1, Reel 23.

if the vessel 'dont arrive at the Coast [of Africa]'. According to Hedges, a letter from Brown to Francis on 27 June 1759 reported that the vessel had indeed arrived on the African coast. Hedges then postulates, based upon an entry in Brown's Insurance Book, that the *Wheel of Fortune* was taken in Africa by French privateers. This seems to be an incorrect conclusion.[20]

On 27 January 1761, according to the Risk Book, Brown underwrote £200 on the *Wheel of Fortune*, Stone Howell, Master, for 12%. The voyage details were not recorded, but the notation 'Taken' indicates that the vessel was captured by privateers or an enemy navy. Brown recorded as paid, in the final pages of his risk book, a £200 claim. Although the claim is not identified specifically as relating to the *Wheel* policy, Brown's line on the vessel was the only £200 risk he had noted in the Book as having yielded a total loss. This entry is the evidence taken by Hedges and others to indicate that the *Wheel of Fortune* was lost during her slaving voyage.[21] However, as Hedges notes, the *Wheel's* master for the Africa voyage was Captain William Earle. The Risk Book lists the captain during the insured voyage as Stone Howell. The rate of 12% may be appropriate for a return voyage from Africa, and it is not inconceivable that Brown had insured the return voyage to cover the interests aboard of a third party, perhaps the master. However, it seems extremely unlikely that the *Wheel* would have still been on that voyage in late January 1761. The Middle Passage generally required six to eight weeks, although it occasionally took longer. Brown's letter to Francis shows that he knew of the *Wheel*'s arrival in Africa in late June; the vessel must have arrived, at latest, in mid-May, for word to have reached Rhode Island. That leaves at least eight months between the arrival of the *Wheel* in Africa and Brown's insurance of it the following year. If two of those months were required for the return journey, and a generous six for the purchase of the small vessel's human cargo, it would still be back in time for the new voyage, which was made under a different master, which Brown insured, and which was interrupted when it was taken. In short, the evidence of the Risk Book (and the fact the slave voyage was insured elsewhere) indicates not that the *Wheel* was taken by privateers during its slaving voyage, but that it was captured during a subsequent adventure.

Insuring colonial local trade in London

Pares declares that London merchants 'played an important role in the trade between New England and the West Indies' as shipowners and traders in goods on their own accounts. London merchant-insurers played a further role as the insurers of western Atlantic voyages. Yet, despite underwriting by resident merchant-insurers such as Brown, London remained a chief source of marine insurance for much or most of the eighteenth century, at least for voyages beginning or ending in Europe. However, as Pares correctly noted based on his examination of contemporary correspondence by American merchants in the Caribbean trade, 'London underwriters disliked policies for vessels whose condition they could not judge because they had never seen them, on cross voyages whose risks they could not estimate, stuffed with all sorts of contingent additions or returns of premium according that the vessel might touch at this island or not touch at that. They would protect themselves by charging higher premiums than an American underwriter, and they might not touch the policy at all.'[22] In other words, London underwriters did not much like to insure the trade between British North America and the Caribbean. That does not mean that they did not insure such voyages.

Some evidence supports Pares's assertion, although underwriters' records from the era show plainly that London merchant-insurers did insure the distant trade between British North America and the Caribbean. For example, on 7 February 1759 the London wine merchant and prominent marine insurance underwriter William Braund insured the vessel *Sally* on her voyage from New York to the Leeward Islands at the rate of 15%. The following day, however, he insured the slaver *Chesterfield* from Liverpool to Guinea and the West Indies for just 12%. The insurance of the latter culminated in a claim; the vessel was captured by the French (and, by coincidence, one of the co-owners of the *Chesterfield* was William Earle, who had been captain of the Brown's slaver *Wheel of Fortune*). The failed voyage was a Liverpool venture, but US slave vessels were also sometimes insured in London for the triangular voyage. For example, a few months later the owners of the Charleston-registered, New England-built, 16-gun *Bance Island* bought cover from Braund through a broker, William Oswald, for its voyage from London to Africa and the West Indies (Braund was unlucky; this vessel too was seized by the French). European voyages dominate Braund's underwriting record (and reflect the continued great importance of European markets throughout the long eighteenth century), but voyages to or from the western Atlantic were also very commonly insured by Braund, as were East India and slave voyages. Cross-risks in the Americas are far less numerous in the record, but could not be considered unusual. In August 1759, to offer one more example of many, the London merchant-insurer put a line of £100 on the *Mary* for a trading voyage from Boston to the Leeward Islands and Jamaica, for a premium of eight guineas per cent. Even in wartime, the rates Braund charged for these distant voyages were strikingly competitive with those charged by Brown for what were, for the Rhode Island underwriter, relatively local voyages. His underwriting terms offered scope for the insured to ply the British Caribbean without breaching his cover, or requiring 'all sorts of contingent additions or returns of premium'.[23]

London's marine insurance market was thus engaged seriously and directly in the provision of cover for merchants on both the eastern and western borders of the Atlantic World, for their transatlantic voyages and their regional trade. Few extant bundles of merchant correspondence do not include references to transatlantic insurance-buying. For example, the Boston merchant Henry Lloyd, writing to London clients in November 1765, requested that they 'make insurance to the value of the cargo' which was to be shipped westwards to England. Agents representing underwriters at Lloyd's were present in Virginia, Alexandria, Baltimore, and Norfolk at least as early as the 1780s. American merchants appeared at least sometimes to have preferred the inconveniences of using London underwriters to the alternative of insuring at home, despite the challenges of time and distance that accompanied the use of an overseas financial services market. McCusker and Menard observe that 'the greater availability of locally negotiated insurance for ships and cargoes represented a considerable savings for colonial merchants', but it was not one of which they always chose to partake.

Despite uncertainty about the length of time required for correspondence to cross the Atlantic – during the mid-eighteenth century a letter would, at the very best, reach London from Boston in five weeks, but could take more than 12, while eight or ten was the norm – the London market was used regularly to insure North American merchants' vessels. Crothers' work on the commercial correspondence of the Norfolk, Virginia merchant Charles Steuart shows that in the 1750s he engaged a network of London and Bristol merchants to arrange insurance for outbound shipments, and had his agents in London arrange to insure inbound shipments on his behalf. Unsurprisingly, like Brown's local underwriting, Crothers found that Steuart's marine insurance buying patterns were affected by the Seven Years' War. Few or no vessels went uninsured, rates doubled, and local underwriters increased their participation in the market, augmenting (but not replacing) supply from London and elsewhere in Britain, notably Bristol, Liverpool, and Glasgow. Further, the use of various underwriting centres and their institutions went both ways. For example, Glasgow tobacco merchants

Buchanan & Simson, writing to Messers Fraser & Wharton in November 1759, reported that 'We find by this days Lloyds List that the [vessel] Maxwell foundered at Sea, as we have insurance made at Philadelphia, we desire you may by first Paquet to New York, send to Mr George Maxwell Merchant in Potuxant [Patuxent] Maryland a proper certificate of the ship being lost that our insurance may be received'. In this example, it is clear the even when the insurance was done elsewhere, London's institutions of underwriting – in this case *Lloyd's List* – supported marine insurance activity in other centres.[24]

The decline of US private underwriting

American independence released the nascent country from the proscription of corporate underwriting set out in the *Bubble Act* of 1720 (which was to remain in force in England until 1824). A number of marine insurance companies were formed shortly afterwards, as US merchants 'energetically developed domestic sources of marine insurance'. The first was the Insurance Company of North America, established by Philadelphia merchants in 1792 with authorised capital of $600,000, although it began operations with just $40,000 in subscriptions. Many others soon followed. The corporate structure of the new US insurers was not standard; some underwrote with a limitation of shareholders' liability, while others underwrote with explicitly unlimited shareholder liability. Most had explicit protections against joint liability among the shareholders. Other underwriting associations were unincorporated groups of merchant-insurers, constituting simply a syndicate of private underwriters, although they could raise and hold mutual capital. Most companies were launched by merchants, who simply took the traditions of the merchant-insurers of old into a new corporate structure.[25]

In 1810 a British Parliamentary Committee was convened to enquire into the state of marine insurance in London and elsewhere, in answer to petitions calling for the repeal of the *Bubble Act's* prohibition of corporate underwriting. Giving evidence, Jenkin Jones of the Phoenix Fire Insurance Office testified that, on a recent exploratory visit to the US, the British North American provinces, and the West Indies, he had found that at least seven companies were operating in Boston, six in New York, eight in Philadelphia, five in Baltimore, one in Norfolk, two in Charleston, and one in New Orleans. Further, Jones testified that the cost of insurance was higher in America than in London for European and East India voyages, 'generally speaking, as much as one third higher'. He attributed the higher prices to typically higher returns on capital that were expected in America. He later noted that US companies charged generally lower rates than London underwriters on voyages along the US coast, and for those between the US and the islands of the West Indies, since 'they know the voyages better [by] a great deal'. The 'primary cause' of the unseating of London underwriters by US insurance companies was, he stated, 'the difficulty of obtaining insurances in England, so as to snit [*sic*] the time and occasion of the American shipper'. Another key factor, raised by other witnesses, was a wartime prohibition which prevented British underwriters from insuring trade with the enemy (a restriction which was often circumvented), and from paying claims arising from the capture of US vessels by British naval forces.[26]

Another testifier, Alexander Glennie, provided a contrasting account of the difference in insurance costs between London and the US companies. 'Premiums on risks from the ports of America to the East or West Indies, or to the Mediterranean and back to America, have been generally about one half what would have been demanded by or underwriters here', he told the Committee, while 'the premium from America to this country... cannot be stated at more than perhaps one third less premium, and in many instances, not quite that; in many instances, the same'. He estimated that his company annually arranged insurance cover worth about £1,000 for US correspondents, where previously the total had been 'twenty, thirty, or forty thousand pounds'. The decline, he assumed, had begun between 1790 and 1795, 'since the establishment of the companies in America'. Again in contrast, Samuel Williams, an American merchant established in London who regularly placed insurances amongst the private underwriters at Lloyd's, testified that he believed 95% of consignments of goods shipped from the US to 'the Continent' were insured at their origin, instead of by him, as an agent, in London. If Williams' experience is accurate and representative, it indicates a significant relative decline in London's share of the US marine insurance market.[27]

Jones reported also that individuals continued to underwrite in the Southern states and Baltimore, and noted that 'private underwriters have nearly disappeared'. In this observation he called the end far too early, however. Private underwriting was to last well into the twentieth century. A 1962 report by the US Commission on Money and Credit described the ongoing operation of organisations it described as 'Domestic Lloyds', comprising unincorporated individuals 'associated together and authorised to operate an insurance business'. Another testifier to the 1810 Committee, Samuel Williams said he believed the number of private underwriters in the US to be roughly the same as it had been before independence and the formation of numerous companies, but noted that the rise in supply was probably due to 'a larger sum insured now in America than there was formerly'. Kingston estimates that in the period April 1798 to March 1799, 'perhaps fifty private underwriters' were active in the Philadelphia market. However, the perceived financial security of corporations encouraged insurance buyers to shift towards them as providers, and away from private underwriters in the 1790s, when wartime capture by privateers was again a serious threat, and premium levels were accordingly high. This wartime threat, Kingston argues, caused at least one private underwriter to increase rates from levels which were, on average, lower than those of the new companies, to levels which were higher. Together such shifts resulted in the collapse of the private underwriters' market share, Kingston concludes.[28]

Nonetheless, the private insurance market in Rhode Island appears to have been operational in 1794. That year Charles DeWolf (rendered in the policy D'Wolfe), who was a member of a prominent Rhode Island merchant family and brother of the better-known merchant, slaver, and later senator James DeWolf, insured his vessel *Sally* and its cargo with private underwriters for £600 'Lawful Money' to cover a voyage from Havana to his home port of Bristol, Rhode Island. The risk was divided between four underwriters, including the partnership Gibbs & Channing, a Rhode Island merchant firm headed by Walter Channing and George Gibbs. On the eve of the arrival of US insurance companies, merchants in Rhode Island were still trading marine insurance amongst themselves in the traditional way, as Obadiah Brown had done two generations earlier. One element of continuity is recourse to arbitration: a clause in the printed policy states 'in Case of any Dispute arising hereupon, the Matter in Controversy shall be submitted to, and decided by Referees, chosen by each party'. The De Wolf policy bears the name of no intermediary, but another, issued in Boston the same year, was 'Underwritten in the Office kept by *Peter Chardon Brooks*', a Boston merchant and insurer who in 1789 reportedly 'engaged in the business of marine insurance, and accumulated a large fortune'. Brooks went on in 1806 to become president of the New England Insurance Company. The 1794 policy grants cover of £900 on the schooner *Nancy* and her cargo for a voyage from Boston to Baltimore and back, and was underwritten by five private individuals.[29]

Meanwhile underwriting Atlantic World trade to the Caribbean continued in London. Extant policies issued by the brokers Sanderson, Brothers in 1717 and 1718 show that ships originating in Britain and sailing for the

Caribbean were insured at Lloyd's. The three brothers Sanderson (Hugh James, William Wood, and John), who conducted business as 'Merchants and Insurance-Brokers' from their office in Lombard Street, became members of Lloyd's in various years between 1800 and 1814. Among others, they arranged for insurance of £200 on the vessel *Grace* for a voyage from 'Liverpool to Barbados or her port or ports place or places of discharge in the West India Islands or Colonies (Jamaica and St Domingo excepted)', on behalf of clients 'Mr. Tho. Dawson and Mrs. Mary Middleton'. The entire risk was underwritten by the Lloyd's member James Carnigie at the rate of 25 shillings per cent. Jamaica was not always a 'no-go' zone for vessels, though. In June 1818 the same brokers arranged cover of £300 for William Pearson on the vessel *Donald* for a voyage from Liverpool to Jamaica and back, at the rate of four guineas per cent. Other policies in the same cache covered vessels heading to 'the Brazils', 'British America', 'Savannah' in Georgia, and from Quebec to Waterford.[30]

Circuits of capital

Whether policies were underwritten in London or the United States, the mechanics of private underwriting were identical, and were based upon those invented by Italian merchants centuries earlier, then refined in London. Knowledge of the practices of marine insurers quickly permeated the Atlantic World, and served to underpin its trade. Merchants of the highest standing actively participated in established and fledgling insurance markets, underpinning the trade of the Atlantic World, enabling the balancing of payments between regions, and maximising traders' investments in cargoes and ships. Merchant-insurers used the instrument to share the grave risks of Atlantic World commerce amongst themselves by establishing a common and virtual pool of contingent capital which provided an important foundation upon which the integrated and expanding Atlantic World economy rested.

Notes

1. For a comprehensive exploration of the Mediterranean origins and early spread of marine insurance, see Leonard, A.B. (ed.): *Marine insurance: international development and evolution*, Palgrave History of Finance Series, Basingstoke: Palgrave Macmillan, 2015.
2. For the identity of some very early Italian underwriters, see Leone, Alfonso: 'Maritime insurance as a source for the history of international credit in the

Middle Ages', *Journal of European Economic History*, Vol. XII (1993), pp. 363–9. Bensa, Enrico: *Il contratto di assicurazione nel medio evo: studi e recerche*, 1884. Translated to French as Valéry, Jules, (trans.), *Histoire du contrat d'assurance au moyen age*, Paris: Ancienne Librairie Thorin et Fis, 1897, p. 28; Muldrew, Craig: 'Atlantic World 1760–1820, economic impact', in Canny, N. And Morgan, P.: *The Oxford History of the Atlantic World, 1450–1850*, Oxford: Oxford University Press, 2011.

3. For Ferrantyn's nationality, see TNA SC 8/111/5523. For his insurance suit, see Thomas, A.H. (ed.): *Calendar of plea & memoranda rolls of the City of London preserved among the archives of the Corporation of London at the Guildhall, AD 1413–1437*, Cambridge: Cambridge University Press, 1943, pp. 208–10. See also Lewin, C.G.: *Pensions and insurance before 1800: A social history*, East Linton: Tuckwell Press, 2003, pp. 85–8.
4. For a discussion of early insurance dispute resolution mechanisms in London, see Leonard, A.B.: *The origins and development of London marine insurance, 1547–1824*, unpublished PhD thesis, Faculty of History, University of Cambridge, 2013. For the Law Merchant and codification in English marine insurance, see Rossi, Guido: 'The Book of Orders of Assurances: a civil law code in 16th century London', *Maastricht Journal*, Vol. 19, No. 2 (2012), pp. 240–61.
5. *The special report from the committee appointed to inquire into, and examine the several subscriptions for fisheries, insurances, annuities for lives, and all other projects carried on by subscription...* London: House of Commons, printed by Tonson, J., Goodwin, T., Lintot, B., and Taylor, W., 1720, Lechemere's report to the King, various affidavits, pp. 40, 44–5; HALSC DE/R/B293/1–47, Business records of the Radcliffe family, policies underwritten for Radcliffe and others; BL Add. Ms. 43,731 f. 58, policy issued to John Fletcher, 5 Dec. 1716 (transcription).
6. *The special report of 1720*, various affidavits; BL GRC 816.m.12, 16, *Reasons humbly offered for the passing of a bill...* undated, 1693/4; A subscriber to Lloyd's: *A letter to Jasper Vaux*, London: Printed for J. M. Richardson, 1810, p. 44; Index, *House of Commons Sessional Papers of the Eighteenth Century 1715–1800*, p. 90.
7. Nash, R.C.: 'The organization of trade and finance in the British Atlantic economy, 1600–1830', in Coclanis, Peter A.: *The Atlantic economy during the seventeenth and eighteenth centuries: organisation, operation, practice and personnel*. Columbia, SC: University of South Carolina Press, 2005, p. 133; For a full discussion of the impact of the efficacy of the Navigation Acts, see Zahedieh, Nuala: *The capital and the colonies: London and the Atlantic economy 1660–1700*, Cambridge: Cambridge University Press, 2010. McCusker and Menard list among the invisibles freight charges, insurance premiums, and the costs of short-term credit. McCusker, J.J. and Menard, R.R.: *The Economy of British America, 1607–1790*, Chapel Hill: University of North Carolina Press, 1985, pp. 36, 91–2.
8. Michael Atkins to John Reynell, July 1751, cited in Pares, Richard: *Yankees and Creoles: The trade between North America and the West Indies before the American Revolution*, London: Longmans, Green & Co., 1956, p. 8; see that volume, p. 13, for the rise in ship ownership; McCusker and Menard, *Economy of British America*, p. 145; Richardson, David: 'Slavery, trade, and economic growth in

eighteenth-century New England', in Solow, Barbara (ed.): *Slavery and the rise of the Atlantic system*, Cambridge: Cambridge University Press, 1991, p. 244.

9. Richardson, *Slavery, trade, and economic growth*, pp. 253–7; Shepherd, J.F. and Walton, G.M.: *Shipping, maritime trade, and the economic development of colonial America*, Cambridge: Cambridge University Press, 1972, p. 115.
10. A heavily edited selection of Claypoole's letters has been published. Balderston, Marion (ed.): *James Claypoole's letter book, London and Philadelphia, 1681–1684*, San Marino, CA: Huntington Library, 1967, p. 136 ff. Morriss, Margaret Shove: 'Colonial trade of Maryland, 1689–1715', *Johns Hopkins University studies in historical & political science, vol. XXXII*, Baltimore: Johns Hopkins Press, 1914, pp. 446–599; Roseveare, Henry (ed.): *Markets and merchants of the late seventeenth century: the Marescoe–David letters, 1668–1680*, Oxford: Oxford University Press, 1987. See especially Appendix E, pp. 582–8; Letterbook of Joseph Cruttenden, letter 17 June 1710, to Habijah Savage, Boston; letter 28 July 1714, to John Nichols, Boston, Bodleian Library, MS Rawl. Lett. 66, pp. 17–18, 158–9.
11. Huebner, Solomon: 'The development and present status of marine insurance in the United States', *Annals of the American Academy of Political and Social Science*, Vol. 26 (Sept., 1905), p. 423; Crothers, A. Glenn: 'Commercial risk and capital formation in early America: Virginia merchants and the rise of American marine insurance, 1750–1815', *Business History Review*, vol. 78, no. 4 (Winter, 2004), p. 611.
12. Diary entry 1 Aug. 1768, Rowe Cunningham, Anne: *Letters and diary of John Rowe, Boston merchant*. Boston: W.B. Clarke, 1903, p. 171; Gillingham, Howard: *Marine Insurance in Philadelphia, 1721–1800*, Philadelphia: no publisher stated, 1933, pp. 42–63; Crothers, *Commercial risk and capital*, p. 610; Pares, *Yankees and Creoles*, pp. 23–4.
13. For a full account of the Browns, see Hedges, James: *The Browns of Providence Plantations: colonial years*, Cambridge, MA: Harvard University Press, 1952.
14. Hedges, *Browns of Providence*, pp. 6, 29; Pares, *Yankees and Creoles*, pp. 61–4.
15. Hedges, *Browns of Providence*, pp. 5–6, 10; Snell, Charles: *A guide to book-keepers, according to the Italian manner: now in general use*, London: 1709, pp. 11, 15.
16. Leyborun, William: *Panarithmologia*, London: for John Dutton *et al.*, 1693, p. Ap. 35.
17. All the references to Brown's underwriting which follow, unless specified otherwise, are from Obadiah Brown Papers, 'Marine insurance book, 1753–1762', RIHS MSS 315, E445/44 Part 1, Reel 23.
18. An average is calculated when losses arise from actions necessary to preserve a vessel (such as the payment of a ransom, or the jettison of cargo in storm conditions). The insurers of all parties with insured interests in the vessel, whether covering the vessel or cargo, participate proportionately, according to the sum insured under each policy, to meet the cost of the claim.
19. For a discussion of convoys and marine insurance, see Leonard. A.B.: *Marine insurance and the rise of British merchant influence, 1649–1748*, unpublished Master's dissertation, University of Cambridge, 2011.
20. Hedges, *Browns of Providence*, p. 72. I am grateful to Christopher Kingston for his insights.
21. Ibid. Brown University (founded by Obadiah's father) has included this tale on the document repository pages of the web site of the *Brown University*

Steering Committee on Slavery & Justice, referring to 'the Wheel of Fortune, a slave ship dispatched to Africa by Obadiah and his nephews, Nicholas and John, that was apparently "taken" by a French privateer'.

22. For the overall dominance of London in the insurance of Atlantic commerce, see, for example, Huebner, *Development and present status*, p. 433. Crothers, *Commercial risk and capital*, p. 612; Pares, *Yankees and Creoles*, pp. 4, 23.
23. ECRO D/DRu B7, William Braund's *Journal of Risks, 1759–1765*; *Trans-Atlantic Slave Trade Database*, www.slavevoyages.org. For the life and business interests of Braund, see Sutherland, Lucy: *A London merchant, 1695–1744*. London: Oxford University Press, 1933.
24. NAS CS96/507 f. 16, *Letterbook of Buchanan & Simson*, letter to Fraser & Wharton, 30 Nov. 1759; Cole, Arthur H.: 'Tempo of mercantile life in colonial America', *Business History Review*, Autumn 1959, 33, pp. 285–7; Crothers, *Commercial risk and capital*, pp. 611–14; McCusker and Menard, *Economy of British America*, pp. 347–8.
25. Crothers, *Commercial risk and capital*, pp. 615–16; Testimony of Jones, Jenkin: '*Report from the Select Committee on Marine Insurance (Sess. 1810), 5 March 1810*'. House of Commons, BPP (226) 1810 IV 247, reprinted 11.5.1824, pp. 36–9.
26. Testimony of Jones, *Report from the Select Committee*, pp. 36–9.
27. Testimonies of Glennie, Alexander and Williams, Samuel: *Report from the Select Committee*, pp. 33–4, 51–5.
28. Testimony of Jones, *Report from the Select Committee*, p. 38; Commission on Money and Credit: *Property and casualty insurance companies: their role as financial intermediaries*, Englewood Cliffs: Prentice-Hall Inc., 1962, p. 13; Testimony of Williams, Samuel: '*Report from the Select Committee on Marine Insurance (Sess. 1810), 5 March 1810*'. House of Commons, BPP (226) 1810 IV 247, reprinted 11.5.1824, pp. 33–4; Kingston, Christopher: 'Marine Insurance in Philadelphia During the Quasi-War with France, 1795–1801', *Journal of Economic History*, Vol. 71, No. 1 (March 2011), pp. 173, 175, 178.
29. Author's collection: policy underwritten for Charles D'Wolfe, 29 Mar. 1794, policy underwritten for Wales and Field, 17 Sep. 1794, other policies; Wilson, J.G. and Fiske, John (eds): 'BROOKS, Peter Chardon', in *Appleton's cyclopaedia of American biography*, vol. 1, New York: Appleton & Co., 1887, p. 389.
30. Author's collection: policy underwritten for Dawson and Middleton, 10 Oct. 1818; policy underwritten for William Pearson, 10 Oct. 1818; *Gazette*, 4 July 1818, p. 1202; Dawson, Warren R. (Honorary Librarian to Lloyd's): *Roll of Lloyd's, 1771–1930*, London: Lloyd's of London, 1931, pp. 48, 243.

8

Slavery, the British Atlantic Economy, and the Industrial Revolution[1]

Knick Harley

Modern economic growth first emerged in Britain at about the time of the first cotton textile factories of the industrial revolution. Urban and increasingly export-orientated industrialisation occurred. The British economy had already experienced economic growth, industrial diversification, and export orientation. Trade with the Americas was central to this development, and the slave colonies of the West Indies were key to Britain's American trade. Eric Williams' *Capitalism and Slavery* emphasised the central role that slavery played in developments leading up to the industrial revolution.[2]

Americanisation transformed Britain's trade in the eighteenth century. In 1700 Britain overwhelmingly exported woollen textiles to Europe, but by the eve of the American Revolution (and the industrial revolution) the Americas had surpassed Europe as a destination for manufactured exports. Probably even more important, exports to Europe remained overwhelmingly woollen textiles, while those to the Americas were overwhelmingly other manufactured products. This diversification seems an important feature of eighteenth-century growth leading to the industrial revolution.

The diversification of British exports, and consequently of its economy, rested heavily on the Atlantic economy, which depended on the slave-based sugar colonies of the Caribbean. However, it is impossible to attribute Britain's success as the first European economy to experience modern economic growth to a unique position in the sugar colonies. Britain's colonies were less productive than those of its rivals, so much so that the prosperity of the British Caribbean depended on mercantile protection which prevented the sale of cheaper French sugar in British markets. In effect, British consumers subsidised British slave plantations, hardly a recipe for differential advancement. Nonetheless Atlantic trade had a differential impact on Britain. What set the British colonial empire apart from those of its rivals was the involvement of the temperate colonies on the North American mainland. The slave colonies had come into being because they presented opportunities for exceptional profits to those who could mobilise labour and capital sufficient to exploit their staple exports.

In contrast, the northern colonies were settled by English emigrants whose primary objective was to establish independent settlement. The resulting colonies in New England and on the middle Atlantic coast did not possess staple products that could be sold at profit in Europe. Instead their residents exploited the opportunities presented by the eighteenth-century Atlantic World as a whole. They became important providers of maritime services in the form of shipping and merchandising. In addition, while there was no significant European market for the timber or temperate agricultural products of the region because transportation costs were too high, the profitability of sugar in the West Indies provided incentives to concentrate resources there on the production of the export staple. The northern mainland colonies' economies evolved in such a way that the residents' purchases of European products were financed by the sale of services, timber, and foodstuffs to the West Indies. In this regard the British empire was unique, and its development provided a growing, diversified, and relatively wealthy market for British manufactured goods, a market which other empires lacked.

Williams' work emphasised questions about the extent to which the British export economy based on West Indian slavery contributed to the coming of the industrial revolution. His own answer, that the profits from the slave trade were crucial to the industrial revolution, has not stood up to critical evaluation. Nonetheless, modern speculations regarding endogenous growth frequently plausibly postulate that manufacturing, urbanisation, and a powerful merchant class all have a favourable impact upon growth. Historians have found statistical support for such propositions. The British Atlantic economy of the eighteenth century enhanced manufacturing, urbanisation, and the mercantile class.

Globalisation, the Americas, and slavery

The industrial revolution capped economic change in the eighteenth century. Prior to the last decades of the century, however, expanding international trade led Britain's economic dynamics. In turn, expansion of trade was firmly situated in a world of mercantilism and colonial rivalry. The competition between European states arose from early modern state-building, and took many forms. Importantly, the state-building process occurred in a time of globalisation, and marked the early stage of a two-centuries-long European political and economic dominance of world affairs. As Williams pointed out as a young scholar, incorporation of the Americas into the dominant Eurasian economy was central to the process. From a Eurasian perspective the Americas that Columbus's voyages brought into contact with the old world were regions of land abundance and labour scarcity (reinforced, of course, by the catastrophic demographic consequences of contact for indigenous American populations). It is now commonplace to point out that if an elite is going to extract a surplus from land abundance, slavery, rather than free labour, is almost certainly involved.

Williams' documentation of this process in the West Indies highlighted key relationships among staple extraction, the slave trade from Africa, and British eighteenth-century prosperity. Relationships between the British sugar colonies in the West Indies and the industrial revolution in Britain were central to his work, but in considering Williams' seminal book, I think it is misleading to overemphasise the famous phrase in the preface, in which he describes the volume as 'strictly an economic study of the role of Negro slavery and the slave trade in providing the capital which financed the Industrial Revolution in England' (p. vii). This thought lies at the basis of the literature on the 'Williams thesis', but the book is more about the political economy of the relationship between British economic policy and the interests of West Indian planters and traders (including slave traders), and how that relationship changed over time. It is worth citing the rest of the famous sentence partially quoted above: 'and of mature capitalism in destroying the slave system.'

By the mid-seventeenth century northern European perception of the gains that could be extracted from the new world had come to focus on the profits of sugar cultivation in the Caribbean. The great early source of American gain, the silver (and to a lesser extent gold) deposits, remained important, but they were firmly in the hands of the Spanish and to a lesser extent the Portuguese. New deposits elsewhere did not appear to be forthcoming. In the seventeenth century the Portuguese transferred sugar cultivation from their Atlantic island possessions to the new world. The potential for sugar seemed nearly unlimited in Brazil and the Caribbean, but its exploitation required labour and capital. Capital and capitalists (the elite) were able to dominate sugar production not only because they received imperial support, but more fundamentally because successful exploitation of the sugar's potential required partial refining of the cane immediately it was cut. Initial refining required fixed capital, and significant economies of scale were realised in processing. As Williams emphasised, the cheapest (but not the only) labour supply came from the African slave trade (which had already been tapped by the Portuguese in the Atlantic Islands). Thus the infamous triangular trade emerged. European powers engaged in intense political rivalry over control and exploitation of the sugar islands. In addition to military confrontation, mercantile legislation attempted to channel surpluses from the expansion of sugar production to home countries. Thus the colonies were required to export and import solely from the home imperial power. Imperial powers controlled the slave trade and the sale of the staple, and monopolised profits from the sale of European goods in the colonies.

As Williams emphasised, the British sugar plantations, first in Barbados and then in the other islands, created vested interests in Britain.[3] Some planters became very wealthy from the trade, particularly in the era of expansion during the late seventeenth century. Shipping interests in London, Bristol, and eventually Liverpool invested heavily in trading links to Africa and the middle passage that transported slaves to the West Indies. In addition, sugar refiners and suppliers of export goods benefited from the trade. In the late seventeenth century the British islands led sugar development. They not only supplied the rapidly growing British market for sugar, but also provided the basis of a substantial re-export trade to the rest of Europe. That changed, however, in the early eighteenth century after the French gained control of Saint-Domingue (today's Haiti). By the end of the 1720s Britain's re-export trade in sugar had disappeared,[4] but high differential tariffs on sugar preserved the large British market. Davis notes that the British consumed a third of all the sugar imported into Europe in the eighteenth century.[5]

As the English islands lost their comparative advantage in sugar production to the larger islands, the prosperity of the vested West Indian interest, resting as it did on protective tariffs that kept British sugar prices substantially higher than prices elsewhere in Europe, depended on its political position. The extent of the sugar interests, reaching from the planters to the ports and into the manufacturing districts, combined with wealthy planters' ability to purchase seats in the unreformed House of Commons, preserved the interests, and supported slavery in the islands throughout the eighteenth century.

Williams argued that this protected West Indian vested interest found its position becoming unsustainable as the eighteenth century ended.

The loss of the mainland American colonies created problems of supply to the islands. The extent of subsidy that British consumers were providing to support a special interest became increasingly obvious. The climate of public opinion was moving away from mercantile policy and the support of special interests. The 'old corruption' that underpinned the political position of the West Indian interest was increasingly under attack. As the industrial revolution proceeded, the main focus of economic attention shifted to the new industries created by Britain's technological prominence. These industries looked not for protection, but for an opening of export markets. As the political economy shifted, the West Indian interest became vulnerable to its opponents. The slave trade was abolished in 1807, and slavery itself in 1833. In Williams' narrative, the key to these changes was not the moral case against slavery and slave-produced products – the British were still happy to purchase slave-produced cotton from the American South – but the West Indian interest's loss of economic and political influence.

Historical importance of slavery

None of the preceding narrative should be taken, however, to negate the importance of slavery and the slave trade in the evolution of the British economy in the eighteenth century. Over the century Britain became more industrial, and exports made major contributions to that industrialisation. Exports to America played a particularly important role, not only in the growth of trade, but also in the development of export trades in manufactured goods. Some basic statistics, presented in Figure 8.1 and Table 8.1, illustrate the process.

At the Restoration, England's exports went almost entirely to Europe. They were overwhelmingly manufactured goods, but also almost exclusively woollen goods. By the eve of the American Revolution (and the industrial revolution) the nature of trade had changed. Exports remained overwhelmingly manufactured goods, but the share of woollen goods had decreased to less than half of the total. Trade to Europe, however, had changed little. The shift in trading patterns emerged from the development of a trade with the American colonies. By the 1770s trade with the Americas rivalled that with Europe, and Britain exported a wide range of manufactured goods other than woollen cloth to these American markets. The American trade had assumed major prominence in British life, and the demand of the colonies provided major markets for diversified manufactured goods. Expansion of the slave economies drove the transformation, but there was another important dimension.

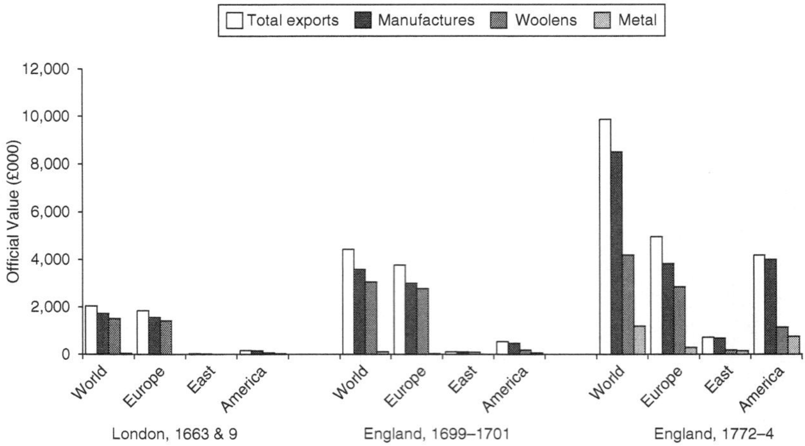

Figure 8.1 English Exports, 1660s to 1770s
Source: Table 8.1.

The mainland colonies of America played a major role. The trade of the mainland colonies, however, was hardly independent, since it depended on the colonists' own triangular trade with the staple colonies.

The sugar colonies and the southern mainland colonies, with exports of tobacco and rice, were classical staple colonies. They had grown to exploit previously unknown opportunities for the production of staples for sale in Europe, under conditions of land abundance and labour shortage. Sales in Europe financed the purchase of African slaves and European consumption goods. The northern mainland colonies differed, however. They financed imports from Europe by providing intermediate goods (food, timber, and so on) and international transactions services (such as shipping) to the staple colonies.[6] Nonetheless, their economies were linked to the slave-based staple trades. Both their trade, and that of the staple producers, was dependent upon the success of the staples themselves. There were, however, important differences.

Basic staple approach

It is useful to conceptualise the staple colonies and the northern mainland colonies in the old British Empire. In their masterful 1985 *The Economy of British America, 1607–1789* John McCusker and Russell Menard introduce their first chapter by noting that historians of

Table 8.1 English trade, 1660s to 1770s (£000s)

	1663 & 9 (London only)				***1699–1701***				***1772–4***			
	World	***Europe***	***East***	***Americas***	***World***	***Europe***	***East***	***Americas***	***World***	***Europe***	***East***	***Americas***
Exports	**2039**	**1846**	**30**	**163**	**4433**	**3772**	**122**	**539**	**9853**	**4960**	**717**	**4176**
Manufactures	1734	1562	19	153	3583	2997	111	475	8487	3816	690	3981
Woollens	1512	1423	19	70	3045	2771	89	185	4186	2849	189	1148
Metal	44	15		29	114	31	10	73	1198	295	148	755
Imports	**3495**	**2665**	**409**	**421**	**5849**	**3986**	**756**	**1107**	**12735**	**8122**	**1929**	**2684**
Manufactures	1292	1077	215		1844	1292	552		2157	1364	792	1
Pepper	80		80		103		103		33		33	
Tea	0				8		8		848		848	
Sugar	292	36		256	630			630	2360			2360
Tobacco	70	1		69	249			249	519	1		518
Re-exports					**1986**	**1660**	**14**	**312**	**5818**	**4783**	**63**	**972**
Manufactures					746	491	3	252	1562	959	7	596
Sugar					287	287			429	428		1
Tobacco					422	421	1		904	884	1	19

Source: Davis, Ralph: 'English foreign trade, 1660–1700', *Economic History Review*, 7 (1954) p. 160; Davis, Ralph: 'English foreign trade 1700–1774', *Economic History Review*, 15 (1962), pp. 300–3.

the economies of colonial America have tended to adopt one of two approaches.[7] One is the 'staples or vent for surplus theory'; the other is Malthusian. The staples approach emphasises the expansion of a land-abundant region driven by potential rents from the cultivation of a staple for sale in the metropolitan economy. The expansion draws capital and labour to the new world. The Malthusian (so termed because it is driven by population growth, although in the absence of diminishing agricultural returns), in McCusker and Menard's words, 'locates the central dynamics of American history in internal demographic processes that account for the principal characteristics of the colonial economy: the rapid and extensive growth of population, of settled area and of aggregate output combined with the absence of major structural change' (p. 18). Although both the slave colonies and the northern mainland economies became major trading economies it is enlightening to think of the trade of the slave economies in the context of the staple approach, and that of the northern mainland economies in the Malthusian context.

For the slave economies it is useful to follow Findlay and O'Rourke's advice and think of the Atlantic economy in the framework of a three-region general equilibrium model.[8] Findlay laid out a useful and simple model of this type, used here with a few modifications, to characterise the slave colonies and their trade.[9] The model explores relationships in an economy comprising a metropolitan manufacturing sector which uses colonial raw materials, the production of which in turn depends on slave labour. The equilibrium of the model simultaneously determines the size of the slave labour force (and, given an exogenous slave mortality, the size of the slave trade), the output of manufactured goods, and the trade in raw materials, as well as the relative price raw materials and of slaves.

Figure 8.2 presents a slightly modified, graphical version of the model of the raw-material–staple–slave economy. In Findlay's version of the model, a unit of manufacturing output required a specific amount of raw materials, with all such materials produced under colonial slavery. That specification tied manufacturing very closely to the slave economy, but can hardly be fully accepted as a reasonable representation of historical reality. After about 1800 the British cotton textile industry depended on slave-produced cotton, but until that time colonial products were heavily dominated by sugar and tobacco, with rice and fish as other exports. None of these were industrial raw materials. Nor was British industry very dependent on imported raw materials. For this reason, and to provide a framework to discuss the initial expansion of the staple colonies, I have modified Findlay's diagram as indicated by the

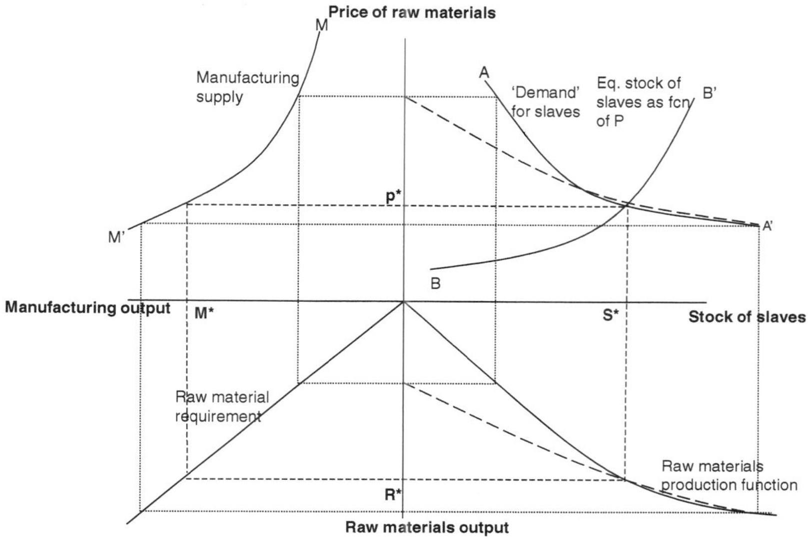

Figure 8.2 Atlantic economy, Findlay model

heavy dashed lines, to allow a non-colonial source of raw materials.[10] The steady-state equilibrium (say, approximately the situation in the mid-eighteenth century) with colonies is qualitatively identical to Findlay's, and indicated by the broken lines. Equilibrium levels of manufacturing output (M*), raw material inputs (R*), slaves (S*), and the relative price of raw materials (p*), and Findlay's comparative statics, also carry through without change.

This model provides a good framework to emphasise the earlier history of the staple colonies. Prior to Columbus, the European economy was constrained by its internal raw material supply. Thus the equilibrium was at R^0, M^0, and P^0 in Figure 8.3. The high price of raw materials in this economy relative to supply opportunities in the colonies generated colonial profits and induced slave imports. The ensuing fall in raw material prices induced expansion of manufacturing. Eventually, the economy attained its post-Columbus equilibrium at R*, M*, S*, and P*.

Trade and the Malthusian northern mainland colonies of British America

The history of the northern mainland colonies did not share the staple-driven dynamic where potential profits from exploitation of

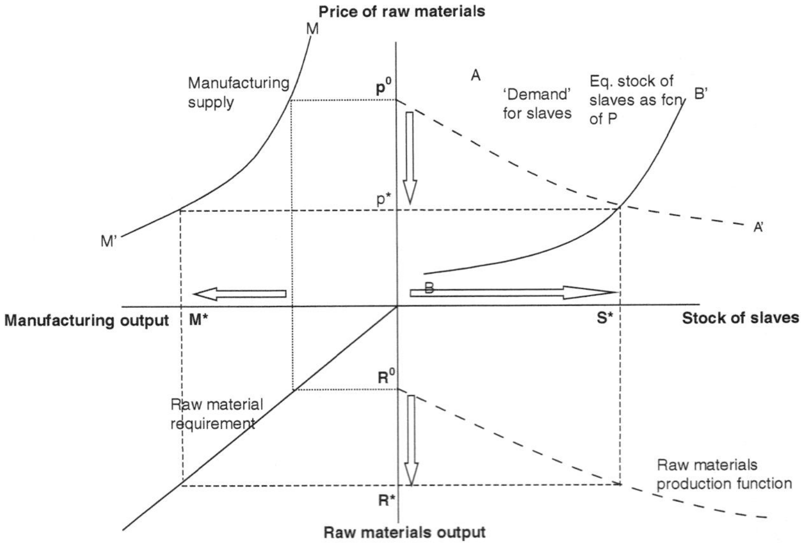

Figure 8.3 Discovery disequilibrium

staples drew labour – through the slave trade – and capital from the Old World. Instead New England and the middle colonies evolved from the migration and subsequent demographic growth of groups interested in creating an independent existence in the new world. These colonies' involvement in Atlantic trade was thus different. Figure 8.4 provides a way of visualising their trading experience. These colonists and colonies represented a potential demand for European commodities; the continent was their main source of manufactured goods and luxuries. The demand curve in the diagram slopes downward in terms of the price of imported goods relative to the price of domestic goods (and wages) in the colonies. Demand may have been quite inelastic, given the difficulties of providing colonially produced substitutes for European goods and the relatively small part these goods played in these intentionally self-sufficient colonies' consumption. The equilibrium colonial price (and wage) level and import quantity are determined by the intersection of this curve with a second curve representing foreign exchange earnings for the colonies at different real exchange rates (determined by colonial price levels). The curve has been drawn with two distinct segments – one nearly horizontal, and one with a clear upward slope. The upward-sloping segment represents the sale of colonial produce in

Europe. Transportation costs precluded the sale of abundant temperate agricultural products and timber, except at very low prices in the colonies themselves, since carriage costs across the Atlantic often exceeded European prices. Some valuable American goods were exported to Europe, such as furs and fish, but revenue from them was modest relative to the demands of the substantial colonial population.

However, the Atlantic economy itself presented a substantial demand for primary products such as foodstuffs and timber in the West Indies, and for shipping services in the maritime economy more generally. At an appropriate level of colonial prices and wages this demand was extremely elastic, since the colonies were marginal suppliers in a much larger imperial market, and could sell at the prices prevailing in the broader market. In the diagram this demand shows up as the nearly horizontal segment of the supply curve for foreign exchange. This horizontal demand determined the colonial price level and the volume of imports.

Shepherd and Walton's estimate of the American balance of payments just before the Revolutionary War (Table 8.2) demonstrates the importance to the northern colonies of this elastic demand in the imperial economy.[11] Exports to Britain were small, at less than ten per cent of estimated foreign exchange receipts. Exports of temperate staples to the West Indies were important, providing about a third of receipts. These exports supported the islands' specialisation in staple production, and without them the islands

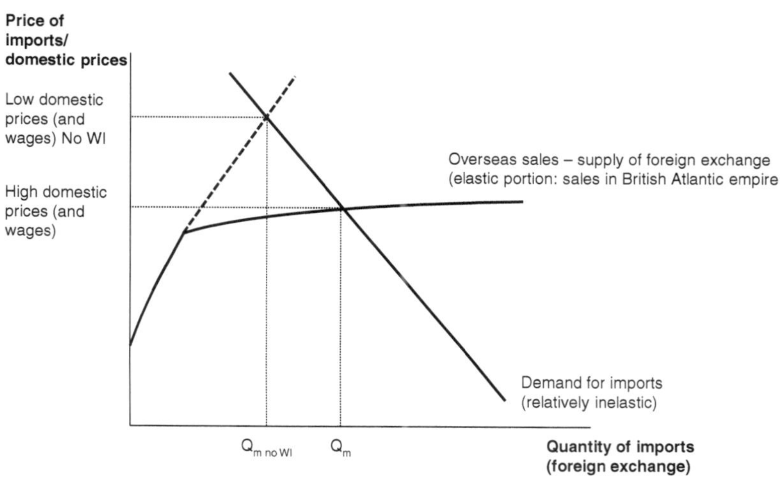

Figure 8.4 Malthusian colonies' trade

would have found these temperate commodities much more expensive. Importantly, nearly 40% of the northern mainland colonies' receipts came from shipping and mercantile services. These services characterised the Malthusian economies that expanded exports into a large market where, as a small supplier, they faced an almost infinitely elastic demand.

Table 8.2 Balance of payments of the New England and Middle Colonies, average 1768–72, £000

Imports		
New England	1,054	
Middle Colonies	1,202	
Combined	2,256	
Earnings		
Commodity exports all destinations		
		%
New England	477	53
Middle Colonies	559	69
Combined	1,036	60
Of which to West Indies		
New England	303	34
Middle Colonies	244	30
Combined	548	32
Of which to Britain		
New England	87	10
Middle Colonies	75	9
Combined	162	9
Shipping earnings		
New England	327	36
Middle Colonies	177	22
Combined	504	29
Other invisibles		
New England	100	11
Middle Colonies	74	9
Combined	174	10
Total Earnings		
New England	904	
Middle Colonies	810	
Combined	1,714	

Source: Shepherd, James F. and Walton, Gary M.: *Shipping, maritime trade, and the economic development of colonial North America*, Cambridge: Cambridge University Press, 1972, pp. 115, 128, 134.

Implications for British growth

There is no question that the growth of British trade and industrialisation was heavily intertwined with the British Atlantic economy of the old imperial system, and with its mercantilist basis. The trade of the Americas rested on the slave-produced staples of the West Indies and, to a lesser extent, the southern mainland colonies. The northern mainland colonies participated by utilising the opportunities that the growing staple trades presented to trade temperate foodstuffs and raw materials to the staple colonies, and to exploit niches in the shipping and mercantile activities that were vital to its success. From Britain's domestic point of view the American colonies within the protected mercantile empire became important markets. This was particularly true for manufacturing industries. The Americas were almost solely responsible for the diversification of Britain's exports, to the point where other manufactured exports exceeded the value of woollen exports. Even though this was clearly the actual historical case, we still do not really know the extent to which the slave-based empire contributed to the coming of the industrial revolution in Britain. We can approach this in various ways.

One starting point is to ask if Britain benefited from slave-based empire more than its European rivals. Here scepticism seems appropriate. To be sure, the English sugar colonies enjoyed a period of impressive expansion and prosperity in the late seventeenth century, but the eighteenth-century picture was much more mixed. Britain's staple colonies were not particularly dominant.[12] As I stated in my survey of British trade,

> The eighteenth century British Empire was not exceptionally large or prosperous. The Spanish, French and English sugar islands in the West Indies all had about the same population (300,000 to 350,000 around 1750). The British islands were high cost producers, unable to compete with the rapidly growing output of French Saint Dominique without protection. During the eighteenth century, French trade to the West Indies grew more rapidly than British and merchants in the French Atlantic ports dominated the re-export Europe of sugar and coffee to Northern Europe (Crouzet 1996). Even in 1750, Spanish America's 10.5 million provided a much larger market than the British America's 1.5 million. Portugal's colony in Brazil had a population equal to that of all of British America. The British Empire's size or trading contribution can hardly have made the decisive contribution to Britain's lead in the emergence of modern economic growth.[13]

If colonial trade did make a crucial contribution, it probably did so by expanding the market for British manufactures, transforming in ways that stimulated innovation the environment in which eighteenth-century British inventors and entrepreneurs made decisions. Here Britain differed from her European rivals, but it is important to be aware that the difference arose not from the staple economies, but from the 'Malthusian' economies of northern mainland America. The principal growing markets for diversified manufactured exports from Britain were not the slave-based staple colonies, but rather the northern mainland. These colonies' population growth was almost entirely internal; New England received no immigration between the Great Migration prior to the English Civil War and the Irish Famine in the 1840s, but was a centre of emigration within America. The demand for manufactured goods was generated by this growing population, which remained relatively prosperous in the absence of a binding land constraint. They satisfied their import demands by trading within the slave–staple-dominated British Atlantic trading network, but it was not the staple economies that distinguished the British Empire from its rivals. It was the presence of these rapidly growing Malthusian colonies.

As Findlay and O'Rourke comment, questions like how important were the slave–staple economies in the growth and diversification of British exports invite counterfactual thought experiments.[14] To what extent would the growth of manufactured imports into the northern mainland colonies been curtailed without the slave colonies? Of course, such questions cannot be fully answered, but they need to be considered if we are to understand the dynamics of economic change. In this context it is perhaps useful to consider Figure 8.4 again. Certainly we can be confident that the absence of the slave colonies would have constrained the supply of foreign exchange to the northern colonies (shifting the curve to the left). It is possible, however, that it would have had no effect on the equilibrium. Table 8.2 indicates that the importance of the horizontal portion of the supply curve of foreign exchange consisted of the shipping and mercantile services the colonist sold in a wider British market, one in which they were relatively small players. This suggests that these sales might have expanded at very little cost to the colonies or to British exporters. This conclusion, of course, is undoubtedly too optimistic. Many of the services that the northern colonies sold were connected with the West Indian trade, and would have been diminished by these colonies' absence. If the supply of foreign exchange were shifted left (say, to the broken line in the diagram) the price of imports in terms of American goods and labour would have risen. How much would that have affected British exports to the colonies? It seems unlikely that it would have had an important impact on the underlying growth in the land-abundant continent. Some expansion of colonial manufacturing production might have occurred if imports had become more expensive, but my own conjecture (and it is no more than that) is that the northern colonists' demand for European manufactured goods was price inelastic, and therefore that the absence of the West Indies would have had little effect on British exports to her northern Atlantic colonies.

It is also the case that simple consideration of the static impact of trade on the British economy does not support the contention that slave-dependent trade generated by the West Indies made a crucial contribution to the British economy. Findlay and O'Rourke, in their spirited argument for the importance of the trade, take me to task for 'play[ing] the old trick of multiplying two fractions by each other to obtain an even smaller fraction' and concluding that trade had a minor impact on the British economy.[15] In response, I contend that this is no trick, but is arithmetic, and a good place to begin, even if a satisfactory analysis needs to go farther. Theoretical speculations demonstrate logical possibilities, but historical explanation needs to connect to extant evidence which can provide indications of relevant importance.

Crafts calculates that Britain's exports were about 16% of national income in 1801.[16] Since most were manufactured goods, exports were a much higher portion of manufacturing output. He calculates net exports were about 45% of the output of manufacturing, mining, and building. About 55% of that went to Africa and the Americas. However, nearly 60% of that went to the United States and British North America.[17] But, as I have just conjectured, much of that, although connected with the slave economies, did not depend on them. If exports to the Americas disappeared, and the resources used in their manufacture were left idle, the reduction of British income would be on the order of eight per cent of national income, and close to a quarter of manufacturing output. Of course, this overstates the static impact after the economy adjusted to the removal of American trade, since British resources no longer employed in the trade would have found some alternative gainful employment. If these alternatives were, say, 25% less productive than the export industries, the loss of national income would have been only two per cent of British income. These calculations have limitations, but they provide the order of magnitude from which discussion should begin. I cannot resist here quoting Samuel Johnson on the value of simple calculations: 'That, Sir, is the good of counting. It brings everything to a certainty, which before floated in the mind indefinitely.'[18] Findlay and O'Rourke quite properly comment on the limitations of similar calculations:

> However, comparative static trade models cannot, by definition, say anything about the impact of trade on growth, and to show that the British economy as it stood in 1860, with the Industrial Revolution already firmly entrenched, would have suffered a small welfare loss had it not been able to trade at all is not only unconvincing on its own terms [whatever this means] but evidently raised the question of what forces were required to bring the economy to that state in the first place.[19]

We all agree that the dynamic questions are the ones in which we are most interested. Unfortunately, we lack a clear understanding of the dynamics of economic growth even in current economies, much less in the transition to modern economic growth. We are trying to tease it from the historical record. In that task, it is important to try to 'bring... to a certainty, which before floated in the mind indefinitely'.

Rather strangely, Findlay and O'Rourke commend (two pages on) the general equilibrium model Findlay proffered in 1990 as a tool to construct counterfactual thought experiments to consider the dynamic issues of trade and British growth. The Findlay model is a useful device to conceptualise the relationships in the British Atlantic economy in the period of old imperialism, and I have already used if for that purpose. However, it can hardly be taken seriously as providing reliable orders of magnitude of various influences. The model is quite straightforward. There are three sectors. Britain (or Europe) produces manufactured goods with domestic labour and mobile capital, and requires a fixed amount of raw materials per unit of output. America produces raw materials using slaves and land. Africa produces slaves at an increasing marginal cost. Capital can be used to produce manufactured goods or to own slaves. An equality of the return on capital in various uses is part of the equilibrium of the system. Now in this system manufacturing depends on the supply of slaves. No manufacturing output is possible without raw materials, and raw materials are produced only in America and require slaves (in terms of the use I made of the model earlier, before the discovery of the Americas everything would be at the origin except the price of raw materials). Useful as this model is in conceptualising the relationships involved in the Atlantic World economy, at least in a comparative static way, it is unconvincing as a guide to assessing the contributions of various factors to British growth overall, or the growth of British manufacturing. In passing I should note that it is a comparative static model, and suffers from the limitations already discussed regarding explanations of growth.

Of course, Findlay and O'Rourke are correct that comparative statics exercises have very limited leverage when it comes to explaining long-run economic growth. Unfortunately, we lack any convincing models of the process of economic growth, although a range of ideas provide background for such a model. One aspect of exploring these ideas is the construction of endogenous growth models. A second is careful examination of the historical record.[20] There is general agreement that technological change lies behind historical economic growth, and that the creation of knowledge and technology must be seen as a part of the economy, that is, as endogenous. Furthermore, knowledge is a 'good' in which market failure is pervasive because it is characterised by externalities and non-exclusivity. This suggests that we should attempt to identify the historical workings of purposeful knowledge creation and the externalities that it involves. Such ideas suggest that the process of technological change is path dependent.

My colleague Bob Allen has recently argued that the classical industrial revolution in Britain at the end of the eighteenth century arose from a process of industrial research and development, work which was directed by the high wages and cheap energy (compared to other economies) that prevailed in Britain in the eighteenth century. This gave British entrepreneurs and capitalists incentives to search for manufacturing techniques that substituted fuel and capital for labour, and their efforts were eventually fruitful. This search in turn created knowledge externalities that further enhanced the process of technological change.[21]

This view of the industrial revolution has the somewhat uncomfortable feature of suggesting that Britain got rich because it was already rich, thus pushing the basic question farther into history. Allen's view, however, is consistent with other work on the emergence of modern economic growth which suggests that understanding of the process needs to extend far earlier than the classical industrial revolution.[22] Allen has investigated data on long-run growth in Europe and concluded that Britain's growth emerged from commerce and the urbanisation that it generated from the early sixteenth century. Urbanisation created incentives and externalities that led to productivity growth.[23] In his view, urbanisation played a key role in stimulating technological change in both the agricultural and non-agricultural sectors of the economy. This technological advance created the high-wage, coal-using economy that lay behind the industrial revolution. The expansion of trade to the Americas fits into this schema in a general sense, although Allen concludes that it occurred too late to have been the trigger that initiated the process of divergence. On the impact of empire he concludes that

> the empire established in the seventeenth and eighteenth centuries also contributed to growth. The greatest impact was on city size. Over half of England's urban expansion is attributed to empire in these simulations.[24]

The view that urbanisation and industrialisation provided both incentives and externalities that contributed crucially to Britain's economic growth is extremely attractive, even if Allen's simulations – based on quite simple regression – are not powerful enough evidence to be conclusive. As I have discussed above, there is no question that the slave-based British Atlantic Empire contributed strongly to both the rise of Britain's port cities and to the expansion of industrial activity. However, by the late seventeenth century the bulk of the trade-based stimulus to industrialisation came not from the slave economies, but from the northern Malthusian economies.

Of course, the industrial revolution was crucially about cotton (Findlay and O'Rourke comment: 'Rostow's original characterisation of cotton textiles as the leading sector of the British industrial revolution appears to have been well-founded'[25]). Cotton depended on a slave-produced raw material. As such, it seems to be the stimulus for Findlay's model. However, several problems arise when building an argument on the importance of slavery for the emergence of modern economic growth on the British cotton textile industry.

The first and most obvious is that the cotton industry emerged too late. A cotton industry existed in Britain from at least the late seventeenth century. However, it was a small industry, initially at least, and dependent upon protection against competition from imports from India. As is well known, that protection was incidental to protection of England's woollen industry, but it helped the cotton industry to become established. This early cotton industry had only limited connections to the slave-based Atlantic. Its raw material came from Ottoman territories in the Levant, although the West Indies became a significant source of supply after the middle of the eighteenth century. Cotton textiles were a part of the cargoes sent to the west coast of Africa to finance slave purchases. These were, however, principally Indian cottons re-exported from Europe, although Inikori has shown that by the second half of the seventeenth century British 'cottons' (cloth of linen weft and cotton warp referred to often, if inaccurately, in the literature as 'fustians') had become significant.[26]

The great expansion of cotton only occurred after Arkwright's innovations at the end of the 1760s. The industry grew spectacularly from the mid-1770s, but remained fairly small until near the end of the century. The United States did not become an important supplier of cotton until Eli Whitney's cotton gin came into widespread use in the 1790s. The expansion of cotton-growing in the United States may have influenced a surge in slave imports in the final years of the legality of slave importation there from 1800 to 1807, but as far as US cotton production is concerned, its expansion occurred with a native-born, although un-free labour force.

How do slavery and cotton fit into endogenous growth models of the emergence of modern economic growth? Not very well. Allen uses Arkwright and the cotton textile innovations as an example of the importance of research and development in the emergence of nineteenth-century technology.[27] The story is persuasive, but the benefits of market size and prospects for market penetration do not really play a role. Although cotton textiles became the British factory industry par excellence in the nineteenth century, it was small until after Arkwright's innovations. The incentives were insufficient to concentrate on cotton. Wool would have appeared to have a much higher payoff. Cotton fibre, however, proved easier to manipulate by machine.

Counterfactual: no slavery

Findlay and O'Rourke invited us to consider counterfactuals.[28] They particularly ask 'what would have happened to the Lancashire cotton industry if there had not been any British colonies or slavery in the New World'. They imply that the effect would have been devastating. Any attempt to answer such a question is inevitably largely speculation, but I am sceptical that the absence of slavery would have had a devastating impact on the cotton industry of the industrial revolution. Slavery and sugar were very closely connected. Sugar technology required large units and concomitant capital resources. An alternative of free white labour would probably have been achievable only at considerably higher cost. We should recall, however, that seventeenth-century Barbados initially attracted white indentured servants. For the Lancashire cotton industry, the labour force on the North American mainland was relevant. Slavery there was used for tobacco cultivation in the Chesapeake and for the rice plantations of the lower south. In 1780 there were about 300,000 blacks in the Chesapeake and 200,000 in the lower south.[29] The labour force in tobacco had become predominantly slave during the eighteenth century, but there were few economies of scale in tobacco production. The Chesapeake had been an attractive destination for indentured servants through most of its history, and it seems likely that the development of the region would have been only slightly retarded in the absence of slavery. The lower south was much less attractive to whites. Here the workforce would undoubtedly have been smaller in the absence of slavery. The population of the cotton-producing states of the US in the nineteenth century arose mainly from natural growth of the population already in place when the country became independent. A somewhat smaller labour force in 1780 would have generated a smaller subsequent labour force. This would have generated higher cotton prices. This, however, would have increased the attractiveness of the region to yeomen farmers, and labour would have moved there. In the absence of slavery, presumably Southern society would have been more attractive to outside labour, and more migration would have occurred.

Cotton was produced in the antebellum American south on both plantations and on yeomen farms. Plantations appear to have had a cost advantage arising largely from the ability of plantation operators to extract harder labour in unattractive circumstances than free men would have tolerated. Even if the labour force had reached the same level that actually prevailed, cotton output presumably would have been lower, since the cost of production would have been higher. What would have been the impact on Lancashire? Modestly smaller output. On British growth? Negligible.

The more interesting, but much more difficult counterfactual question on which to speculate relates to the impact of the sugar-slave economy on the underlying dynamics of growth. The rise of commerce and industry that were stimulated in England by the Atlantic economy may have been important in creating the underlying dynamics of technological change that drove modern economic growth. Unfortunately, we have no way really to estimate the magnitude of possible dynamic forces. It is easier to speculate about the effects on trade of the absence of slavery. Certainly the sugar economy would have been smaller, but it would hardly have disappeared.

I have already argued that if a decisive impact of American trade existed in differentiating the growth of the British economy from that of her European rivals, it lay in the trade with the mainland colonies. It was the mainland colonies that provided the overseas demand for manufactured goods. I have just speculated that the absence of slavery would probably not have decisively slowed the development of the tobacco colonies. The northern colonies were a key market for British exports. These colonies financed their imports primarily through the sale of goods and services to the sugar colonies. How much different would their development have been in the absence of these slave colonies? Here, thinking about their colony's trade in the 'Malthusian' framework is helpful. If we look at Figure 8.4 two features seem relevant.

To what extent would the horizontal portion of the foreign exchange supply curve have been affected? It seems almost certain that there would have been an effect if the sugar trade disappeared completely, since in historical fact most of the temperate agricultural goods and the maritime services that this horizontal curve represents were sold to the West Indies or in West Indian trade. However, after independence American shippers were excluded from the British West Indies by the Navigation Acts. They found new trades, including the Far East.

The second key element in thinking about the impact of the absence of slavery on British trade is the elasticity of the American demand curve and the extent to which it moved over time. First, the growth of population in the northern colonies was largely independent of trading opportunities. Between 1700 and 1780 the population of New England increased from just over 90,000 to over 700,000, and that of the middle colonies from just over 50,000 to over 700,000.[30] It seems unlikely that these numbers would have been much different in the face of reduced export markets. New England's population grew exclusively on its natural increase, and even lost population through migration to other colonies. The attraction of the middle colonies consisted of good agricultural lands at very low prices. Imports made up only a small part of the colonies' yeoman farmers' consumption. It is likely also that the demand for European goods was relatively price inelastic. If this were true, British sales to the mainland colonies of North America would have been only modestly decreased if the sugar colonies had never existed.

Slavery, the British Atlantic and the industrial revolution

Eric Williams was certainly right to bring interaction between industrialising Britain, slavery, and the Atlantic economy into the centre of discussion of British change in the eighteenth century. The Atlantic economy provided the focus of expanding and diversifying trade, and trade contributed greatly to the expansion of manufacturing. The sugar colonies of the West Indies provided the focus of the Atlantic economy, and were fuelled by the trade in African slaves. However, if we believe, as Williams did, that the Atlantic economy made a central contribution to the subsequent industrial revolution, it seems likely that the route through which this contribution came was the trade to the colonies of the northern mainland. Here the story becomes rather more complex. These colonies were not created on the basis of slave-based sugar plantations, but primarily as refuges in the new world. They grew rapidly because they had abundant agricultural land into which the settlers could expand rapidly. This rapidly growing population demanded industrial goods that were imported from Britain. The northern colonies, in turn, financed their imports by sales of agricultural and forest commodities, and, crucially, shipping and mercantile services to the West Indies. In this way the entire American trade rested on the slave colonies. However, this statement almost certainly overemphasises the role of the slave colonies. In the absence of slavery, the northern settlements would have found alternative goods to sell into the Atlantic economy and their growth, and their demand for British manufactures, seems unlikely to have been stifled.

Notes

1. Presented at *A Centenary Conference: New Perspectives on the Life and Work of Eric Williams*, St Catherine's College, Oxford University, 24–5 September 2011.
2. Williams, Eric: *Capitalism and slavery*, Chapel Hill: University of North Carolina Press, 1944.
3. Davis, Ralph: *The rise of the Atlantic economies*, Ithaca: Cornell University Press, 1973, ch. 15.
4. Davis, Ralph: 'English foreign trade 1700–1774', *Economic History Review*, 15 (1962), p. 294.
5. Davis, Ralph: *The rise*, p. 255.
6. See Leonard, this volume.
7. McCusker, John J. and Menard, Russell P.: *The economy of British America, 1607–1789*, Chapel Hill: University of North Carolina Press, 1985, p. 18.
8. Findlay, Ronald and O'Rourke, Kevin H.: *Power and plenty: trade, war, and the world economy in the second millennium*, Princeton: Princeton University Press, 2007, p. 339.
9. Findlay, Ronald: *The triangular trade and the Atlantic Economy of the eighteenth century: a simple general-equilibrium model*, Essays in International Finance, No. 177, 1990, Princeton: Department of Economics, Princeton University.
10. Ibid., p. 9.
11. Shepherd, James F. and Walton, Gary M.: *Shipping, maritime trade, and the economic development of colonial North America*, Cambridge: Cambridge University Press, 1972.

12. Inikori, Joseph E.: *Africans and the industrial revolution in England: a study in international trade and economic development*, Cambridge: Cambridge University Press, 2002, p. 101.
13. Harley, C. Knick: 'Trade: discovery, mercantilism, and technology', in Floud, R. and Johnson, P. (eds.), *The Cambridge economic history of modern Britain, Vol. 1: Industrialisation, 1700–1860*, Cambridge: Cambridge University Press, 2004, p. 196.
14. Findlay and O'Rourke: *Power and plenty*, p. 339.
15. Ibid., p. 337.
16. Crafts, Nicolas F. R.: *British economic growth during the industrial revolution*, Oxford: Oxford University Press, 1985, p. 127.
17. Davis, Ralph: *The industrial revolution and British overseas trade*, London: Leicester University Press, 1979, p. 89.
18. Cited in McCluskey, D. N.: 'The industrial revolution, 1780–1860: a survey', in McCloskey, D. N. and Floud, R. (eds): *The economic history of Britain since 1700, Vol. I. 1700–1860*, Cambridge: Cambridge University Press, 1981, p. 105.
19. Findlay and O'Rourke: *Power and Plenty*, p. 337.
20. Harley, *Growth theory*.
21. Allen, Robert C.: *The British industrial revolution in global perspective*, Cambridge: Cambridge University Press, 2009.
22. Such as Crafts, N.F.R. and Harley, C. K.: 'Output growth and the British Industrial Revolution: a restatement of the Crafts-Harley view', *Economic History Review* 44 (1992); van Zanden, Jan Luiten: *The long road to the industrial revolution: the European economy in a global perspective, 1000–1800*, Leiden: Brill, 2010.
23. Allen, Robert C.: 'Progress and poverty in early modern Europe', *Economic History Review*, vol. LVI (2003), 403–43.
24. Ibid., p. 431.
25. Findlay and O'Rourke, *Power and plenty*, p. 320.
26. Inikori, *Africans and the Industrial Revolution*, p. 444.
27. Allen, *British Industrial Revolution*.
28. Findlay and O'Rourke: *Power and Plenty*, p. 339.
29. McCusker and Menard, *Economy of British America*, pp. 136, 172.
30. McCusker and Menard, *Economy of British America*, pp. 103, 203.

9

Commodity Frontiers, Spatial Economy, and Technological Innovation in the Caribbean Sugar Industry, 1783–1878

*Dale Tomich**

Debate over the relationship between slavery and technology has a long history. Broadly speaking, two approaches to the question have been advanced. One school of thought has regarded slavery as an archaic, pre-capitalist system of production relations, and has emphasised the incapacity of slaves to utilise sophisticated technologies and the backwardness of planters. From this perspective, either slave relations suppress technological innovation, or technological advance spells the end of slavery. The other school, by contrast, has regarded slavery as modern, and has emphasised the productivity, profitability, and technological efficiency of slave economies. Despite this opposition, both approaches share a common view: technology, particularly steam power, is presumed to have a transformative effect on economies and societies. The debate turns on whether or not technology is adopted. No matter what the position, it is treated as a variable independent of the social relations of slavery and is regarded as virtually a free-floating signifier of modernity.

Generally overlooked in such discussions are the geographical and physical conditions of production in which technology is deployed.[1] Economies operate in specific historical geographical configurations. Instruments of production, including modern industrial technologies,

---

* This chapter builds upon a previous approach to the crisis and reconstitution of the nineteenth century Caribbean sugar economy that appears in 'Commodity frontiers, conjuncture and crisis: the remaking of the Caribbean sugar industry, 1783–1866', in Laviña, Javier and Zeuske, Michael (eds): *Second slaveries and the Atlantization of the Americas*, Münster: LIT Verlag, 2013. The author thanks Ina Brownridge and the staff of The Binghamton University Multimedia Information Technology Services for preparing the images.

mediate between human labour and nature. Together land, labour, and technology form an interdependent unity. Their interaction creates historical-geographical complexes whose 'geometry' is shaped by material, social, and technical conditions.[2] The creation and ordering of productive space and the transformation of the natural environment at once both enable and constrain technological innovation.

Sugar imposes its own spatial order on the production process. The material processes of sugar require both agricultural and industrial operations. Land, labour, and technology must be combined in the proper proportions to secure the optimal conditions of production. Each step in the production process must be performed sequentially. No more cane can be planted than can be milled, and no more syrup can be extracted than can be processed into crystallised sugar. Further, cane has to be processed within 72 hours of being cut, since otherwise fermentation impedes crystallisation. Consequently, the material and economic requirements of sugar production impose a particular scale of activity and economy of time and space on producers. They involve the production of space and transformation of nature, and, at the same time, are subject to specific physical-material as well as social-historical limits.

This chapter examines the production of plantation space in three major Caribbean sugar colonies during the world-economic conjuncture of the first half of the nineteenth century. Each site represents a particular local historical-geographical complex of sugar production formed within the world-economic division of labour. In each, the articulation of environment, material processes of production, technology, and forms and scales of collective labour shape the productive capacity of the sugar industry and determine its position in the world-economy. From this perspective the specific geographies of sugar production in Jamaica, British Guiana, and Cuba disclose the processes of world-economic expansion and differentiation that shaped the crisis of the British West Indian sugar industry, and formed a new Atlantic division of labour.

Two concepts are central to this investigation. The concept of commodity frontier calls attention to the ways that the production and distribution of specific primary products restructures geographic space at the margins of the world-economic system. Such zones have geographical and environmental conditions that are favourable to the production of specific commodities, particularly agricultural products and raw materials. They are zones in which further expansion is possible, as long as uncommodified land, and, to a lesser extent, labour, are present. Incorporation and exploitation of land beyond the frontier is driven by the demand for the product. The movement of population and the transformation of nature follow upon the demand for the commodity, and the open frontier permits dramatic restructuring of land and labour relations. This concept calls attention to the transnational migratory character of primary production and of slave labour, and enables conceptualisation of the relation between specific ecologies, place, and the geographical expansion of the capitalist economy.[3] On the other hand, Barry Higman utilises the concept of spatial economy to examine the ways that location, environment, topography, and the material requirements for the production of the principal crop, together with the planter's changing notions of profitability and control, affect the size, shape, and physical organisation of the plantation. This concept permits analysis of the ways in which spatial configuration orders productive activities and social life within the plantation unit.[4]

Shifting patterns of production

The expansion and restructuring of world sugar production after 1815 accentuated the difference between old sugar zones and new commodity frontiers. World sugar production and consumption expanded dramatically during the first half of the nineteenth century. Old producing regions increased their output, and new regions emerged. British West Indian sugar production expanded during this period, but it was unable to keep pace with the changing conditions of world production. From 1791 to 1815 British West Indian sugar production increased more rapidly than at any other time in its history. Peace in 1814 brought about a boom in sugar production as pent-up demand was released, and planters throughout the Americas rushed to fill the void left by the destruction of the French colony of Saint-Domingue, the world's leading sugar producer.[5] The British West Indian colonies were best positioned to take advantage of this situation. They had been least disrupted by war and revolution, had regular access to shipping, and benefitted from the acquisition of new territories, particularly Guiana. Their output fell below that of the previous period, but between 1815 and 1819 they still accounted for nearly half of world supply, which had risen by 38% despite the destruction of Saint-Domingue.[6]

By 1820 the post-war boom ended and prices fell. Markets became more stable and increasingly synchronised with one another. Competition between producing regions became more acute. This contraction disclosed considerable geographical differentiation among the British West Indian colonies, even as the British West Indies as a whole declined in relation to world sugar production. In each historical-geographical complex the specific 'geometry' of land, labour, and technology produced a distinct pattern of development of the sugar industry. After 1814–18 sugar production ceased to expand in the small islands, which accounted for a progressively smaller share of British output. Jamaica remained an important centre of production, although it no longer attained pre-1807 levels. In 1820 it still produced nearly 98,000 metric tons of sugar, but production gradually declined thereafter, and during the 1820s the island was replaced by Cuba as the world's largest single supplier. Led by Demerara, British Guiana's sugar production grew rapidly, and it became a valuable addition to the empire. Production increased nearly fivefold between 1814–18 and 1829–33, reaching a pre-emancipation high of nearly 60,000 metric tons in 1830.

Mauritius in the Indian Ocean was also acquired by Britain during the Napoleonic Wars. It represented an open commodity frontier: it enjoyed relatively continuous access to labour either through the clandestine slave trade from nearby Africa and, after 1834, by the importation of 20,000 indentured Indian contract labourers. Annual average sugar production in Mauritius reached 32,700 tons by 1834–8 and 113,000 tons by 1854–8. However, the cost of transportation always put its sugar at a disadvantage. By the late 1820s and early 1830s, sugar production stagnated and declined in the Lesser Antilles and Jamaica. The losses in these old colonies were offset by increases in Guiana, Mauritius, and Trinidad, but the British colonies' share of world production fell from nearly 50% in 1815–19 to just under 25% in 1838–42. During the same period sugar production also fell in the French West Indian colonies, where the slave trade remained active. Land was available in Brazil, and sugar cultivation spread to areas outside of the northeast. The Atlantic slave trade continued to supply labour. However, attempts at innovation were sporadic, and production techniques remained largely unchanged. The expansion of sugar cultivation was extensive rather than intensive, with production tripling between 1820 and 1850, before levelling off.[7]

Changes in sugar markets and sugar consumption also undermined the British West Indian colonies. Producers there continued to enjoy a protected position in the British national market as long as the Navigation Acts remained in effect, but British territorial expansion in the Caribbean ended in 1815, and planters there were faced with rising costs. With the end of the Napoleonic Wars Britain gained command of the Continental market. British refiners, shippers, warehousemen, and commercial and financial interests were becoming involved in trading large quantities of both British and foreign sugar in Europe. As cheaper Cuban and Brazilian sugar found markets in Europe, British West Indian sugar was excluded from the Continent. In order for British West Indian planters to remain competitive with countries that were customers for British manufactures, re-exports of British West Indian sugar had to be subsidised by government drawbacks and bounties. At the same time Britain either warehoused cheap foreign sugar for re-sale on the Continent, or British merchants or carried it directly to Europe for sale in 'spot markets'.

If mercantilism was intended to secure sources of overseas production and trade for the metropolis, by the nineteenth century it had become the means for West Indian planters to protect themselves from competition. Britain continued to pay higher prices for sugar without the benefits that mercantilism had previously provided for the national economy. This problem was especially acute in the re-export market. Huskisson wanted to make London the 'depot of the merchandise passing between the two worlds'. In 1823 his Warehousing Bill allowed foreign goods, including sugar, to be re-exported free of duty. This policy was based on notions of 'comparative advantage' and the allocative efficiency of the market system. Nonetheless, until the preferential duties were repealed in 1846, Britain had a dual market system: one that subsidised colonial producers and kept domestic prices high, and another that attempted to control Continental markets by managing the unrestricted flow of produce from American producers. British West Indian re-exports dropped from about 100,000 tons in 1802 to 27,000 tons in 1827. As British sugar producers became increasingly dependent on preferential duties and the home market, British domestic consumption doubled between 1815 and 1840. By the 1830s the West Indian colonies were no longer able to supply British demand.[8]

If the British West Indies remained significant centres of sugar production, and continued to yield profits to those who invested in them into the 1830s and beyond, this does not mean that they did not decline. The condition of a capitalist market economy is not simply to make profit. Rather, it is to make profits on an ever-renewed and ever-increasing scale. This process of capital accumulation continually establishes new material and social conditions of production. The decline of the British West Indian sugar industry is apparent if we look beyond the British colonial economy to the dramatic rise of new producing zones, especially Cuba, and the restructuring of the Atlantic division of labour as a whole. Sugar production in Jamaica and British Guiana, the two most important British West Indian sugar colonies, was eclipsed by Cuba (Figure 9.1). Jamaican production declined gradually after the post-war boom, while British Guiana's output increased sharply between 1807 and about 1827. Its production then remained relatively stable until the abolition of

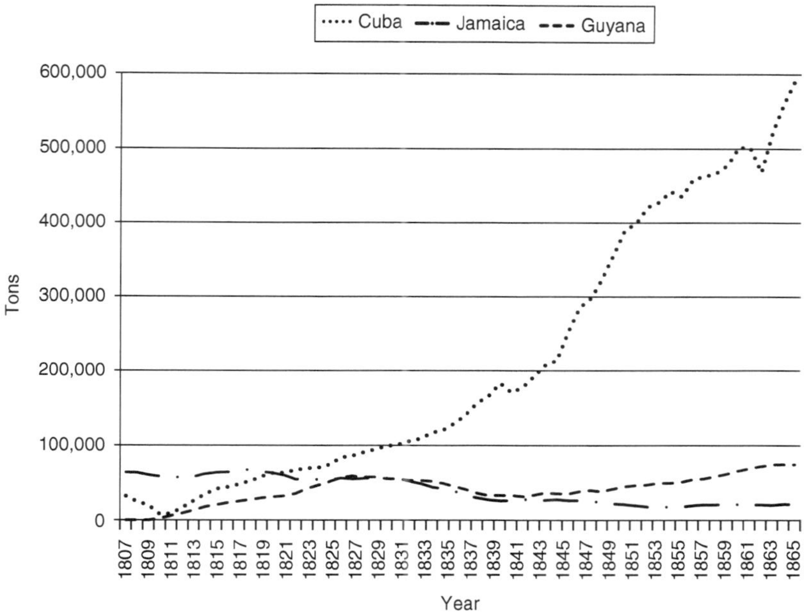

Figure 9.1 Sugar production in Jamaica, British Guiana, and Cuba, 1807–1866
Source: Deerr, Noel: *The history of sugar*, London: Chapman and Hall, 2 vols, 1949: I, pp. 126, 131, 193–204.

slavery, when it declined slowly until the 1840s, as indenture resolved the labour question. After that it climbed to a little beyond its previous high levels. Cuban sugar production, in contrast, climbed steadily and sharply upward. It surpassed both Jamaica and British Guiana in the early 1820s to become the world's leading producer, and increased nearly eightfold between 1820 and 1860 to dominate world sugar production. In contrast to British West Indian sugar planters, Cuban planters had to place their sugar in highly competitive continental European markets, and increasingly in the United States. After independence, US economic interests became steadily more active in Cuba. The US was destined to become the primary consumer market for Cuban sugar, and Cuba enjoyed exceptional access to US goods and capital. From this perspective the production curves for Jamaica, British Guiana, and Cuba attest to the expansion and differentiation of Caribbean sugar production during the first half of the nineteenth century, and the formation of a new Atlantic division of labour.

Jamaica: a mature production zone

Jamaica was a settled plantation zone in which the size and geography of the island, as well as the existing pattern of land tenure and cultivation, restricted possibilities for expansion, technological innovation, and greater productivity during the first half of the nineteenth century. It had been the leading sugar producer in the British West Indies during the eighteenth century. The plantation zone was concentrated on the prime sugar lands in the southern coastal lowlands, the inland alluvial plains extending from St Thomas to Clarendon, across the north coast, and into the Westmoreland plain. At the time of the Haitian Revolution, room remained for new investment and territorial expansion. Jamaican sugar production nearly doubled between 1792 and 1805, and surpassed the output of Saint-Domingue at its peak. New plantations were constructed, but these were often located in regions with less favourable soil and climatic conditions, or in the interior of the island where transportation costs were higher. Eighty-four new sugar estates were established between 1792 and 1799, over half of them in the northern districts of St Ann, St James, and Trelawny. James Robertson's map, drawn in 1804 at the highpoint of Jamaican sugar production, documents 830 sugar estates for the entire island, densely concentrated on the northern and western coastal plains and the southern coastal lowlands.

Jamaica's response to the new conditions of sugar production after 1815 was that of a mature production zone, rather than that of a new sugar frontier. The island's sugar industry had reached its geographical, environmental, and socioeconomic limits. The response was contraction and consolidation rather than expansion. Jamaican sugar production declined during the years of war, blockade, and embargo between 1810 and 1816. However, it recovered during the boom at the end of the Napoleonic Wars, and was the largest single sugar producer in the world until the mid-1820s, although it never again reached the level of 1805. Output declined after 1821, but remained relatively stable until slave emancipation in 1834.[9] Throughout this period the number of estates declined as properties were abandoned, amalgamated, or converted to livestock. (Figure 9.2). This contraction of productive land was especially pronounced in Portland, where climatic conditions were unfavourable, and among those marginal estates that had been established in the interior of the island at the end of the eighteenth century. By 1832 the number of estates had dropped to about 670. The least-affected parishes were those that had been occupied longest. Production was reconsolidated in the original core sugar zone. This process of

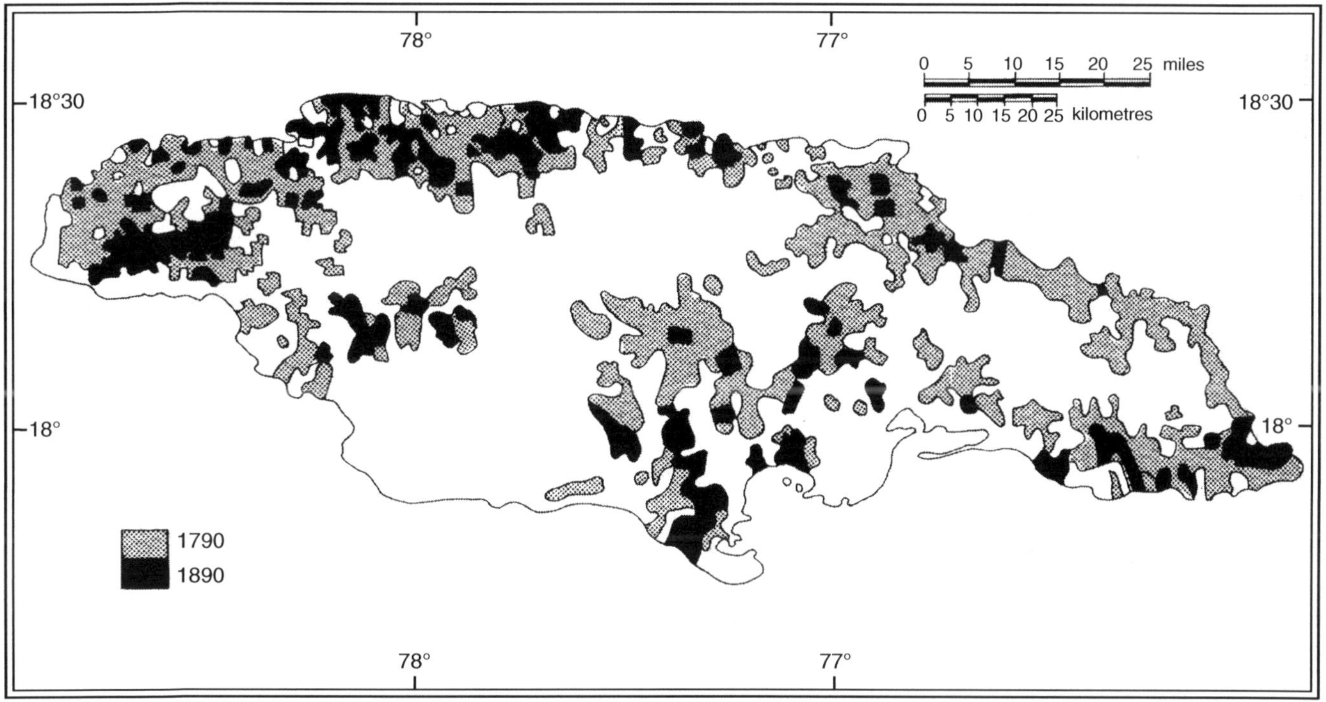

Figure 9.2 Distribution of sugar estates in Jamaica in 1790 and 1890
Source: Higman (2001, p. 14).

contraction of the plantation zone and concentration of individual sugar estates continued well into the nineteenth century. Between 1834 and 1846, 157 sugar estates were abandoned, and, following the passage of free trade legislation, another 86 estates were abandoned.[10]

The end of the slave trade in 1807, of course, meant the end of the forced importation of new labour to Jamaica's plantations. It did not, however, mean an absolute shortage of labour. The labour supply mirrored sugar production. More slaves were employed in sugar in 1817 than had been in 1804. The labour force then gradually declined until emancipation in 1834 (although it should be noted that the decline of labour was more gradual than the decline of sugar production. There was no sharp reduction of workers coinciding with the decline in production after 1821.). With abolition of the slave trade, the labour force underwent a process of 'creolisation'. Slaves of African origin, who had swollen the cohorts of adults of working age during the period of the slave trade, gradually disappeared from the slave population. In their place were slaves born in the West Indies. The percentage of females slowly increased, and children and the elderly accounted for a larger proportion of the enslaved population on the estates.[11] The medium-term consequence of the abolition of the Atlantic slave trade was not a general shortage of labour, but the shortage of *optimal* labour. It seems probable the slaveholding units were becoming larger, as many planters consolidated properties in the coastal sugar zone to secure favourable conditions for production. The enslaved population moved from towns to rural parishes, and from the coffee zone to coastal sugar estates. In 1830 almost half of Jamaica's enslaved population lived on units of over 150 slaves, and the average number of slaves on a sugar estate was 223 in 1832.[12]

Jamaican planters were actively concerned with increasing their efficiency and with technological improvement of their operations. Beginning around 1800 those planters who were able to do so were quick to take advantage of new industrial technologies. Both the steam engine and horizontal three-roller iron mills became common in Jamaica between 1810 and 1830. Veront Satchell has documented the purchase of fifty-one Bolton and Watt steam mills in Jamaica between 1809 and 1830; the majority were low-pressure engines of ten horsepower or less. The number of purchases of Bolton and Watt engines fell after 1818 when more powerful and efficient high-pressure engines of American manufacture became available. These presumably dominated the market after 1818, but there is no systematic data on the number purchased, or their distribution.[13] By 1854, 108 of the island's 330 sugar estates were powered by steam.[14] Satchell further records that the firm of Fawcett and Preston sold 63 horizontal iron mills to Jamaica between 1813 and 1817, as opposed to only eleven vertical mills. The sturdier horizontal mills were better able to withstand the stress of steam power, increased the grinding surface, and could be operated in combination with one another. According to Satchell, horizontal iron mills were in common use in Jamaica by the 1830s despite their higher cost. In 1854, 140 such devices were in operation there.[15]

No significant improvements in sugar boiling were made during the first half of the nineteenth century; the vacuum pan was slow to appear in Jamaica. By reducing air pressure, it allowed evaporation to take place at a low temperature, eliminating the problems of excessive heat and exposure to open air and humidity inherent in open-kettle boiling. Further, it yielded more and higher-quality sugar from a given quantity of syrup. However, the vacuum pan was an extremely expensive apparatus, and had to be employed on a large scale to be effective. In Guiana, where it was more common, planters felt that an estate had to produce a minimum of 500 hogsheads per year in order to install it profitably.[16] In Jamaica, the reduced scale of planting inhibited its introduction, and planters felt that the higher-quality sugar obtained from it did not compensate for its expense. Only five or six were in operation in 1846.[17] In 1870s, according to Higman, only one vacuum pan was in operation in Jamaica, and in the 1890s it was upgraded to a triple-effect apparatus.[18] The centrifuge, which more effectively drained molasses from muscovado sugar, was more widely adopted in Jamaica than the vacuum pan. Better drainage of molasses encouraged rum production, and after 1872 lower-priced unrefined muscovado sugar found a market in the United States.[19]

Despite the efforts of a significant portion of the Jamaican planter class to incorporate these new technologies during the first part of the nineteenth century these innovations did not create a major breakthrough in sugar production. Both output and productivity were stable between 1812 and 1820, and declined gradually thereafter.[20] Steam and water mills were superior to wind and cattle mills. Steam-powered mills offered a savings in livestock, and perhaps permitted more regular production and marginal improvement in yield. However, by itself the steam engine, especially the small, lower pressure engine, did not offer substantial improvement over the water mill. In 1854 just over two-thirds of all sugar mills in Jamaica were powered by steam or water, with 108 steam mills and 125 water mills. In Westmoreland, Metcalfe, and St Thomas in the east, the parishes with the highest production per estate, water mills outnumbered steam mills 36 to 21. These three parishes are located in river delta areas with abundant water supply and superior sugar soils.[21]

Horizontal iron grinding mills used the entire surface of their cylinders, and allowed more cane to be processed than the old vertical mills. They were effective when used in combination with steam power, but were also compatible with water mills. Data for 1854 indicate that there were 140 horizontal mills employed in Jamaica, but only 108 estates utilising steam power. Presumably a significant number of the horizontal mills were installed on estates using waterpower.

New milling technologies could process more cane and extract more juice from a given quantity of cane. However, to be successful innovations had to operate within the constraints created by material requirements of sugar cane production, geography, and planter strategies for profitability and social control. These restrictions ordered the factors of production, and shaped the spatial economy of the plantation.[22] Time, distance, and the quantity of material to be processed regulated spatial arrangements, and became increasingly important as the scale of production increased. Planters sought to maximise efficiency and profit by organising the plantation in such a way as to minimise movement, and secure the material integration of production as the sugar moved through sequential steps. Ideally, the works were centrally located, and cane fields were planted in close proximity to them to facilitate timely transportation of cut cane to the mill. Fields were laid out in a systematic manner that permitted calculation of the amount of material to be processed over the course of the harvest, and the apportionment of labour. The lanes between cane pieces allowed carts to carry the cane to the mill, and served as firebreaks. Transport capacity determined the location and extent of cane fields, and the capacity of the factory limited their size.

Though few Jamaican estates actually conformed to this ideal, their spatial configuration effected the material economic integration of the production process. Irregular topography, slope of the terrain, soil quality, access to water, the disposition of pre-existing buildings, fields, and roads, and the proximity of neighbouring properties could all influence the spatial configuration of a plantation and the movement of materials and workers through the various phases of the production process. Estates with windmills or watermills had to locate the factory where adequate power was available. The shape of both cane pieces and the estate itself, and the arrangement of fields, varied with topography, slope, and with the irregularities that followed from the amalgamation of estates. Irregular shapes did not affect the quantity of cane planted, but did affect the speed and continuity with which it was transported to the mill, and therefore the integration of field and factory.

The full potential of the technological innovations of the first half of the nineteenth century could only be realised if the area under cultivation could be increased in proportion to the increased capacity of the mill (and boiling house), and if more cane could be supplied to the mill. However, conditions in Jamaica did not permit significant expansion of the area cultivated in cane: as marginal properties were eliminated, and sugar production was contracted geographically to the original core plantation zones on the coastal plains and in inland valleys. Plantations in these core zones were larger than before. Many planters either brought previously unused land into cultivation, or amalgamated several properties, in particular to take greater advantage of the new milling technologies. The slave population in these zones was increasingly concentrated. However, it seems reasonably clear that if the Atlantic slave trade had continued, Jamaica could only marginally absorb more labour. The geometry of the sugar zone inhibited further reorganisation of the sugar estates, and the physical appropriation of labourers as slaves meant that the labour force could not be easily adapted to new conditions. Additional labour could not be productively employed on the available land.

Within the core plantation zones, geography and the built environment were obstacles to the amelioration of production. In these settled districts production was organised on a scale adequate to eighteenth-century technical conditions, that is, to wind-, water-, or animal-powered mills. Dense occupation of arable land and the existence of contiguous properties made it difficult for estates to expand or to establish new properties. Expansion might be possible within individual properties, or estates might be amalgamated. Already at the end of the eighteenth

Table 9.1 Average total acreage and acreage in sugar cane for selected Jamaican estates, 1810–1859

Years	*Total area (acres)*	*Acres in cane*	*No. of estates*
1810–1819	768	269	14
1820–1829	984	240	14
1830–1839	884	202	25
1840–1849	1,217	273	15
1850–1859	1,147	312	11

Source: Higman (1987, p. 26).

century Bryan Edwards wrote that it was impossible to find in Jamaica a block of land of even 300 uncleared acres with more or less uniform soil and topography.[23]

The extent of estates varied with their location. Plantations on the coastal plains were more compact and had relatively larger cane fields, while those in the interior were spread across more irregular terrain and had smaller and more dispersed fields.[24] Relatively little information is available on the amount of acreage in planted cane on Jamaican estates.[25] In 1774 Edward Long recorded that a medium-sized estate had 600 acres, of which 266 were planted in cane. In his analysis of Jamaican survey maps, Higman provides a sample of the acreage in cane on selected estates.[26] In general terms, this sample indicates increasing acreage in cane between 1810 and 1859, as sugar plantations became larger. However, if these averages may be regarded as representative of Jamaican sugar plantations, the area planted in cane was relatively small in comparison to that of Guiana and Cuba.

Thomas Harrison's 1872 plan of Greenwich Estate in Clarendon (Figure 9.3) provides an example of a sugar estate on the alluvial coastal plain. Its scale and spatial ordering express its productive capacity and the conditions under which the material processes of production are integrated. Soil and slope are uniform, and the property is relatively compact and symmetrical. Cane fields are laid out in more or less regular squares, with the exception of those that were adapted to the changing course of the Rio Minho. The factory and houses are located at the centre of the plantation. Greenwich is surrounded on all sides by other estates, and Vizzard's Run, which is subdivided into ten fields, is on the other side of the neighbouring Amity Hall Estate. Greenwich covers 886 acres, of which 348 are planted in cane. Vizzard's has an additional 274 acres, but there is no indication about what is produced there. Greenwich had a cattle mill around 1800, but subsequently a steam mill replaced it. In 1872 it produced 209 hogsheads of sugar and 155 puncheons of rum.[27]

Under the circumstances prevailing in Jamaica it could be difficult to maintain the optimal proportions between the capacity of the slave gangs, the size of the cane fields, and the mill. In 1843 John Biggs, an engineer resident in Jamaica, noted that planters who installed steam engines were often disappointed because they had not proportionally increased the capacity of the boiling house, or they did not have a sufficient supply of cane. Further, even if they could afford it, planters with capital already sunk in slaves and equipment might be reluctant to invest in new plant that could offer only marginal increases in output.

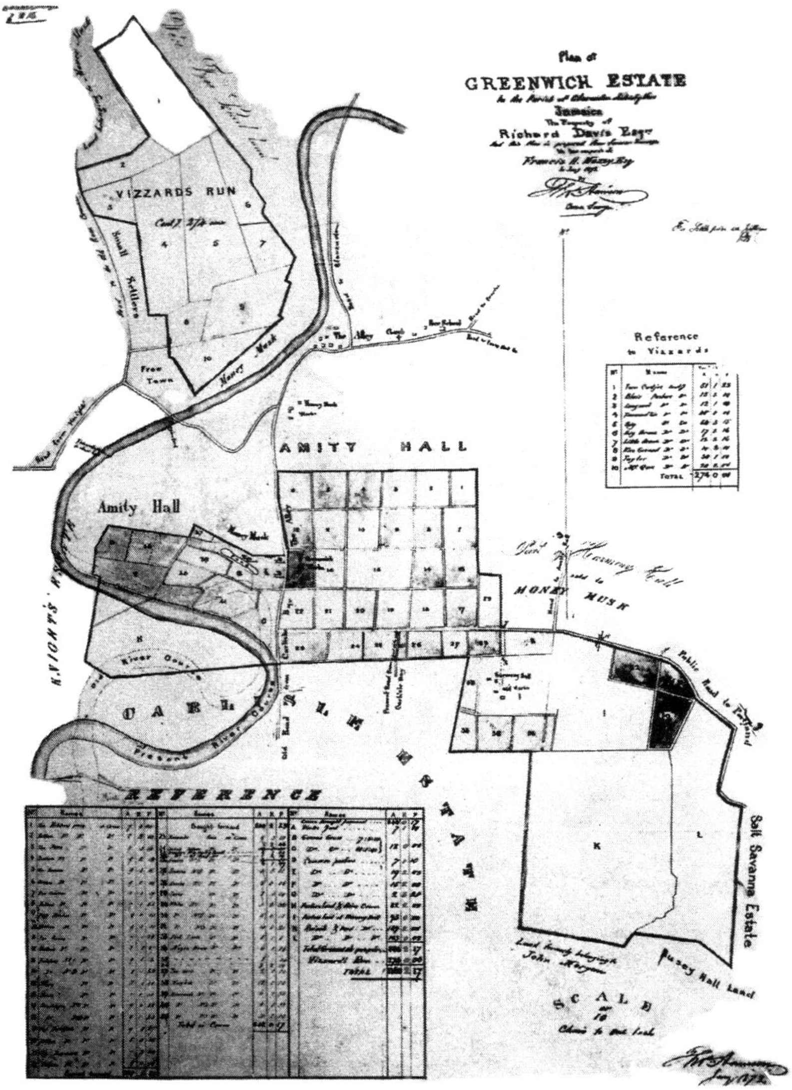

Figure 9.3 Plan of the Greenwich Estate, Clarendon, 1872, by Thomas Harrison
Source: Higman, 'Spatial economy', p. 96.

Historian Douglas Hall notes that a lack of cane made planters reluctant to install improved boiling-house equipment.[28]

After 1820 sugar plantations in Jamaica were bigger, had more slaves, and in many cases employed new milling technologies. However, in the absence of substantial new territory for expansion technological innovation did not have a transformative effect. Planters could seldom achieve an optimal combination of land, labour, and technology. The sugar estates were, by formal measure, productive and presumably profitable, but after the 1820s, productivity, like production, gradually declined.[29] Inability to transform productive space consigned the Jamaican sugar industry to slow decline. At best, technological innovation facilitated the consolidation of large estates and slowed the decline in output as the sugar industry contracted.

Guiana: a commodity frontier

In contrast to Jamaica, both Guiana and Cuba were commodity frontiers. In each, environment and material conditions of production shaped distinct sugar frontiers that offered different responses to the conjuncture of the nineteenth-century sugar economy. In each, highly specialised and rationalised forms of production combined with coercive and indeed virtually regimented forms of labour organisation.[30]

Guiana was the most important addition to Britain's West Indian colonies after the Napoleonic Wars. Attracted by the rich soil and favourable climatic conditions, English planters began moving to the Dutch colonies of Berbice, Demerara, and Essequibo as early as the 1760s. By 1813, 95% of the European settlers were British. Guiana was late in developing compared to the rest of the Caribbean. At the beginning of the nineteenth century it still possessed large tracts of virgin land and excellent conditions for cotton, coffee, and sugar cultivation. However, the colony's unusual ecology imposed distinctive conditions on the development of the plantation system there. The Guianese plantation belt was restricted to the coastal plain, an area of 1,750 square miles. This land was at or below sea level, and subject to alternating periods of heavy erosion and substantial accretion. The plantations were constructed on land reclaimed from the sea using Dutch techniques of polder construction. Plantation agriculture in Guiana depended on the construction and maintenance of an elaborate system of dams, canals, and sluices for drainage, irrigation, and transportation. Walter Rodney estimates that the original construction of this hydraulic system required moving at least one hundred million tons of soil by hand.[31]

Hence a heavy capital investment was required even before sugar production was undertaken.

Poldering and the need to control the movement of both water and land created a particular agricultural landscape. Plantations in Guiana were large, contiguous, and relatively uniform in their internal organisation. Flood and water control was centralised on each estate. Collectively, they safeguarded the plantation zone from the sea and back-land flooding, and controlled the water supply. They were connected with one another through the network of canals and waterways, and the coastal road to Georgetown (Figure 9.4).[32] Although sugar was grown under the Dutch, cotton and coffee predominated, and much land remained uncultivated. Plantations were already large when the colony was under Dutch control. Initial allotments along the coast were 250 acres, and planters had the right to a second grant extending into the interior. A limit of 1,000 acres was established for sugar plantations, and 500 for coffee.[33]

British reconquest of Guiana in 1796 initiated a massive shift of British capital and slaves to the colony. The enslaved population doubled from 1792 to 1802, and between 1803 and 1805 20,000 Africans were imported.[34] Cotton, sugar, and coffee exports increased dramatically between 1789 and 1802.[35] After 1810 cotton and coffee, faced with foreign competition, began to decline gradually, while sugar production rose significantly, particularly during the boom period between 1815 and the early 1820s. Many planters sowed all three crops, and shifted gradually to sugar. By the 1820s the decline of cotton and coffee became more pronounced; sugar production more than tripled between 1814–18 and 1829–33. The need to invest in land, new steam technology, and slaves made Guianese sugar planters vulnerable to falling prices after 1820. Cotton and coffee gave way to sugar, and small sugar plantations were unable to compete under the new conditions. Land was gradually concentrated in the hands of British merchants with enough capital to maintain production, and planters grew more cane and produced more sugar in order to offset declining prices. Between 1810 and 1834 output per slave tripled, and was the highest in the British West Indies. By 1830 most East Coast Demerara plantations had converted to sugar. Steam mills replaced wind- and cattle-driven mills. Several plantations had merged, and some passed into the hands of British corporations.[36]

Land was abundant and there was room for expansion of the area under cultivation, but Guiana was chronically short of labour.[37] Guianese plantations were already big in 1815. Eight per cent had more than 300 slaves, 40% between 200 and 300, and 46% had between 100

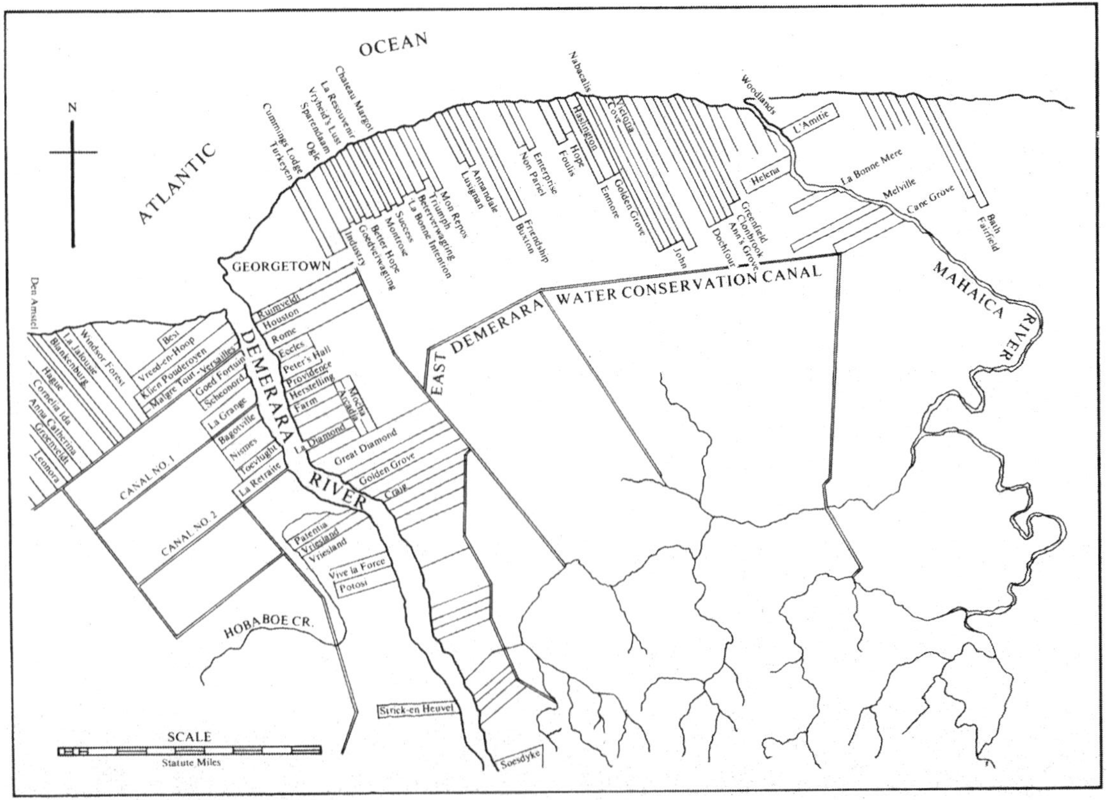

Figure 9.4 East Coast Demerara sugar zone, *c.* 1880
Source: Rodney (1981, p. 8).

and 200.[38] However, the expansion of sugar production and increased demand for labour coincided with the end of the slave trade. As in Jamaica, abolition of the slave trade resulted in the slow decline and gradual creolisation of the enslaved population. The intercolonial slave trade brought some relief to Guianese planters' demand for labour. Between 1808 and 1821 approximately 8,000 slaves were brought to Guiana from the neighbouring Caribbean. However, from 1817 to 1829 the number of those enslaved declined from 77,173 to 69,358, while the number of slaves of working age declined. The demographic structure of this population changed appreciably. The share of slaves of African origin fell from 55% to 38%. The percentage of females and young people increased, and the percentage of old people grew appreciably. Slave prices rose with the decline in the size of the labour force. Nonetheless, the demand for labour grew sharply. Mechanisation of the sugar industry required more field labour. This demand was particularly acute in the expanding sugar districts of Demerara and Essequibo, where the enslaved populations became concentrated. In 1813 only 32.6% of slaves in these districts lived on sugar estates. By 1832 this rose to 78.5%, when there was an average of 233 slaves on a Demerara-Essequibo sugar estate.[39]

Guianese sugar plantations were larger, more capital intensive, and more technologically advanced than those the rest of the British West Indies. The largest plantations had 400–700 slaves. John Gladstone owned 2,000 slaves on several plantations at the time of emancipation.[40] Climate, topography, and hydrology combined with political-economic factors to create a distinctive spatial economy on Guianese plantations that required an extraordinary amount of human labour (Figure 9.5). Cane could be harvested year round in Guiana, which required elaborate strategies for staggering planting, harvesting, and the distribution of labour over the course of the agricultural year. Because of the numerous drainage ditches, draft animals could not be employed for fieldwork, and human beings had to be used in their place. Cane was moved from field to factory by punts traversing the drainage canals. Considerable labour was required to maintain the extensive hydraulic system necessary to control the movement of water and land. Before the adoption of the steam engine tidal flow powered the sugar mills. The high productivity of the soil, economies of scale, and technological innovation compensated for the relative shortage of labour. The size of plantations, the amount of capital required to make them profitable, and the shortage of labour all encouraged technological improvement. At the same time, the size of the estates and the scale of operations

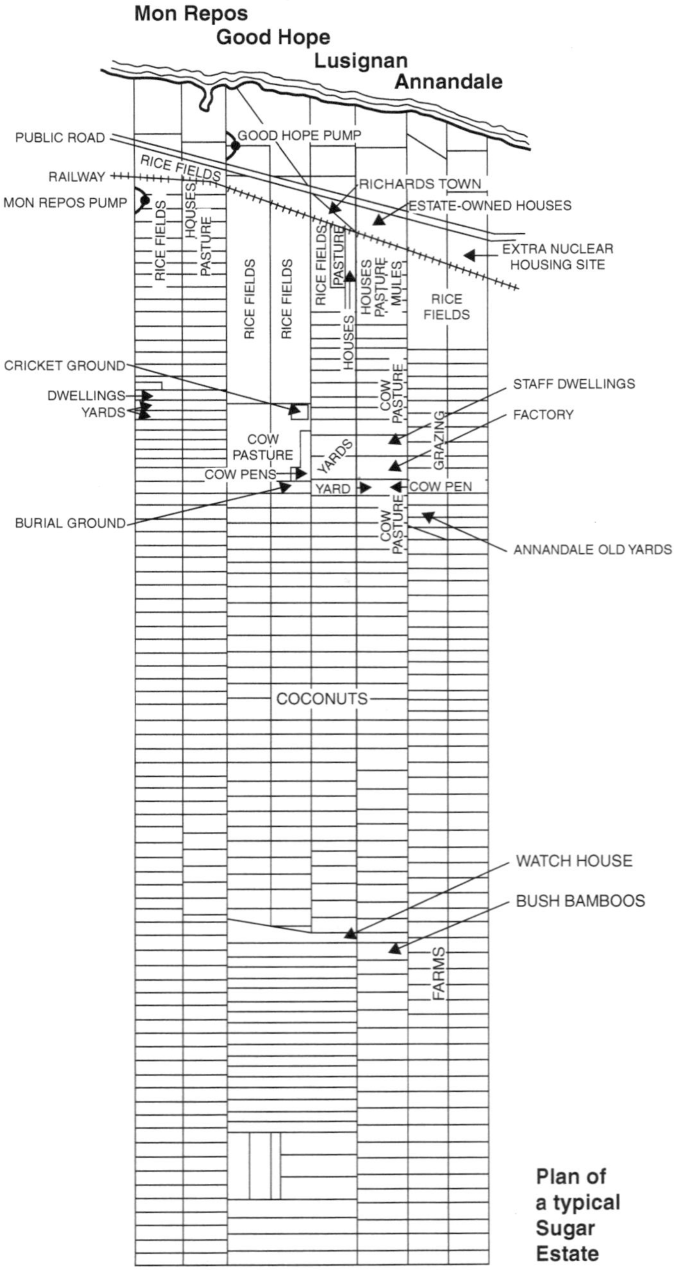

Figure 9.5 A typical plantation

Source: *Michael Swan, British Guiana: The land of six peoples*, London: HMSO, 1957, p. 77, licensed under the Open Government Licence[OGL].

meant that planters could take advantage of the increased capacity of new machinery. That estates were contiguous to one another and relatively uniform in their layout meant that consolidation of estates was a possibility, as new refining technologies required an increased area under cultivation in order to achieve their maximum efficiency.

Guianese planters innovated in both agricultural and manufacturing aspects of sugar production, and they invested heavily in machinery.[41] Boulton & Watt sold 114 steam engines to Guianese planters between 1803 and 1820. The number fell to 32 between 1820 and 1852, but by 1830 most estates had steam power. It is likely that after 1820 these were higher-capacity and more efficient high-pressure engines, and not Boulton & Watt machines. The other major innovation was the vacuum pan, which revolutionised sugar boiling. Various historians have complained that that the vacuum pan was invented in 1813, but only deployed in the Caribbean decades later.[42] However, a great difference exists between the early models and the vacuum pans and multiple effect boilers that appeared in the 1830s-1840s. The latter devices were extremely expensive, costing between £40,000 and £50,000. They increased both the quantity and quality of the sugar obtained from a given amount of syrup, but they required a far more extensive area of cane to be cultivated than was necessary for non-mechanised mills in order to make full use of their capacity, and to compensate for the costs of purchase and maintenance. The first vacuum pans were successfully employed in Guiana in 1830, 16 years before they were first installed in Jamaica. By 1832 vacuum pans were operating on six estates in Guiana.[43] A shortage of skilled 'pan boilers' slowed their adoption in the 1830s and early 1840s, but they became more common afterwards.[44]

From 1838, the year of end of apprenticeship, and 1853, the transition to free labour, increasing costs, the abolition of protective sugar duties, and a credit shortage provoked a period of crisis for the Guianese economy. Cotton, coffee, and small sugar estates were abandoned. Some 175 estates, 60% of plantations under cultivation in 1838, were sold at auction, and property values fell sharply. With the end of apprenticeship newly emancipated labour struggled over wages and working conditions, and in 1842 and 1848 engaged in protracted strikes against the sugar industry. Sugar production dropped drastically. During last five years of slavery, average annual sugar production was 57,197 tons; from the end of apprenticeship to 1846 it fell to an annual average of 31,685 tons. 1847 was the peak year of crisis, when the abolition of protection for sugar in the British market compounded the commercial crisis, but from that point on the sugar industry began to recuperate, encouraged by the low price of property and the importation of Asian indentured labour, which allowed wage rates to be stabilised, and control over the labour force to be reasserted. Between 1847 and 1852 annual average sugar production rose to 39,278 tons.[45]

The recovery of the Guianese sugar industry was characterised by the consolidation of estates, the extension of cultivation, technological innovation, and the trend toward corporate ownership. Small properties were amalgamated with larger ones. In 1841, 173 estates produced less than 250 hogsheads of sugar. They accounted for 64% of the crop, in contrast to 24 medium estates producing 250–499 hogsheads that accounted for 25% of the crop and 11 estates producing over 500 hogsheads accounting for 11% of the crop. By 1851 the number of small estates had fallen to 84, and they account for only 24% of the crop, while 58 medium-sized estates contributed 38%, and 28 large estates 38%. Eleven of the medium-sized properties and twelve of the large ones employed the vacuum pan. The average output of each large estate was 624 hogsheads. The process of expansion and consolidation continued after 1851, and indeed into the twentieth century. Total acreage in cane rose from 48,087 acres in 1855–57 to 75,944 in 1870–72. The number of sugar estates fell from 173 in 1853 to 105 in 1884, but the average acreage in cane per estate rose from 256 to 757. Of the 66 estates that disappeared during this period, at least 45 were absorbed into larger enterprises. While an annual crop of 500 hogsheads was regarded as large in 1851, the average for 1882–4 was 1,291 hogsheads, and the giant mills far surpassed this number.[46]

Consolidation allowed Guianese planters to realise economies of scale, and to take full advantage of technological innovation. They produced more and better sugar per plantation than the other British West Indian colonies. By 1852 the steam engine had become commonplace. There were 173 active estates employing 208 steam engines, most of which powered sugar-grinding mills. The greater part of the growth in acreage in cane took place before 1870, while output per acre increased slowly. The adoption of the vacuum pan was inhibited by its high cost and the unfavourable duties charged on its superior-quality sugar, in order to protect British refiners. By 1852 the number of plantations employing vacuum pans had risen to only 25, which produced 16% of the sugar crop. While no data have come to light on the number of vacuum pans installed after 1852, it is reasonable to assume that some estates adopted it as the process of amalgamation of properties proceeded. Before 1884, when the first triple-effect vacuum was installed in Guiana, single-effect vacuum pans were used in 'mixed trains', that is, open-kettle boiling was used until the last phase of crystallisation, when vacuum pans were used to complete the process.[47]

Figure 9.5 is a map of a consolidated sugar estate in British Guiana, one most probably made during the twentieth century. It gives an indication both of the uniform layout of Guianese plantations, and of the way in which their contiguous location facilitated amalgamation. In a series entitled 'Our Sugar Estates', the Guianese newspaper *The Argosy* published descriptions of plantations indicated on the map as they existed at the end of the crop year 1882–3, before the properties were consolidated. The descriptions of the sugar estates depicted in Figure 9.5 give an idea of the scale of production in Guiana during that period. *Mon Repos* was the only property in East Coast Demerara where open-kettle boiling was retained. Yet the plantation was renowned for the high quality of its muscovado sugar and its rum, which fetched superior prices. It possessed 909 acres in cane, and employed 111 creoles, 60 workers of 'other lands', and 236 indentured and 447 unindentured Asians. Annual production was 1,260 tons of sugar. *Good Hope* was a failed cotton plantation that was being consolidated with *Lusignan*, which was regarded as one of the best estates on the East Coast. It had 1,136 acres in cane, and produced 2,070 tons of sugar. Its labour force included 138 Creoles, 56 workers from 'other lands', and 440 indentured and 374 unindentured Asian workers. It had recently erected an 'entire new sugar works with many improvements in sugar machinery and special appliances'. It was regarded as one of the pioneer estates of the colony. The neighbouring *Annandale* estate had 1,000 acres planted in cane and produced about 2,000 tons of sugar. It employed 74 Creoles, 19 workers of 'other lands', and 247 indentured and 173 unindentured Asian labourers. It had doubled the capacity of its mill, and utilised steam evaporators in conjunction with double vacuum pans.[48]

After 1884 the process of concentration of estates accelerated, and the multiple-effect vacuum pan became the industry standard. The number of sugar plantations fell from 105 in 1884 to 84 in 1890, as estates were consolidated or abandoned. The number of multiple-effect vacuum pans rose from three in 1885 to sixty in 1892. Acreage in cane and output also increased. The price of vacuum-pan produced sugar increased appreciably over that of muscovado, and rum and molasses ceased to be significant products. Indeed, 'Demarara crystals' set the standard in international sugar markets.[49]

Despite the high overhead imposed by the colony's peculiar environment and chronic shortage of labour, the Guianese sugar industry was able to increase output and productivity through the amalgamation of properties, economies of scale, technological innovation, and control over wage rates and the labour force. Yet while the size, productivity, and output of individual estates increased, sugar production could not expand beyond the geographical constraints of the commodity frontier. In a manner distinct from that of Jamaica, Guiana too reached a material and economic limit beyond which it could not go, and its production remained stable.

Cuban exceptionalism

This chapter is not making the case for geographical determinism. Rather, its concern is with the historical conditions under which land, labour, and technology are combined to transform natural environments in accordance with the requirements of sugar production. In contrast to both Jamaica and British Guiana, Cuba offered exceptional conditions for the development of sugar production from the 1790s onward. The Cuban sugar frontier encompassed approximately 30,000 square miles. This zone was virgin prairie with ideal soils and climate for sugar production. However, the extent of the Cuban frontier was an obstacle to its development until the problem of overland transportation was solved by the development of the railroad network, from 1837.

Slavery and sugar production expanded slowly in the region around Havana from the time of the English occupation in 1763 until 1792. The outbreak of the Haitian Revolution and the destruction of the Saint-Domingue sugar industry provided Havana planters with their great opportunity, and they took full advantage. Between 1792 and 1818 they established the conditions necessary to develop the Cuban sugar industry. First they freed themselves from the restraints of Spanish mercantilism by securing the rights both to free trade in sugar and to import slaves without restriction. (The sugar boom signalled an unprecedented expansion of the slave trade to Cuba. David Eltis estimates that over 780,000 slaves were imported to Cuba between 1791 and 1867.)[50] Further, Cuban planters were competing with protected sugars, especially in European markets. They were constantly concerned with the scale and productivity of their operations, and sought to promote systematically technological innovation and the application of scientific methods to production. Finally, in 1818 they won the right to absolute private property in land, including the right to clear the forest, which had been the preserve of the Royal Navy.[51] In addition, they enjoyed exceptional access to the rapidly expanding US consumer market, and to US goods and capital.

Between 1792 and 1800 the number of sugar mills increased from 237 to 350, with another fifty under construction in regions around the ports of Havana and Matanzas. Total production increased from 14,500 tons in 1791 to 41,000 tons in 1807, while average production per mill rose from 60 tons 1791 to 136 tons in 1804. Significant technological innovations were not adopted during this period, but the first giant mills made their appearance, employing over 300 slaves and producing over 300 tons of sugar each. Even though the majority of Cuban sugar mills were still similar in size and composition to their foreign competitors, contemporaries commented on the higher yields in Cuba, the product of virgin soils and the intensive exploitation of slave labour.[52] By 1820 there were 625 plantations throughout the entire province of Havana, predominantly located near the coast to facilitate transport, and total production climbed to 55,000 tons. The 1827 census registered 1,000 sugar mills in Cuba. Only 2.5% of these had steam power, but Cuba produced nearly 77,000 tons of sugar, and surpassed Jamaica's peak year. Three years later, its output reached 105,000 tons, and it became the world's leading producer, placing its sugar not only in European markets, but especially in the United States, which had emerged as its major trading partner.[53]

Cuba's extensive sugar frontier permitted both the virtually unlimited expansion of sugar cultivation, and the continuous reconfiguration of the sugar mill. The virgin prairie landscape encouraged the formation of spatial economies that closely approximated the ideal arrangement of fields and factory. More and larger mills could be constructed, and economies of scale realised. Increasing scales of production more closely integrated field and factory. Technological innovations could be adopted with fewer obstacles than elsewhere. The steam engine, all-metal horizontal grinding mills, the railroad, the vacuum pan, and the centrifuge were successively introduced in Cuba between the 1820s and 1840s. Each of these technologies restructured the productive space of the Cuban frontier, and shaped new models of plantation organisation.

The first successful application of steam power to sugar production in Cuba took place in 1817; during the 1820s there were various experiments with the technology. Early low-compression steam engines of between eight and 12 horsepower did not transform production, particularly when used in conjunction with pre-existing grinding mills. An advance was made with the introduction of all-metal grinding mills. Both vertical and horizontal models were produced, and could be used with either animal or steam power. (Water mills were rare in Cuba, which possessed few suitable rivers to power them.) Horizontal mills enjoyed the greatest success, and were soon employed in conjunction with steam power. These mills proffered savings in terms of both oxen and slaves, and could process more cane more rapidly, but required more area under cultivation to be efficient. After 1817 the British firm Fawcett and Preston sold 16 horizontal mills and fifty steam engines of between eight and 12 horsepower. Steam technology advanced rapidly. More powerful high-compression engines soon became the norm. By the 1840s improved steam engines and all-iron horizontal mills increased their capacity and established their superiority over other types of mills. By the 1850s steam mills could be as large as 100 horsepower.[54]

Before 1830 increased production owed more to the multiplication of the number of sugar mills than to technological transformation. The majority of Cuban mills were still clustered around the ports of Havana and Matanzas. Transportation was the key obstacle to full exploitation of Cuba's sugar frontier, particularly as the volume of sugar to be moved increased so dramatically. Havana's planter elite was increasingly concerned with road construction, and developed a plan to build a canal from Havana to Güines. The key breakthrough was the construction of the Havana–Güines railroad in 1837, the first in South America. The railroad was built by and for sugar. Rail transport provided cheap and efficient transportation which opened Cuba's interior to sugar cultivation.[55] The railroad network permitted the rapid extension of the sugar frontier, and the construction of sugar estates whose spatial economies were optimally suited to the new milling and refining technologies. José María de la Torre's map of railroads, steamship routes, and telegraph stations (Figure 9.6) shows the close association between the rail system and sugar production. The railroad network coincides with the sugar zone and the projected extension of the sugar frontier. The sugar mills were constructed in close proximity to the rail lines, and the railroads carried sugar directly to modern warehouses and improved port facilities. The map presents not only the distance, but also the time between stations. It conceives the island as an abstract time-space grid, and makes possible continuous calculation of time-distance and price through which the Cuban countryside is integrated into world markets.

As the rail network extended into the interior, the number of sugar mills multiplied. Cuban planters were able to extend the sugar frontier, and to establish new plantations on a scale that made optimal use of each new technological innovation, as the Atlantic slave trade

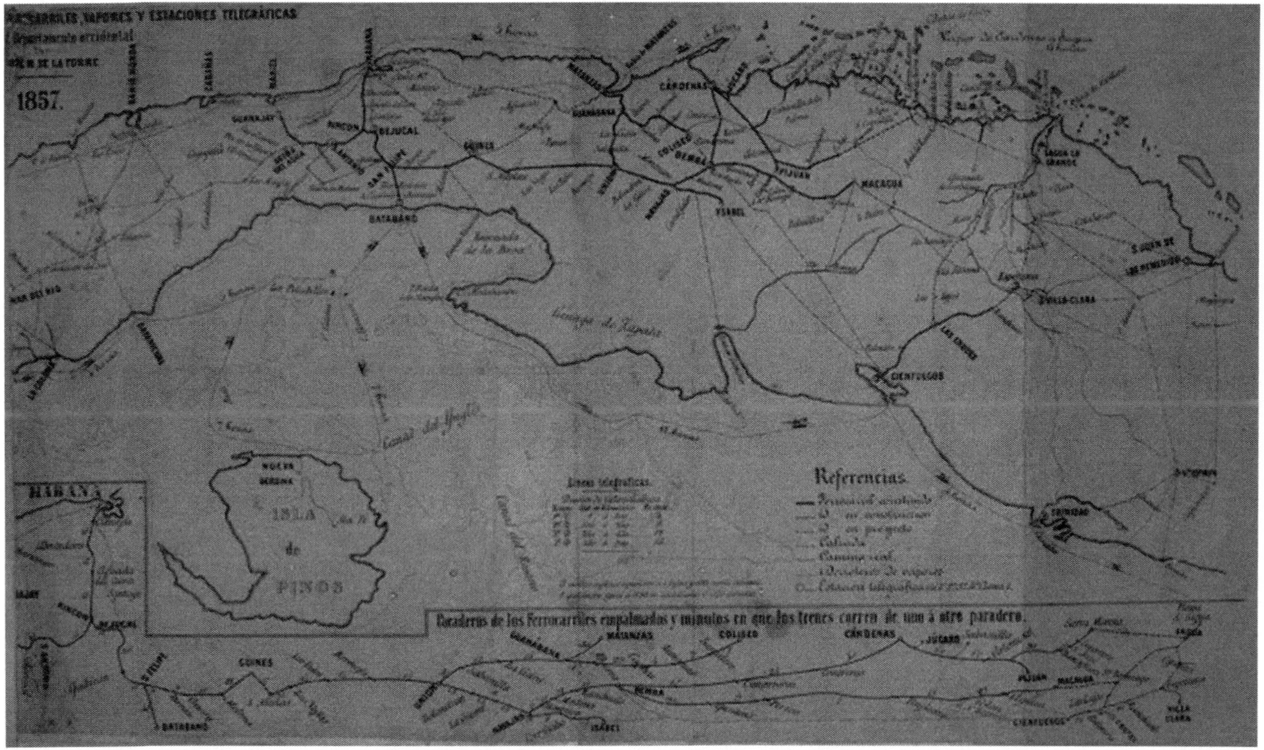

Figure 9.6 José María de la Torre, Map of the Western Department in 1857
Source: Fundación Antonio Núñez y Jiménez de la Naturaleza y el Hombre. Fondo de Agrímensores Serafín Sánchez Govín.

continued to supply necessary labour. In 1827, 26 sugar mills in Havana and Matanzas used steam engines. In 1846, 251 out of 735 sugar mills in the same two regions were employing steam, while in Cárdenas, to the east of Matanzas, 115 out 199 mills had steam power.[56] The centre of gravity of the sugar zone moved eastward, the scale of production increased significantly, and many older and smaller mills were no longer viable. The number of sugar mills in the Havana region fell from 439 in 1820 to 228 in 1860. In contrast, the number of mills in Matanzas rose from 95 to 442, and in Villa Clara, Cienfuegos, and Sancti Spíritus, from 91 to 395 during the same period.[57]

The new sugar zones had more virgin land and forest to provide fuel. Larger plantations with more extensive acreage in cane were established there, to take advantage of the new, more powerful and efficient steam engines and iron horizontal grinding mills, which became available after 1840. In order to process the increased quantity of cane juice produced by these mills, planters multiplied the number of trains of open kettles used to boil the juice down. These semi-mechanised mills could have four, six, or even ten sets of kettles. The vacuum pan also became available in Cuba in the 1840s. Its great expense inhibited its adoption, but by the 1850s multiple-effect vacuum pans of enormous capacity, in conjunction with the most advanced milling technology, were concentrated in the new zones of the sugar frontier: Colón, and the provinces of Villa Clara, Cienfuegos, and Sancti Spíritus. They required a much more extensive scale of operations and obtained more and better sugar from a given quantity of syrup. These fully mechanised mills reconfigured the spatial and technical organisation of sugar production.[58]

The map of the Merced Sugar Mill (Figure 9.7) documents the scale and spatial economy typical of the sugar mills being formed on the Colón prairies. The landscape of Merced approximates the ideal plantation discussed at the beginning of this chapter. The great majority of the property was given over to cane cultivation. Merced covered 1,376 acres, of which 1,007 were planted in cane. Its regular rectangular shape was organised symmetrically. The works and living quarters were located at the centre of the property to minimise the movement of workers to the fields, and of cane to the factory. Fields of more or less equal size were located on each side of the works. They were subdivided into uniform and numbered cane pieces to facilitate the calculation of the amount of cane to be cut each day, and the distribution of labour during the harvest. As the scale of production increased, rapid transport of large quantities of cut cane to the mill became a growing concern for planters and managers. Strategies of landscape and labour management

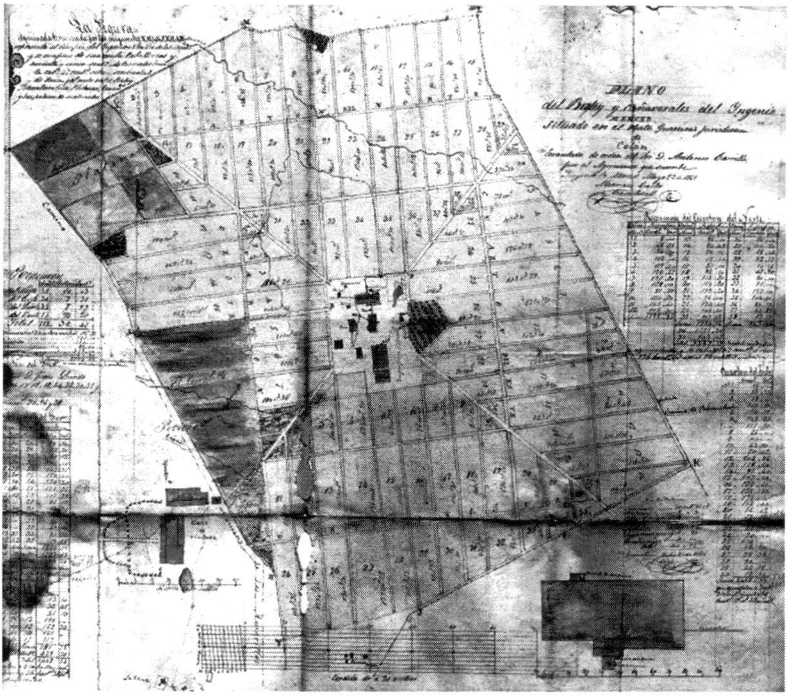

Figure 9.7 Ingenio Merced, Colón, Cuba, 1868
Source: Fundación Antonio Núñez y Jiménez de la Naturaleza y el Hombre. Fondo de Agrímensores Serafín Sánchez Govín.

emerged that were organised around the movement of ox carts to supply a continuous flow of cane to the mill.[59] As shown on the Merced map, diagonal lanes, or *guadarayamas*, were cut through the cane fields in order rapidly to move the great quantities of cane over extended distances, and to secure the integration of field and factory.

Carlos Rebello's 1860 enumeration of sugar production shows both the increasing scale of production and technological innovation on Cuba's sugar mills, and the shift of the centre of production toward the interior of the island. There were 359 animal-powered mills with an average capacity of 113 tons; 889 semi-mechanised mills with an average output of 411 tons; and 64 fully mechanised mills with an average production capacity of 1,176 tons. The latter accounted for five per cent of the plantations, but 15 per cent production. A productive, semi-mechanised mill could utilise a labour force of over 350 slaves, while a fully mechanised mill could have as many as 500, or even 600. The average size of an estate for the entire island was 1,452 acres, with 604 acres in cane. In Colón, where the greatest number of mechanised mills were located, the average size holding was 1,817 acres, with 916 acres in cane.[60] The extensive Cuban sugar frontier permitted the rapid multiplication of large and productive semi-mechanised mills, and the establishment of fully mechanised mills and spatial economies adequate to optimal utilisation of the most advanced milling and refining technologies available. Output and productivity increased steadily and dramatically. By 1840, Cuba produced over 160,000 tons of sugar, more than 19% of the world market. Ten years later its output reached nearly 295,000 metric tons, one-quarter of the world's supply, and by 1868 Cuban production climbed to 720,000 tons, or 30% of a growing world market.[61]

Cuba achieved domination of the world sugar market through the extensive development of its sugar frontier. Its natural environment combined with railroad transportation to redefine productive space in ways that allowed Cuban planters to build more and bigger plantations, to create economies of scale, and to make optimal use of available technological innovations. They achieved unprecedented levels of production and productivity. They were able to open new markets, and to establish greater synchronisation between production and market demand. The enormous mass of commodities they produced brought them greater returns, even as they drove down the price of sugar. They thus transformed the conditions of world sugar production, and created a new hierarchy among production zones which redefined the world division of labour.

In Jamaica, Guiana, and Cuba, planters struggled to adapt their activities to the new conditions of sugar production and labour in the world-economic conjuncture of the nineteenth century. At each site production was reorganised, new technologies were adopted, and larger and more productive plantations emerged. The success of innovation depended upon the capacity of the sugar industry to restructure the natural environment, and to reorder productive space in ways that were compatible with the optimal utilisation of new technologies. However, in each local instance the close interdependence of geography, environment, material processes of sugar production, labour, and technology formed specific historical-geographical complexes. Despite apparently common features, land, labour, and technology were constituted differently in each commodity frontier, and created distinct spatial economies. The reorganisation of production and technological innovation had different consequences in each zone, and resulted in contrasting historical trajectories. These different outcomes disclose the uneven processes of expansion and differentiation that shaped the crisis of the British West Indian sugar industry, and the formation of a new world-economic division of labour.

Notes

1. Although their emphasis is on geography and environment, historical geographers have produced an important body of scholarship on the technological development of the Caribbean sugar industry. Watts, David: *The West Indies: patterns of development, culture and environmental change since 1492*, Cambridge: Cambridge University Press, 1987, and Galloway, J.H.: *The sugar cane industry: an historical geography from its origins to 1914*, Cambridge: Cambridge University Press, 1989, are examples of recent works that present broad syntheses of the development and diffusion of technology in the Caribbean sugar industry from the perspective of historical geography.
2. Magalháes Godinho, Vitorino: 'Complexo histórico-geográfico', in Serrão, João (ed.): *Dicionário de história de Portugal*, Porto: Livraria Figueirinhas, 1961, pp. 130–1.
3. Moore, J. W.: 'Sugar and the expansion of the early-modern world economy: commodity frontiers, ecological transformation, and industrialization', *Review (Fernand Braudel Center)*, Vol. 23, No. 3 (2000) , pp. 409–33; Barros de Castro, Antônio: *7 ensaios sobre a economia brasileira*, (2 vols.), Rio de Janeiro: Forense 1971, II, 48–83.
4. Higman, Barry W.: 'The spatial economy of Jamaican sugar plantations: cartographic evidence from the eighteenth and nineteenth centuries', *Journal of historical geography*, 13, (1987), pp. 17–39.
5. Marrero, Levi: *Cuba: economía y sociedad. Azucar, ilustracion y conciencia 1763–1868* (14 vols.), Madrid: Editorial Playor, 1985, 12, IV, p. 108.
6. Tomich, Dale: *Slavery in the circuit of sugar: Martinique in the world economy, 1830–1848*, Baltimore: Johns Hopkins University Press, 1990, p. 23.
7. Deerr, N.: *The history of sugar*, (2 vols). London: Chapman & Hall, 1949, I, pp. 193–201; Drescher, Seymour: *The mighty experiment: free labour versus slavery in British emancipation*, Oxford: Oxford University Press, 2002, pp. 154–6; Moreno Fraginals, Manuel: *El ingenio: Complejo económico social cubano del azúcar*, 3 vols, Havana: Ciencias Sociales, 1978), I, pp. 40–2; II, p. 173; Instituto Brasileiro de Geografia e Estatística, *Estatísticas históricas do Brasil*, Rio de Janeiro, 1987, p. 342.
8. Marrero, *Cuba*, 12, IV, p. 108; Tomich, *Slavery in the circuit of sugar*, p. 28; Williams, Eric: *Capitalism & slavery*, Chapel Hill: University of North Carolina Press, 1944, pp. 133–54; Drescher, Seymour M.: *Econocide: British slavery in the era of abolition*, Pittsburgh: University of Pittsburgh Press, 1977, pp. 155–7; Ingham, G.: *Capitalism divided? the city and industry in British social development*, New York: Schocken, 1984, pp. 113–15; Moreno Fraginals, *El ingenio*, II, pp. 157–61; Ragatz, Lowell: *The fall of the planter class in the British West Indies, 1763–1833*, New York: The Century Co., 1971, pp. 434–5.

9. Satchell, Veront: *Sugar, slavery, and technical change: Jamaica, 1760–1830,* Saarbrücken: VDM Publishing, 2010), pp. 178–80.
10. Deerr, *History of sugar,* I, pp. 193–202; Drescher, *Econocide,* pp. 93–103; Moreno Fraginals, *El ingenio,* II, pp. 106–30, 156–65; Higman, 'Spatial economy', p. 14; Higman, Barry W.: *Jamaica surveyed: plantation maps and plans of the eighteenth and nineteenth century,* Kingston : University of West Indies Press, 2001, pp. 10–16; Satchell, *Sugar, slavery, and technical change,* pp. 102, 166; Watts, *The West Indies,* p. 497.
11. Higman 'Spatial economy,' pp. 45–99. 'Creolisation' may also be understood more broadly as a process of transformation of social relations (Brathwaite, Kamau: *The development of creole society in Jamaica, 1770–1820,* Oxford: Clarendon Press, 1971). Particularly relevant in the context of this argument is the practice, which had become more widespread during the period following US independence, of providing provision grounds to the enslaved population. They offered the enslaved population an effective resource to contest conditions of labour, social reproduction, and social meaning, and to form themselves as a 'proto-peasantry' (Mintz, Sidney W.: *Caribbean transformations,* Chicago: Aldine Publishing Co., 1974, pp. 131–250; Mintz, Sidney W.: 'Slavery and the rise of peasantries', in Michael Craton (ed.), *Roots and branches: current directions in slave studies* [*Historical reflections/Réflections historiques,* Directions Series, 1], Willowdale: Pergamon Press, 1979, pp. 213–42).
12. Higman, 'Spatial economy', pp. 68–71; Higman, Barry W.: *Slave population of the British Caribbean, 1807–1834,* Baltimore: Johns Hopkins University Press, 1984, p. 106.
13. Satchell, *Sugar, slavery, and technical change,* pp. 132–64, 165.
14. Hall, Douglas: *Free Jamaica, 1838–1865: an economic history,* New Haven: Yale University Press, 1959), pp. 70–1.
15. Satchell, *Sugar, slavery, and technical change,* pp. 91–5; Hall, *Free Jamaica,* p. 71.
16. Hall, *Free Jamaica,* p. 78.
17. Hall, *Free Jamaica,* p. 75; Eisner, Gisela: *Jamaica, 1830–1930: A study in economic growth,* Manchester: Manchester University Press, 1961, p. 297.
18. Higman, *Jamaica surveyed,* p. 141.
19. Galloway, *The sugar cane industry,* pp. 137–8, 148.
20. Satchell, *Sugar, slavery, and technical change,* p. 182.
21. Hall, *Free Jamaica,* pp. 69–71.
22. Higman, 'Spatial economy,' pp. 17-39.
23. Higman, 'Spatial economy', p. 25.
24. Higman, 'Spatial economy', pp. 17, 33.
25. Satchell, *Sugar, slavery, and technical change,* p. 177.
26. Higman, 'Spatial economy', p. 26.
27. Higman, *Jamaica surveyed,* p. 96.
28. Hall, *Free Jamaica,* pp. 75–6.
29. Satchell, *Sugar, slavery, and technical change.*
30. Scott, James C.: *Seeing like a state: how certain schemes to have improved the human condition have failed,* New Haven: Yale University Press 1998, pp. 18–52; Thompson, Edgar: *The Plantation,* Columbia: University of South Carolina Press, 2010.
31. Rodney, Walter: *A history of the Guyanese working people, 1881–1905,* Baltimore: Johns Hopkins University Press, 1981, pp. 2–3; Viotti da Costa,

Emilia: *Crowns of glory, tears of blood: the Demerara slave rebellion of 1823*, New York: Oxford University Press, 1994, p. 9; Mohamed, Wazir: 'Frustrated peasants, marginalized workers: Free African villages in British Guiana, 1838–1885', unpublished PhD dissertation, Binghamton University, 2008, pp. 93–110.
32. Mohamed, 'Frustrated peasants,' p. 99; Costa, *Crowns of glory*, p. 44; Adamson, Alan H.: *Sugar without slaves: the political economy of British Guiana, 1838–1904*, New Haven: Yale University Press, 1972, pp. 167–9.
33. Costa, *Crowns of glory*, p. 44; Mohamed, 'Frustrated peasants,' pp. 93–110.
34. Costa, *Crowns of glory*, p. 46.
35. Costa, *Crowns of glory*, pp. 46–7; Adamson, *Sugar without slaves*, p. 24.
36. Deerr, *History of sugar*, I, pp. 193–201; Moreno Fraginals, *El ingenio*, II, 106, 156–65; Costa, *Crowns of glory*, pp. 46–8; Higman, *Slave population*, p. 63.
37. Mohamed, 'Frustrated peasants,' p. 133; Adamson, *Sugar without slaves*, p. 24.
38. Costa, *Crowns of glory*, p. 47; Adamson, *Sugar without slaves*, pp. 24–5.
39. Costa, *Crowns of glory*, pp. 48–54, 57; Higman, *Slave population*, pp. 63–4, 79–85, 91–2, 106.
40. Costa, *Crowns of glory*, pp. 9, 47–8; Mohamed, 'Frustrated peasants,' pp. 116, 127–33.
41. Stoll, Allyson: 'Not for "want of go-aheadism" in field and factory: the technological trajectory of the British Guiana (British British Guiana) sugar industry from 1800 to the 1930s', unpublished PhD dissertation, Cornell University, 2011; Stoll, Allyson: 'Thoroughly tested and tried: Cane culture, technology, and environmental change in nineteenth century British Guiana', *Journal of Caribbean History*, 45, (2011), pp. 92–127.
42. Adamson, *Sugar without slaves*, pp. 26–7; Satchell, *Sugar, slavery, and technical change*.
43. Stoll, 'Not for "want of go-aheadism"', p. 164; Mohamed, 'Frustrated peasants', pp. 127–33; Satchell, *Sugar, slavery, and technical change*, p. 99.
44. Tomich, *Slavery in the circuit of sugar*, pp. 195–7; Stoll, 'Not for "want of go-aheadism"', pp. 162ff.
45. Adamson, *Sugar without slaves*, pp. 160–7; Rodney, *A history of the Guyanese working people*, pp. 31–59.
46. Rodney, *A history of the Guyanese working people*, pp. 175–9, 202–6; Stoll, 'Not for "want of go-aheadism"', p. 21.
47. Adamson, *Sugar without slaves*, pp. 171–3, 179, 188–90, 204–6.
48. Rodney, Walter: *Guyanese sugar plantations in the late nineteenth century: a contemporary description from the* Argosy, Georgetown, British Guiana: Release Publishers, 1979, pp. 59–62.
49. Adamson, *Sugar without slaves*, pp. 187–92, 202, 204–6, 209.
50. Eltis, David: *Economic growth and the ending of the transatlantic slave trade*, New York: Oxford University Press, 1987, p. 245.
51. Funes Monzote, Reinaldo: *From rainforest to cane field in Cuba: an environmental history since 1492*, Chapel Hill: University of North Carolina Press, 2008.
52. Funes Monzote, *From rainforest to cane field*, pp. 43–5.
53. Moreno Fraginals, *El ingenio*, I, 67–71; Funes Monzote, Reinaldo and Tomich, Dale: 'Naturaleza, tecnología y esclavitud en Cuba. Frontera azucarera y revolución industrial', in José Antonio Piqueras (ed.), *Trabajo libre y coactivo*

en sociedades de plantación, México: Siglo XXI de España, 2009, pp. 75–117; Knight, Franklin W.: *Slave society in Cuba during the nineteenth century*, Madison: University of Wisconsin Press, 1970, pp. 30–1.

54. Moreno Fraginals, *El ingenio*, I, 203–14; Funes Monzote, *From rainforest to cane field*, pp. 128–31, 306–7n.
55. Zanetti, Oscar and García, Alejandro: *Sugar and railroads: A Cuban history, 1837–1959*, Chapel Hill: University of North Carolina Press, 1998.
56. Funes Monzote, *From rainforest to cane field*, p. 129.
57. Funes Monzote, *From rainforest to cane field*, p. 130.
58. Moreno Fraginals, *El ingenio*, I, pp. 170–3; Funes Monzote, *From rainforest to cane field*, pp. 129–31.
59. Tomich, Dale: 'The invention of the Cuban sugarmill: space, time, and labour management, 1820–1860', in González-Ripoll, Mª Dolores and Álvarez Cuartero, Izaskun (eds), *Francisco Arango y la invención de la Cuba azucarera*, Salamanca: University of Salamanca, 2010, pp. 133–49.
60. Cantero, Justo G.: *Los ingenios: Colección de vistas de los principales ingenios de azúcar de la isla de Cuba* [1857], reprinted Madrid: Centro Estudios y Experimentacíon de Obras Públicas, 2005 (García Mora, Luis Miguel and Santamaria García, Antonio eds.), p. 44; Marrero, *Cuba*, 12, IV, p. 249; Funes Monzote, *From rainforest to cane field*, pp. 129–31.
61. Moreno Fraginals, *El ingenio*, I, 46–7, 67–71, 95–102, 167–255; II, 93–7, 106–74; III, 35–6; Knight, *Slave society in Cuba*, pp. 14–18, 40–1.

10

From Periphery to Centre: Transatlantic Capital Flows, 1830–1890

Martín Rodrigo y Alharilla[1]

One of the major factors in the configuration of the 'first globalisation' observed in the Atlantic economy, as analysed by Kevin O'Rourke and Jeffrey Williamson, was the international flow of capital. Both authors argue that the first wave of the globalisation process took place before the First World War, no earlier than 1870. In this regard their impeccable work merits at least two critiques. First, O'Rourke and Williamson barely consider the presence of international flows of capital before that period – in particular, during the second and third quarters of the nineteenth century. Second, and most importantly, their analysis centres on, in their own words, 'capital exports from the centre to the periphery'.[2]

This chapter's main objective is to provide evidence that a longer time-frame is required to grasp a full understanding of the complexity of the configuration of the global capital markets and their *fin de siècle* integration, a time-frame that includes the middle decades of the nineteenth century. The evidence presented here also suggests that any analysis of transnational flows of capital in the Atlantic World must be, necessarily, bidirectional: researchers must take into account not only the process of capital *exportation* from a European centre (Great Britain, France, or Germany, for example) to an American periphery, but also the existence of a process of a previous, but also parallel, transfer of capital from the American periphery (in the case considered here, the Hispanic Caribbean) to the centre. Any study that does not integrate both ends of this process is markedly short-sighted.

This analysis demonstrates that during the entire length of the nineteenth century, capital accumulated initially in the largest of the Antilles, Cuba, was transferred to Europe in tireless transatlantic waves. What started as agrarian or commercial capital was transformed into industrial and financial capital or real estate in a process that culminated in the exportation of capital *from the periphery to the centre*. The cities within that centre which are addressed in this chapter are Barcelona and Paris. The first is considered because Barcelona, and Catalonia more generally, had intense relations with the island of Cuba during this period at various levels, including trade, migration, and transatlantic capital flows; the second because, despite having less direct and less intense exchanges with the island, Paris, like Barcelona, became an important recipient of Cuban capital during the nineteenth century. Focusing on the Cuba–Barcelona and Cuba–Paris axes, this chapter will explore flows of capital that were initially accumulated in the Caribbean for ulterior investment in Europe.

The first section explores the key, defining features of Cuba's economic growth across the nineteenth century. The second traces the historical and economic development of the wealthiest entrepreneur in the island, and through him addresses a further defining feature of the Cuban economy: the capacity of its fields to generate vast amounts of profit which were seldom reinvested in the island. The third focuses on a particular type of entrepreneur, the so-called *indianos* ('from the Indies') or *americanos*, who successfully enriched themselves in the Americas, then returned to Europe with their fortunes. Finally, the chapter provides details of the mechanisms and instruments that made possible the transfer of capital which *indianos* carried out from the Caribbean periphery into the European centre.

Cuba – a growing economy

Numerous scholars of contemporary Cuban economic history point to the last third of the eighteenth century as the initial moment of a period of extraordinary growth that continued up to the first decades of the twentieth century. Certain authors have reached a general consensus regarding the main engine behind that process: sugarcane. The production and export of sugarcane products (raw sugar, rum, molasses, etc.) led the Cuban economy to register a long cycle of growth which was only mildly affected by various small crises, all of them brief and circumstantial.[3] Impelled by sugar, Cuba's productive capacity registered an unstoppable growth from 1762, when the English took Havana, until the end of the period known as 'the Dance of the Millions' in 1920.

Cuba's relationship with its metropole, Spain, was radically different from those that other Caribbean sugar islands had with their metropoles, such as Barbados and Jamaica vis-à-vis Great Britain, Martinique and Guadalupe vis-à-vis France, or even Surinam vis-à-vis the Netherlands.

Three main differences stand out: (1) Cuba was never a plantation society or plantation economy, in the strict sense of the term, even though its economy was plantation-driven, especially in some regions. The island's history records no moment, for example, when the slave population surpassed the free population. (2) Cuba's transformation into a plantation-driven economy was produced by the island's own oligarchy, and not by individuals, companies, or capital arrived from Europe, or more concretely, from Spain. (3) Unlike the Netherlands, France, or England, Spain did not have sugar refineries where the industrial transformation of sugar could be completed. On the other hand, the weak Spanish market could not generate enough demand to absorb Cuban exports, so the Spanish authorities did not establish a commercial monopoly to ensure that Cuban sugar products went exclusively to or through Spain.

Cuba's productive and export capacity was multiplied by the convergence of various internal and external factors. The island's oligarchy was already betting on sugar, but the collapse of the Haitian sugar economy after 1791 encouraged them to intensify their activities in this sector. Still, the continued increase in the demand for sugar in European and US markets ensured that the island's production could continue growing. Internally, the availability of land and capital facilitated the process, and the need for labour led to the massive importation of African slaves, despite the fact that the slave trade in the Spanish possessions was banned after June of 1820. Indeed, encouraged by the cane industry, an increasing number of slaves was imported each year. The slave population in Cuba rose from 25.8% of the population in 1744 (nearly 44,000 individuals) to 43.3% in 1841 (close to 430,000). Recent estimates calculate that between June of 1820 and 1866 – the year of the final slaving expedition to the island – 547,288 enslaved Africans arrived at Cuba illegally. When British pressure over the illegal trade intensified, the island's plantation owners found other ways to fill their cane fields with labourers. These included the importation of *colonos* (settlers) from Yucatan (Mexico), the promotion of European immigration, and, most significantly, the importation of Chinese indentured workers. More than 124,000 Asian 'colonists' arrived in Cuba between 1847 and 1874, the majority ending up working in the island's expanding cane-fields.

Accumulation with little reinvestment or diversification

All of these internal and external vectors which configured and encouraged Cuban economic growth have been the subject of ample study, but a significant result of this same process has not been sufficiently analysed. Transformed into a plantation-driven economy, Cuba became an ideal space for the accumulation of capital, but not necessarily for its reinvestment. Most plantation owners and traders sought to relocate to outside of Cuba a significant portion of the profit they had obtained from the exploitation and exportation of sugar products. This process of capital relocation, most evident in the second half of the nineteenth century, was carried out by both the youngest generations of prominent Cuban *criollo* families (whose forebears had launched the intensification of the sugar industry at the end of the eighteenth century and throughout the first third of the nineteenth), but also, and more significantly, by immigrants, especially from Europe, who had been attracted to the island by the dynamism and promise of the Cuban economy, and had succeeded in striking it rich.

Both these groups used part of the capital accumulated in the sugar industry to boost, extend, and modernise their sugar mills, transforming some of them into modern sugar *centrales* at the end of the period. However, the Cuban sugar boom beneficiaries not only engaged in sumptuous consumption and a relatively small, but sufficient reinvestment in their Cuban businesses. They also located vast amounts of their capital outside the island, establishing new businesses in faraway latitudes.

The process of capital export (whether transfer, relocation, or flight) was one of the most problematic effects of Cuba's economic growth, especially during the second half of the twentieth century. This phenomenon was well-known (in some cases, even denounced) by powerful sectors of Cuban society. For example, Vicente Galarza, vice-president of the Cuban party *Unión Constitucional,* was highly critical. In the Spanish Senate's session of 11 February 1889, he lamented the capital flight that bled the greatest of the Antilles, denouncing that 'Floating capital disappears [from Cuba], it flees because it does not see sufficient guarantees, and not because it is less productive than here [in Spain], for in that island, even under the present circumstances, capital always produces more than in the peninsula.'[4] It is ironic that it was Galarza himself who said this, for he had invested part of his fortune outside the island. Merely three months after uttering these words in the Senate, the Cuban politician and businessman assumed a limited partnership of 440,000 pesetas in Barcelona's most important printing press, Heinrich & Cía., to become its major shareholder.[5] Beyond this anecdote, the long-drawn capital flight described by Galarza did indeed bleed the Cuban economy, particularly during the second half of the nineteenth century.

An analysis of the investment patterns and wealth sources of the richest man in Cuba, trader and plantation owner Tomás Terry Adan, highlights the extent of this process.

Terry first saw the light of day in Caracas, on 24 February 1808, and closed his eyes for the last time in Paris, on 4 July 1886, having spent the greater part of his life in the Antilles. Of illegitimate birth, he had arrived at the Cuban village of Cienfuegos in 1830 at the age of 22, without any assets, after having lived with his father for some time on the Dutch island of Curaçao. In Cienfuegos he worked for the trader Martín Iradi, whom he had probably met in Curaçao.[6] Saving enough to start a business of his own, he invested his money in a mule team with which he traversed the hinterlands of Jagua Bay. By the time he married in 1837, his fortune had reached 20,000 *pesos fuertes*. Using this capital, in January 1839 he joined forces with Bostonian Augustus L. Richardson to create the firm Terry & Richardson. In only three and a half years, he had doubled his capital: when Terry & Richardson was dissolved in July of 1842, he recovered his initial investment and had an additional 25,000 pesos as net profit. From then on, his wealth grew exponentially, rising from 45,000 in the summer of 1842 to nearly 725,000 pesos in 1851, and to almost two million pesos by the end of 1856. Four years later, in 1860, Tomás Terry's assets had already surpassed three million pesos, and ten years after that, at the close of 1870, they had reached nearly eight million *pesos fuertes*, a truly spectacular amount.[7]

Terry's immense fortune was rooted in Cienfuegos' sugarcane plantations. He was involved from the beginning in the illegal importation and sale of African slaves to the Cienfuegos cane fields, often using the nearby Ciénaga de Zapata.[8] Terry also amassed a significant portion of his fortune as *comerciante-refaccionista*, an activity that encompassed the provision of a broad set of goods and services related to trade and finance for the region's plantation owners. He imported and sold food as well as machinery, tools, and equipment related to the sugar industry on credit, he exported sugar products, and he provided loans, amongst other activities. Finally Terry ended up becoming a plantation owner himself, purchasing some mills and raising others. He was not just a plantation owner: with seven mills (Juraguá, Teresa, Palmarola, Reparador, Santo Tomás, Esperanza, and Caracas) he was the planter with the largest number of mills in central Cuba, and perhaps the entire country. Some of these mills would later become important *centrales*. It is therefore evident that Terry's fortune was built upon sugar and plantations in the region of Cienfuegos, a fortune that did not stop growing throughout the latter part of his life, even during the course of the Ten

Year War in Cuba (1868–1878). His assets went from 7,891,784 pesos in 1870 to 13,763,594 pesos in 1880. When he died in July of 1886, Tomás Terry Adán had 20,669,170 pesos, an impressive fortune. Indeed, Roland T. Ely described Terry as the Cuban Croesus, stating that 'in the final years of his life, Tomás Terry was probably the wealthiest man in Cuba'. Manuel Moreno Fraginals went further, arguing that Terry held 'one of the greatest capitals in the world of his day'.[9] He was right.

Where did Terry invest his immense fortune, one of the greatest of his time? The answer can be found in the post-mortem inventory of his estate, carried out in Cuba months after his death. Terry's heirs and executors initially tried to manipulate the inventory by underestimating the deceased's wealth at 14,429,174 pesos, probably in an attempt to elude their fiscal duties. In any case, the analysis of where Terry's assets were invested proves to be very revealing of the process of capital transfer from Cuba, in Terry's case, to the United States.

Beginning in 1851, but to an increasing extent after 1864, Terry began buying government bonds, stocks, and shares from diverse institutions and companies across North America and Europe. In 1870, with the Franco-Prussian War raging across the latter continent, Terry abandoned his investments in the Old World to concentrate in the United States, as shown in Table 10.1. For the Caracas-born businessman, Cuba had been the ideal place to make a fortune, but it was apparently not the ideal place to invest and diversify his assets beyond sugar. He preferred to do this outside the island, especially in the great neighbour to the north. Stocks and shares in North American institutions (mostly government bonds and railroad company shares) comprised nearly three-quarters of his capital at the time of his death. A capital sum accumulated in the

Table 10.1 Estate of Tomás Terry Adán, July 1886

Country	*Asset types*	*Share of total assets*
United States	All assets	77.55 %
	Cash, shares and bonds	76.68 %
	Urban properties, N. York	0.87 %
Cuba	All assets	14.73 %
	Cash, shares, bonds and stocks	3.78 %
	Urban and rural properties	10.95 %
England	Cash, shares and bonds	6.15 %
France	Cash, shares and bonds	1.57 %
TOTAL		100 %

Source: Archivo Provincial de Cienfuegos, Protocolos Notariales, José Joaquín Verdaguer, 18.05.1887, fol. 1286 and ss.

Cuban countryside, on the backs of slaves, was invested in the last third of the nineteenth century in the development of the institutional and transportation networks of the US.

Terry's case may seem exceptional and therefore unrepresentative, given the immensity of his wealth. However, it is simply the extreme form of a phenomenon that was not exceptional at all: the transfer to central economies of capital created in Cuba, a peripheral one. Moreover, Terry's life-story, especially his business trajectory, exemplifies another process that goes beyond his particular case. It shows how the accelerated expansion of sugarcane cultivation (and, by extension, the intense growth of the Cuban economy during the first two-thirds of the nineteenth century) opened the doors to the incorporation of businessmen who came from all across the globe into the world of sugar, men who contributed with their activities to the crowning of king sugar in Cuba. Some became landowners, others traders, and yet others (Terry, for instance) combined both activities, but almost all begat their wealth from the sugar industry. Although the old Havana oligarchy had been the main agent in the transformation of the island into a plantation-driven economy (through a process forged in Havana Bay's hinterlands, but which soon took over the Colon-Matanzas plains), it would soon be joined by foreign businessmen who made a decisive contribution to the growth of the island's sugar economy.

This new generation of entrepreneurs, many of whom arrived without resources in Cuba in the second quarter of the nineteenth century, made a fortune during the sugar boom, and joined their counterparts among the well-established *criollo* oligarchy. The list of names that illustrate this is too long to relate in detail, but it includes, for example, Julián Zulueta (of Anucita, Araba), who owned the mills of Álava, Vizcaya, Habana, España and Zaza; Francisco de Sola Nanclares (of Mondragón, Gipuzkoa), owner of Cieneguita; Manuel Calvo (of Portugalete, Biscay), owner of Portugalete; Agustín Goytisolo Lezarzaburu (of Lekeitio, Biscay), owner of Simpatía, Lequeitio, San Agustín, and Lola; José P. Taltavull García (of Maó, Menorca), owner of Caridad; Joaquín Fábregas Estrada (of Piera, Barcelona), owner of Delta and San Joaquín; José Carbó Martinell (of Sant Feliu de Guíxols, Girona), owner of Santa Catalina; and the brothers from Barcelona, Tomás Ribalta, owner of Santo Tomás and Santa Teresa, and Pablo Luis Ribalta, owner of Rosa and Santa Marta. Together, these men owned 19 sugar plantations.

All were born in the Iberian Peninsula, and had gone to Cuba either with very little money, or with none at all. They all built great fortunes in the island, and many (such as Calvo, Sola, Goytisolo,

Taltavull, Fábregas, Carbó, and Tomás Ribalta) decided to return to Spain – some as middle-aged men, others much older – where they lived quietly as *rentiers* until they died. Terry, as we saw, died in Paris; Carbó, Taltavull, Fábregas, Goytisolo, Ribalta, and de Sola all met their deaths in Barcelona, where they had spent the last years of their lives and invested a significant proportion of their fortunes. Agustín Irizar Declouet, Terry's junior partner in the Cienfuegos-based firm Tomás Terry y Cía., also died in Barcelona. Irizar had left Terry y Cía. in 1881 to move to Barcelona, where he had transferred most of the fortune he had made in Cuba.

During the last two-thirds of the nineteenth century Barcelona became a favoured place of resettlement for businessmen who had made their fortunes in Cuba. Some of them were among the most dynamic entrepreneurs of the city, but all were protagonists in the transatlantic flow of capital which, with Cuba as its starting point, ended up in the main European and North American markets during the latter nineteenth century. Barcelona not only received significant numbers of entrepreneurs who made their fortunes in the Americas (especially in the Hispanic Caribbean); it also received vast amounts of their capital. These men were referred to as *indianos* by their peers, a word with such historical and cultural specificity as to be practically untranslatable.

Indianos: a particular type of nabob

The phenomenon of circular migration to the Americas was so widespread and so important in nineteenth-century Spain that it became part of a shared imaginary among the Spanish, and also, of course, among Catalans, so much so that it was inscribed in everyday expressions and language. Differences and translation difficulties aside, an equivalent concept evolved in British society: the *nabobs*.[10] Like them, the *indianos'* peculiar way of acquiring wealth outside Europe was acknowledged by their peers in the metropole.

The figure of the *indianos*, those who had gone with little or no resources to the Americas only to return, prosperous and respectable, to Spain, is as old as the colonisation of the new world itself. In Sebastián de Covarrubias' *Tesoro de la Lengua Castellana o Española*, originally edited in 1611 and considered the best dictionary of the Siglo de Oro, the term *Indio* is defined as 'he who is native to India', while *indiano* is defined as 'he who has gone to the Indies, who usually returns wealthy'. Note that by the early seventeenth century Covarrubias included in the definition of *indianos* both their status as returnees and the wealth that they had accumulated. The joining together of these two elements is present one way or another in all of the dictionaries consulted. In the last editions of the *Diccionario de la Lengua Española* of the Royal Spanish Academy, the term *indiano* is defined as 'natural of, but not born in, America, that is, the West Indies', with the additional observation that 'it is said also of those who came back rich from America'. The second definition of the term *indiano/a* in María Moliner's *Diccionario de uso del Español* is 'emigrant who returns rich from America'. In Catalan, two different words are used to refer to this archetypical emigrant: *indiano* and *americano*. According to the latest version of the *Diccionari de la Llengua Catalana* edited by the Institut d'Estudis Catalans, an *indiano* is a person 'who has resided for a long time in the West Indies and has returned to his hometown', while the dictionary edited by the *Enciclopedia Catalana* says of *americano*: 'in the Catalan countries and other European countries, an emigrant who has returned from America'. The terms apparently do not connote riches or wealth. In its present use, but also in contemporary nineteenth-century parlance, *indianos* need not be returnees. Some had been born and raised in the West Indies, but resettled in Europe later in their lives, reclaiming in a way their parents' (or grandparents') ties to the land. Another telling colloquial expression, in Spanish as well as in Catalan, reflects just how common, and at the same time significant, this pattern of circular migration to the Americas had become: '*hacer las Americas*' ('making the Americas'). Moliner's *Diccionario* defines this expression as 'establishing oneself in America to make one's fortune'.

Thus, terms that apply specifically to these types of individuals in both Catalan and Spanish, and to the act of migrating and returning, were coined and adopted as early as the 1610s. This points to a generalised and relevant social process which, in the Catalan case, was especially visible along the Mediterranean coast. In this regard the words of the historian Lluís Costa, who analysed the emigration from the town of Begur to Cuba in the nineteenth century, are particularly relevant: 'one cannot explain, nor make any sense of, the history of Begur without taking into account the history of Cuba', for emigration from Begur to Cuba 'determined the entire social system; there was practically no family that did not have some member in the American continent'.[11] With 87% of Begur's emigrants, Cuba, described by Costa as 'the isle of dreams', concentrated most of that circular migratory flow, while the rest of America received the remaining 13%.

Indianos were so important in their day in the towns across the Catalan coast that they still indirectly hold an important economic and cultural role: a number of towns recently institutionalised a network, the *Red de Municipios Indianos*, to promote their character as '*indiano* towns' and to attract tourism. Many of these towns hold so-called *Ferias de Indianos* each year to commemorate and celebrate their particular *indiano* heritage. Tens of volumes of local history analyse, town by town, the historical figure and significance of *indianos* in Catalonia. It is not the objective of this chapter to analyse the social, cultural, or economic effects that *indianos* had in the diverse locales of the Catalan coast, but to look at the investments that they carried out in Barcelona, and not because the city was the capital, but because it concentrated the greater part of the most successful, wealthiest, and enterprising *indianos*, those who were more interested in investing in new businesses and engaging in European-based economic activity than they were in retiring to their original hometowns to live in splendour.

A significant, but by no means complete, list of the *indianos* who settled in Barcelona throughout the nineteenth century compiled for this research employed somewhat restrictive criteria. In the case of resettled married couples, only the husband has been considered *indiano* (for historically relevant reasons), while in the case of resettled families, I have not counted second- or third-generation returnees. One could say that I have included only the *pater familias* in every case – which has yielded 112 *indianos* who resided in the Catalan capital in the nineteenth century. Of these individuals, we know the birthplace of 100, and an initial exploration reveals that only five of them were originally from Barcelona. The rest were born in other Catalan locales (66), elsewhere in Spain (15), or in the Americas (14). This reveals that Barcelona attracted mostly – but not exclusively – Catalonia-born *indianos*. The majority of these wealthy returnees settled in Barcelona (rather than their home towns) when they returned from the Americas to Europe, because of the economic vitality of the city. One can surmise that it was this same vitality that produced so few Barcelona-born emigrants in the first place.

Once in the Catalan capital their presence, and especially the investment of their American capital, contributed to the maintenance and increase of the city's economic dynamism. The Barcelona of the nineteenth century was not only the 'return capital' of those who 'made the Americas'. It was also the 'capital of capital' made in the Americas.[12] This capital underpinned, much more than has previously been acknowledged, the economic modernisation of Barcelona (as well as that of the rest of the Spain) throughout the nineteenth century, and well into the twentieth.[13] But what place did the Caribbean, and Cuba specifically, hold in this process of capital production, transfer, and reinvestment? We know for certain the places in America where 110 of

Table 10.2 Places where *indianos* garnered their wealth, number of people

Caribbean		***South America***	
Cuba – total	91	Argentina	3
Havana	37	Chile	2
Santiago de Cuba	22	Ecuador	1
Matanzas	10	Brazil	1
Cienfuegos	10	**Total South America**	**7**
Sagua la Grande	5		
Gibara	3	***Central America***	
Las Tunas	2	Guatemala	2
Santa Clara	1	Chiapas	1
Manzanillo	1	**Total Central America**	**3**
Puerto Rico	5		
Jamaica	2		
Veracruz	1		
Cartagena De Indias	1		
Total Caribbean	**100**		

the 112 *indianos* who had resettled in the Catalan capital made their fortunes: 91% of them did so in the Caribbean, only 6.3% in South America, and just 2.7% in Central America.

Table 10.2 shows just how significant the Caribbean was as the place where nineteenth-century *indianos* produced their wealth. The fact that 82.7% of Barcelona's *indianos* had lived in Cuba underlines the particular significance of the island. The other Caribbean societies where *indianos* had lived were Puerto Rico (five individuals); Jamaica (two); Veracruz, on the Mexican Caribbean coast (one) and Cartagena de Indias, on the Colombian Caribbean coast (one). Although the list of 112 *indianos* may underestimate the actual number of *indianos* who lived in Barcelona, it is telling. To paraphrase a Spanish saying (*no están todos los que son, pero son todos los que están*), in this case not all who are *indianos* are on the list, but all who are on the list are *indianos*. It shows an undeniable tendency: the great majority of Barcelona's *indianos* had amassed their fortunes in Cuba.

Some of the listed *indianos* were among the wealthiest of the Barcelona elite. Such was the case of José Xifré Casas, quite possibly the wealthiest Catalan businessman in the mid-nineteenth century. His prominent role in the Barcelona bourgeoisie was secured by the fortune that he had made in the Caribbean. By looking at certain aspects of Xifré's life-story, and at those of other *indianos*, the mechanisms and instruments used to transfer their fortunes from the Caribbean to Europe can be explored.

From periphery to centre: Caribbean capital in Catalonia

Born in Arenys de Mar in Catalonia in 1777, José Xifré Casas left for Havana at a very early age. Starting in 1803, he set up a trading business (concentrated on leather) in Cuba. Twenty years later, he moved to New York, and in the Spring of 1830 decided that it was time to return to Europe. Xifré wrote to his representatives in Havana, ordering the following: 'I want 100,000 pesos made available to me in Barcelona no later than half a year from now. Do what is necessary, because I have decided to bury that money there.'[14] When transcribing this letter, his biographer stated that 'When Xifré wrote "bury", he meant invest in real estate'. Xifré became one of the most important real estate developers in the capital in the mid-nineteenth century. He bought various buildings in Barcelona's Old Town (*Ciutat Vella*), and, most importantly, he financed the construction of 11 contiguous structures of various heights in an enormous lot next to the city's port. This lot had been freed up for construction when the old *Muralla de Mar* of Barcelona was torn down. As his biography notes: 'The three wealthiest Catalans in the first half of the Elizabethan period in Spain were Xifré, Safont and Remisa, but when the time came to execute their wills, Xifré turned out to have the largest patrimony.'[15] Thus, the largest fortune in Catalonia in the period had been originally accumulated in the island of Cuba, and was later transferred to Europe.

Cuban-born Manuela Xiqués Romagosa (1807–91) shared some features with José Xifré Casas. Born in Havana, Xiqués had married the wealthy businessman Roque J. Llopart Azúa, and upon his death in 1846 she was left with three daughters and a vast fortune. The four women left Cuba to travel across Europe, and decided to settle in Barcelona. According to an autobiographical document left by Mercedes, the eldest of the Llopart-Xiqués daughters, in 1854 their mother decided that they would not return to the island, 'and so, having determined that it was much more convenient to stay in Barcelona, she sent orders to rent the house in Havana. Funds were sent from Cuba, and soon after the great house of Baron Rocafort was put up for sale, my mother hurried to get the lot where our actual house in la Rambla was built.'[16]

In another fragment of the extremely interesting autobiography, Mercedes Llopart-Xiqués declares that the family survived on money sent from Cuba, generated by the Llopart and Xiqués estates, until the death of the matriarch in Barcelona in 1891. 'Here I must mention the exceptional role of Providence, which conserved all of the material possessions of my mother through many vicissitudes and dangers so well, that she alone passed on to her grandchildren the inheritance of Grandfather Xiqués', that is, Lorenzo Xiqués Godomá, who had died in Havana in 1842.[17] Thanks to this stream of Cuban funds, Manuela Xiqués became an important real estate owner in Barcelona, especially in the area known as the Ensanche. In a period of just 16 years between 1855 and 1871, the wealthy *indiana* invested more than 1,200,000 pesetas in Barcelona real estate alone.[18]

Often Cuban cash arrived in Catalonia directly from the island's countryside, from the sugar mills. Such was the case of José P. Taltavull García, a Menorcan-born entrepreneur who lived for over twenty years in Cuba, where he developed the Caridad sugar mill. He left the Caribbean in 1864 to live in Barcelona, leaving the administration of his properties in the hands of his partners in the firm García Taltavull y Cía., who were to send his annual profit share to Europe. The contract signed before he embarked stated explicitly that 'the liquid benefits of the Ingenio Caridad will be remitted every year to Spain for D. José P. Taltavull, and if that remittance be sent in bills of exchange, as is customary, they shall bear the official company seal without charging a commission for this'.[19] His case was not unique, as is revealed by a conversation held in Cienfuegos by two entrepreneurs. Agustín F. Goytisolo, administrator of the mills of San Agustín and Lequeitio, wrote to his brother Antonio, who lived in Barcelona:

> I tell you that in conversation with Don Juan del Campo, this gentleman repeated to me that on three or four occasions Don Tomás Ribalta has received [in Barcelona] of [his nephew-in-law and representative] Don Juan de Oña one million eight hundred thousand pesos in gold. I guess [this has happened] since he has been in Spain, and with the half a million pesos that he inherited from [his brother] Pablo Luis [Ribalta, owner of the Rosa and Santa Marta sugar mills]... But in the end I showed myself as a doer and told him that I would have done more, that with two mills and half the laborers that Don Tomás' had in his mills, I had made almost as much as that amount. The man was struck dumb, as one who sees things.[20]

The economic trajectories of José Xifré, Manuela Xiques, José P. Taltavull, Agustín Goytisolo, and Tomás Ribalta show that throughout the nineteenth century a significant flow of transatlantic capital left Cuba for Catalonia, and was instrumental in the modernisation of the Catalan capital. This transfer was possible thanks to the generalisation of the use of bills of exchange, as the Taltavull example reveals. In a monographic study on the use of bills of exchange on the Barcelona exchange, Lluís Castañeda saw that as early as 1840 these financial instruments were used widely to transfer capital from Cuba to the Catalan capital. Castañeda states that 'customarily, effects proceeded from the colonies as bills drawn against London pounds sterling'. Indeed, most of the bills of exchange drawn against pounds in the Barcelona market were sent from Cuba. After reviewing an extensive sample, Castañeda argued that in the decade of the 1840s Cuba 'had become the main provider of pounds sterling in the Barcelona market... Cuban exporters created bills against London which were acquired by Spanish trading houses and shipping companies to reap the benefits'.[21] This shows that a partial segmentation divided the markets for bills of exchange and merchandise flows, a phenomenon which continued beyond the second half of the nineteenth century.

Cubans were aware that the drawing of bills of exchange had negative effects on the island's economy. In 1872, in the midst of Cuba's Ten Year War, young Miguel Plana, the administrator of four mills in Cienfuegos, wrote to Barcelona lamenting 'the great extraction of capital that flees the island, as is proven by the fact that so far this year a value of 46 million pesos [bills of exchange] has been drawn'.[22] This capital left Cuba not only for the Barcelona exchange, but also for London, Paris, Boston, and New York. The relationship between such capital flight and cash flow problems on the island, reflected in high annual interest rates of 12 to 18%, was a zero-sum game in which the gains of one centre (the central economies, and in this case the Catalan economy) were the losses of another (the Cuban economy).

Seven years after Miguel Plana's complaint about Cuba's loss of capital, the *Junta de Agricultura, Industria y Comercio* of Barcelona wrote a report in which it analysed the impact of construction in the *Ensanche*. The *Junta* argued in 1879 that 'the insurrection of Cuba encouraged the influx of immense capital to the peninsula, in search of places in industry or urban constructions, and this led to the erection of hundreds of houses in the Barcelona *Ensanche* and the towns around its plains'.[23] The politician and writer Valentí Almirall also addressed this issue when he wrote that 'the most flourishing cities in Spain owe a great part of their prosperity to the capital amassed in America... and imported later into the peninsula'. Speaking specifically of the Barcelona *Ensanche*, Almirall stated in 1886 that its construction was 'almost entirely due to the *indianos* and the *Americanos* who made their fortunes during the war in Cuba and the great crisis which followed it, and then returned to settle among us'.[24] Indeed, as shown in a preceding monograph, the main proprietors in the city's *Ensanche* were *indianos* whose wealth came from Cuba.[25]

Real estate was not the only sector of the Catalan economy targeted for the investment of wealth brought from Cuba. Various studies have shown that, for example, *indianos'* capital (especially Cuban capital) was important in the growth of Catalan industrial activity during the second half of the nineteenth century.[26] Moreover, various *indianos* contributed directly to the country's financial sector, not only through the investment of their capital, but also through the dissemination of their transatlantic knowledge and contacts.[27] A more detailed analysis of the contribution of Antillean capital, both Cuban and Puerto Rican, to the development of steam transportation, remains a pending task. Some of the most important Catalan shipping companies of the second half of the nineteenth century were financed with capital acquired in the Caribbean. The Compañía Catalana de Vapores Costaneros; Martorell, Bofill y Cía.; the Compañía Barcelonesa de Vapores Transatlánticos; and the mighty Compañía Trasatlántica (the main Spanish shipping firm up to 1920) were founded by wealthy *indianos* who had made their fortunes in the Caribbean.

Although it may be impossible to comprehend the full impact of Caribbean capital upon the nineteenth-century Catalan economy solely from an analysis of numbers and percentages, it seems unquestionable that productive and financial activity in that country benefited greatly from the flow of capital from the Caribbean, especially Barcelona. This capital contributed to the intense economic growth registered in Catalonia, directly impacting its modernisation and industrialisation. Over the entire course of the nineteenth century Catalonia was the only Spanish territory to experience a successful industrialisation process.

Cuban capital in Paris

Barcelona was not the only European city capable of attracting Caribbean capital (mostly Cuban and Puerto Rican). Tomás Terry Adán opted for the United States as the main recipient of his money, but also invested in London and Paris, and he was not alone in doing this. Some of the wealthiest and most distinguished Cuban businessmen (such as Sabino Ojero, the Solar family, the Ayala family, the Campuzano family, the Suárez Argudín family, the Count of Casa Lombillo, and Alonso Jiménez and sons, to cite just a few) opened delegations of their Havana-based trading houses in England (some in London, some in Liverpool), or used English trading firms to invest part of their fortunes there. Years ago Ángel Bahamonde and José Cayuela pointed out the significance of capital flight from Cuba to Great Britain: 'The transfer of Antillean capital to Great Britain constitutes a stable flow through time... representing a continuous line that barely suffers any alteration... Great Britain attracted Cuban capital because it was the main European consumer of Antillean sugar,' because it provided Cuban mill owners with industrial inputs, and also because 'the City guaranteed a range of opportunities for investment that was unparalleled by any other stock market at that time'.[28]

In an effort to highlight the importance of London as the main European trading location for Cuban capital in the nineteenth century, Bahamonde and Cayuela did not pay Paris the attention it deserves. According to these authors, in mid-nineteenth century Paris had 'no commercial network as consolidated as that of London, that could ensure connections with Havana'. They acknowledged, however, that in the 1870s there existed 'an important Cuban rentier colony in Paris'. They registered only one name, the Count of San Fernando de Peñalver,[29] but obviously there were many more.

The French capital, like the English one, attracted a great number of entrepreneurs who decided to resettle there after making their fortunes in the Caribbean. Tomás Terry Adán, shown earlier, died in Paris, and is buried in the world-famous cemetery, Père-Lachaise. Many other landowners and businessmen also moved to the French capital during the nineteenth century. Considering only those individuals who belonged to the old Havana oligarchy (transformed into a *sugarchy*) and who lived in Paris, the list includes Juan Montalvo O'Farril and his wife, María Antonia Montalvo Cárdenas; Dolores Pedroso Cárdenas, sister of the first Marquise of San Carlos de Pedroso; José María de Herrera, Count of Fernandina; Concepción de Peñalver Calvo and her husband, Guillermo Julián de Bullet y Desabaije, as well as the brothers Pedro Regalado and Martín Javier Pedroso Pedroso.[30] Yet it was not only Cuban entrepreneurs who resettled in Paris after having struck it rich in Cuba. The city was also home to the Cienfuegos traders Julio Leblanc and Juan Avilés, born in France and Colombia respectively. Another Paris resident was the physician Eugenio Dupierris (son of the wealthy Havana trader Marcial Dupierris), who in April of 1866 inherited more than one hundred thousand francs from his mother, recently deceased in Cuba. Others had been born on the peninsula but settled in Paris when they decided to leave Cuba and head for Europe. This was the case of Cantabrian Venancio de Aldama Gil, formerly a partner of the Havana jewellery store Pacheco y Cía; Burgos-born Gerónimo del Val, who had sold imported clothing in his Havana shop La Bomba; and also the widow of Francisco Labayen, who had been the senior partner in Francisco Labayen y Cía., which had imported machinery in Matanzas.

Other entrepreneurs who moved from Cuba to Paris did so not to live off their rents or real estate investments, but to operate their businesses from the French capital. This was the case of Baltasar Mitjans Ricart, an entrepreneur born in Catalonia who made his fortune in Havana, then opened a trading house in Paris in 1834. Mitjans' Parisian trading and credit business was known precisely for its intense relations with the Antillean world. In 1838 another young Catalan entrepreneur described Mitjans' trading house as a 'firm that works very much, and mainly with Havana'.[31] In time Baltasar Mitjans was to become the wealthiest Spanish banker in the French capital, leaving at his death a fortune of nearly six million francs. His banking house survived him, and continued to operate until the Second World War.

This was also the case of tobacco magnate José de Susini Ruiseco, first Count of Susini Ruiseco, born in Gibraltar in 1822 to Malagueña Concepción Ruiseco and Nicean Luis Susini Sepadari. The family had left for Cuba, then entered the tobacco industry. In 1863, according to Jean Stubbs, 'Susini's factory of La Honradez registered a daily production volume of 3 million cigars, which was ten per cent of the total production of Havana (five per cent of the island's total production), and it supplied the royal houses of Europe. The factory was enormous for its day, employing a total of 2,500 workers. Indeed, Susini strove to create a technical revolution in Cuba, inventing a complex cigar-making machine. This machine was exhibited in the 1867 Paris World's Fair, and as a result, this factory's renown extended throughout Europe'.[32]

In 1863 the Susini family patriarch died, but his son continued to run the company. Hoping to exploit the success that the family firm garnered in 1867, Luis Susini Ruiseco decided to resettle in the French capital. He patented the Susini 'machines to make paper cigars', and created *ex novo* the firm Compañía Francesa de Tabacos, of which he became director. In March of 1870 Susini sold the La Honradez factory to the French firm, and the Compañía Francesa de Tabacos immediately listed 28,000 shares on the Paris bourse. Moreover, from his Paris headquarters he established and increased his business contacts with various European firms and governments. In January of 1874, for example, he signed a 'contract related to the mechanical production of cigarettes in the peninsula' with the Spanish government. Five months later he signed another, this one to transfer the exploitation of his patents and rights 'in the Russian

Empire, excepting Poland' to 'an Anglo-Russian private limited company to be established in England' by Nicolas Gessler.[33]

Like Luis Susini Ruiseco, Asturias-born Gerónimo Castañón and Ramón Menéndez Fernández also moved to Paris after amassing their fortunes in Cuba. In January of 1879 they created the trading company Castañon et Menéndez with a starting capital of 300,000 francs. The partners hoped to secure '*la comisión des marchandises pour tous les pays et principalement pour l'Espagne et les Antilles*' during the next ten years.[34]

It was not only through returning entrepreneurs that Cuban capital found its way to Paris. French citizens operated businesses in Paris and Havana simultaneously. In 1867 young Gustavo Mathias declared his desire to establish 'a trading company with the gentlemen D. Carlos Gerson and D. Eduardo Mathias, in the city of Havana, to be named Mathias Gerson y Cía.', which would have a branch in Paris. It was a rollover of a previous company with the same name, which had capital of 46,292 pesos. Apparently, the firm prospered: in April of 1870 Mathias Gerson y Cía. was again rolled over for another six years, with a capital of 100,000 pesos. Gustavo continued living in Paris and directing the French branch, while his brother Eduardo and their business partner, Carlos Gerson, operated in the Cuban capital.[35]

The individuals listed in this chapter were not the only entrepreneurs or *rentiers* who lived in Paris off wealth that they had accumulated in Cuba. They have been chosen to illustrate the phenomenon because their business trajectories reflect the varied and complex economic and migratory relations that tied together Cuba and Paris throughout the nineteenth century, and reveal the continual flow of capital produced and accumulated in the Caribbean, then transferred to Europe.

Reconsidering the Caribbean role

During the whole of the nineteenth century Barcelona and Paris attracted a share of the benefits generated by the Cuban economy in the heat of the cane fields by the sweat and blood of slaves. London also became a significant destination for Cuban capital during the same period, as Bahamonde and Cayuela have shown.[36] The axes Cuba–Barcelona, Cuba–Paris, and Cuba–London reveal the existence of a process of transference of capital generated in the Hispanic Caribbean that was later reinvested in Europe.

Returning to this chapter's starting point, it is necessary to reconsider the role played by the Caribbean, particularly the prosperous island of Cuba, in the heart of the international financial circuits during the nineteenth century. Contrary to what O'Rourke and Williamson argued in their studies on the international flows of capital, which they considered one of the major components of the first wave of capitalist globalisation, this chapter has demonstrated that centre–periphery financial flows were preceded and accompanied by capital flows from the periphery to the centre.[37] Capital transfer from Cuba to Europe was most visible after the second third of the nineteenth century, and it was instrumental in the expansion and dynamism of financial markets in Paris, London, and, most especially Barcelona.

These financial circuits constituted a significant mechanism of interconnection between the Caribbean world and old Europe during the long nineteenth century. The transfer of capital from Cuba to Spain, France, and England was one of the most intense and complex expressions of the transatlantic relations woven between the Caribbean and Western Europe. Moreover, the remittance of capital from the Antilles contributed to the economic growth of Barcelona, Paris, and London. Cuban capital flows contributed, in particular, to the intense growth of the Catalan economy, particularly during the latter half of the nineteenth century, encouraging the urban development registered by Barcelona during that time. The transfer of capital from the Caribbean to Europe was also behind the development of a set of transatlantic institutions and financial relations participated in by the major trading houses and banks of the cities involved (not just Havana, London, Paris, and Barcelona, but also Philadelphia, Boston, New York, Madrid, etc.). Some of those businessmen and bankers increased their fortunes and multiplied their business opportunities in an effervescent Atlantic World economy, thanks to their dealings with Antillean clients and capital. Finally, this chapter underscores the complexity of these types of relations, which extend well beyond analyses of nation-bound histories, and which can be studied only from transnational perspectives such as those developed in the field of Atlantic history.

Notes

1. This work has been supported by the following research project of the Spanish 'Ministerio de Economía y Competitividad': HAR 2012-39352-C02-01.
2. O'Rourke, Kevin H. and Williamson, Jeffrey G.: *Globalization and History: the evolution of a nineteenth-century Atlantic economy*, Cambridge: MIT Press, 1999.
3. Riverend, Julio: *Historia económica de Cuba*, La Habana: Editorial de Ciencias Sociales, 1985; Santamaría García, Antonio and García Álvarez, Alejandro: *Economía y colonia: la economía cubana y la relación con España, 1765–1902*, Madrid: CSIC, 2004.

4. Galarza, Vicente: *En propia defensa. Contestación al General Polavieja por el Conde de Galarza*, Barcelona: 1898.
5. Archivo Histórico de Protocolos Notariales de Barcelona, Luis G. Soler Pla, 1889, 3, 25.05.1889, fols. 2657–2687. Other Cubans who resided in Barcelona, such as Cienfuegos' Antonio Goytisolo Digat, also participated in the company.
6. García Martínez, Orlando: *Esclavitud y colonización en Cienfuegos, 1819–1877*, Cienfuegos: Ediciones Mecenas, 2008, p. 45.
7. Ely, Roland T.: *Cuando reinaba su majestad el Azúcar*, La Habana: Imagen Contemporánea, 2001 (1963), pp. 393, 406.
8. Rodrigo y Alharilla, Martín. 'De la esclavitud al cosmopolitismo: Tomás Terry Adán y su familia', in Laviña, Javier; Piqueras, Ricardo and Mondéjar, Cristina: *Afroamérica, espacios e identidades*, Barcelona: Icaria Editorial, 2013. References to the fact that Tomás Terry Adán was a slave-trader have been made by scholars including Manuel Moreno Fraginals, Hugh Thomas, and Orlando García, as well as individuals who knew Terry in Cienfuegos, such as Boston businessman Edwin F. Atkins, who nonetheless sweetened this aspect of Terry's life-story in his autobiography: Atkins, Edwin F.: *Sixty years in Cuba*, New York: Arno Press, 1980 (1926), p. 57.
9. Ely, *Cuando reinaba*; Moreno Fraginals, Manuel: *El Ingenio: complejo económico social cubano del azúcar*, La Habana: Editorial de Ciencias Sociales, 1978.
10. Nechtman, Tillman V.: *Nabobs: empire and identity in eighteenth-century Britain*, Cambridge: Cambridge University Press, 2010.
11. Costa i Fernández, Lluís: *L'illa dels somnis: l'emigració de Begur a Cuba al segle XIX*, Begur: Ajuntament de Begur, 1999.
12. Rodrigo y Alharilla, Martín: 'Barcelona, capital del retorn' in *Les bases colonials de Barcelona*, 1765–1968, Barcelona: Museu d'Història de Barcelona, 2012, pp. 78–91.
13. Solà i Parera, Àngels: 'Os "americanos" cataláns e o seu impacto económico en Cataluña ó longo do século XIX', *Estudios Migratorios*, nos 11–12, 2001, pp. 141–68; Solà i Parera, Àngels: 'Els capitals americans i la industrialització de Barcelona al segle XIX', in Rodrigo, Martín (ed.): *Les bases colonials de Barcelona, 1765–1968*, Barcelona: Museu d'Història de Barcelona, 2012, pp. 68–76; Rodrigo y Alharilla, Martín: *Indians a Catalunya: capitals cubans en l'economia catalana*, Barcelona: Fundació Noguera, 2007.
14. Ramón de San-Pedro, José María: *Don José Xifré Casas: pequeña historia decimonónica de un archimillonario 'americano'*, Barcelona and Madrid: Banco Atlántico, 1956, p. 73.
15. Ibid., p. 4.
16. *La Bisabuelita: Manuela Xiqués de Llopart*. Manuscript generously provided by José Ramón Bofarull Sivatte.
17. Ibid.
18. Rodrigo y Alharilla, Martín: 'Trasvase de capitales antillanos: azúcar y transformación urbana en Barcelona en el siglo XIX', in Santamaría, Antonio and Naranjo, Consuelo: *Más allá del azúcar: política, diversificación y prácticas económicas en Cuba, 1878–1930*, Aranjuez: Doce Calles, 2009, pp. 127–58.
19. Archivo Nacional de Cuba, Protocolos Notariales, Gabriel Ramírez O'Brien, 1864, 1, p.107, 09/02/1864.
20. Fundación Luis Goytisolo, Fondo Agustín Goytisolo Lezarzaburu, letter from Agustín F. Goytisolo (Cienfuegos) to Antonio Goytisolo (Barcelona), 9 November 1882.

21. Castañeda Peirón, Lluís: *Esplendor y ocaso del mercado de letras de cambio en la Barcelona del siglo XIX*, PhD thesis, Instituto Universitario Europeo de Florencia, 2001.
22. Fundación Luis Goytisolo, Fondo Agustín Goytisolo Lezarzaburu, box 97, exp. 2: letter from Miguel Plana Iradi (Cienfuegos) to Agustín Goytisolo (Barcelona), 27/05/1872.
23. Cited in Tafunell, Xavier: *La construcció de la Barcelona moderna. La indústria de l'habitatge entre 1854 i 1897*, Barcelona: Ajuntament de Barcelona, 1994, p. 61.
24. Almirall, Valentí: *España, tal como es*, Barcelona: Anthropos, 1983.
25. Rodrigo y Alharilla, *Trasvase de capitals*. See also Tafunell, *La construcció*.
26. Solà i Parera, *Os 'americanos'*; idem. *Els capitals americans*; Dorel Ferré, Gràcia: 'La contribution des indianos à la formation du territoire et du paysage industriel catalans durant la seconde moitié du XIX siècle', in Le Bot, Florent and Perrin, Cédric: *Les chemins de l'industrialisation en Espagne et en France. Les PME et le développement des territoires (XVIII–XXI siècles)*, Peter Lang, 2011, pp. 283–99; Dorel Ferré, Gràcia: *Les colònies industrials a Catalunya: el cas de la Colònia Sedó*, Barcelona: Publicacions de l'Abadia de Montserrat, 1992; Soler Becerro, Raimon: 'Comerciants i fabricants: una reflexió sobre l'origen comercial del capital industrial: el cas de Vilanova i la Geltrú', *Recerques*, no. 36, 1998, pp. 109–37; Soler Becerro, Raimon: 'La Vilanova ultramarina i industrial: el paper dels indianos en el bastiment d'una ciutat industrial', in *Ultramar: política de Víctor Balaguer i progres per a Vilanova*, Biblioteca Museu Víctor Balaguer, 2000, pp. 89–102.
27. Badenes Badenes Escudero, Carles: *Capitales indianos en el sistema financiero catalán, 1833–1882*, Unpublished Master's Dissertation, Universitat Pompeu Fabra, 2011; Castañeda Peirón, Lluís and Rodrigo y Alharilla, Martín: 'Los Vidal Quadras: familia y negocios', in Grau, Ramon (ed.): *La ciutat i les revolucions, 1808–1868. II, El procés d'industrialització*, Barcelona: Quaderns d'Història, no. 11, 2004, pp. 115–44; Rodrigo y Alharilla, *Trasvase de capitales*; Blasco Martel, Yolanda: 'Retornos de América, banca y capital humano: el caso de Jaime Badia', *Historia Social*, no. 59, 2007, pp. 125–50; Blasco Martel, Yolanda: *Epistolari de Jaume Badia: el pensament bancari en el segle XIX*, Torredembarra: Arxiu Municipal de Torredembarra, 2009.
28. Bahamonde, Ángel and Cayuela, José: *Hacer las Américas. Las élites coloniales españolas en el siglo XIX*, Madrid: Alianza, 1992, pp. 76–7.
29. Ibid.
30. Archivo Histórico de Protocolos Notariales de Madrid, Consulado de España en París, 32022, 13.01.1866, fols 20–5 and 12.02.1866, fols 104–6; 32026, 24.01.1870, fols 168–73, 07.03.1870, fols 345–9 and 29.03.1870, fols 470–3; 34790, 28.05.1883, fols. 679–85 and 03.08.1883, fols 969–73.
31. Archivo Privado Leopoldo Gil Nebot, box 6, file: 'Pablo Gil Serra, cartas', letter from Pablo Gil Serra (London) to Pedro Gil Serra (Paris), 11/09/1838.
32. Stubbs, Jean: *Tobacco on the periphery: a case study in Cuban labour history, 1860–1958*, Cambridge: Cambridge University Press, 1985.
33. Ibid.
34. Archives Nationales de France, Paris, Minutier Central, ET CXII, 1376.
35. Archivo Histórico de Protocolos de Madrid, Consulado de España en Paris, 32023, 31.12.1867, fols 1511–15; 32026, 26.02.1870, fols 311–19.
36. Bahamonde and Cayuela, *Hacer las Américas*.
37. O'Rourke and Williamson, *Globalization and History*.

11

Baring Brothers and the Cuban Plantation Economy, 1814–1870

Inés Roldán de Montaud

Historians of the North Atlantic have shown how its economic system mobilised great effort and resources to produce, distribute, and sell commodities such as sugar. This activity helped to weave a trading and financial network which, from the sixteenth century onward, increasingly enmeshed the eastern American ports with European consumers and African enclaves on the Atlantic.[1] Sugar economies and planters required the ready availability of credit; merchant banking houses in Europe and the US provided the financial link between local exporters and their markets throughout the world. The House of Baring emerged as one of the premier entrepreneurial firms on either side of the Atlantic, and was among the first to act on the idea of establishing a permanent transatlantic organisation.[2] By the first decade of the nineteenth century Barings was one of the leading Anglo-American houses providing financial services to sugar merchants, which soon became an important part of its business. This importance drove Barings' involvement in the wealthy Spanish Caribbean colony of Cuba, which, after the collapse of Saint-Domingue, had become the world's greatest sugar producer, at a time when rising demand in European markets seemed unstoppable. Thus, Barings stands out among the creators of the intricate circuit of commerce, money, and capital flows which shaped a single transatlantic economy. Its history illustrates how changing fortunes on either side of the Atlantic World reverberated across and through the entire network.

Several decades ago, in a pioneering work, D.C.M. Platt gave a summary account of trade and financial ties between Cuba and Great Britain. He studied British interests on the American continent, a theme almost completely ignored by Leland Jenks a few years earlier.[3] Many years have passed since this work, but only a few scholars have referred to the British presence and influence in this area, usually citing only the migration of British individuals, mainly machine operators and mechanics,[4] and investments in trade, railways, and the tobacco and mining sectors. Certainly the nature of British interests and presence in the Spanish Caribbean varied over the years. During approximately the first two-thirds of the century British capital financed trade above all through provision of commercial credit, by accepting drafts and discounting bills of exchange. From the middle of the century direct investment in various productive activities had an increasingly important role, and the extension of railway lines was possible thanks to loans placed by British bankers on the London market. At the end of the century British investors owned almost the entire railway system.[5] Recent studies have revealed the mobilisation of mining resources on the eastern part of the island through the creation of several British-owned companies which for years supplied smelters in Swansea with a third or a half of the mineral they used. Those firms fuelled the trade in African slaves on the eastern half of the island, and the migration of miners from Cornwall.[6] The British also had an important presence in the tobacco business.[7] However, the level of overall direct investment was relatively low until the last decades of the century; British investment through public issues amounted to just £1,231,600 in 1880, but grew rapidly to £26,806,000 in 1890.[8]

Based to a large extent on Barings Brothers' business papers in London and documents kept in Cuba's National Archives, this chapter deals with the bank's presence and trade-related activities in Cuba, and the role these merchant bankers played in the process of mobilising the products of the island's agricultural export economy. Existing studies of Barings have overlooked their connections with Cuba, although among the European merchant banks it had probably the most outstanding role in marketing the sugar that was produced in such quantities in the colony. This abundance was observed by Baron Alphonse de Rothschild, who wrote on a visit to Havana in 1849 that 'the sugar business here is the monopoly of the exporters, Drake, Burnham, Picard & Albert. However, they are not doing the most important or weighty business, this being done by Barings, Coutts, Fruehling & Goschen in London, who are making all of the profit from commissions, credits and consignation. The credits are mostly given for account of continental Houses.'[9]

Rather than simply quantifying the relevant commerce and trade, this chapter focuses on the London bank's actual experiences in Cuba, and maps its operations from the 1820s until well into the 1870s. The first section considers the origin of Barings' relationship with Cuba, and the nature of its interests there, which consisted mostly of providing financial services to local sugar merchants. The second part examines the company's working methods through its business with a small set of trading firms considered to have first-class reputations. The circumstances and specific production conditions of the Cuban sugar market are analysed to show the effects of the failure of correspondents on Barings' business. The final section considers the circumstances in which Barings' interests in Cuban sugar faded in the late 1860s, and how its ties to merchant houses lapsed while other British firms – those less involved in trade – continued with their investments on the island.

The origins of Barings' business in Cuba

Originally from Bremen, the Barings had settled in Exeter by the early eighteenth century, where they were engaged in the manufacture and trade of woollen cloth. By 1760 their London firm, John and Francis Baring & Co., had trading links with merchants in the Baltic and Mediterranean Seas, the British North Atlantic colonies, and the Caribbean. The bank perceived the potentially unlimited opportunities emerging on the other side of the Atlantic early on, and sent Alexander Baring to Philadelphia in 1795 to establish an agency there. While Europe was in the throes of the French Revolution and the Napoleonic Wars Barings became leaders in Anglo-American commerce, with an interest in most of the goods traded at the time, especially tea, sugar, coffee, indigo, and cotton. As Barings financed the campaigns of Great Britain's allies against revolutionary and Napoleonic France between 1796 and 1814, the partnership became one of the most powerful merchant banking houses in Europe, and contributed to England's evolution into a great commercial power.[10]

Barings seems to have had an eighteenth-century connection to the ports of Cadiz and Barcelona, and one of the partners, Charles Baring, travelled to the Iberian Peninsula to establish contacts in Spain and Portugal.[11] Later the firm intervened directly in financing the payments the US was required to make to Napoleon pursuant to the 1803 Treaty which outlined the terms of the Louisiana Purchase.[12] Barings' links to Russia were strong, but its interests in the rest of Europe were limited. The bank had only occasional business in mainland Spain in the nineteenth century,[13] but its activities in Cuba were intense. This preceded the Spanish concession of complete free trade to her colony in 1818, which had initially been restricted to Cadiz, and was later extended to additional Spanish ports. The trading monopoly of these ports had been faltering for several years, and now was effectively broken, leaving fewer obstacles to the establishment of foreign merchants in Cuba. Barings' ledgers testify to its dealings with Cuban exporters in 1814. An account was opened that year for James Drake, a British merchant established in Havana in 1792 under a warrant issued by King Charles IV. Drake was one of the first to be permitted to settle in the colony in the late eighteenth century, as the rules restricting trade to merchants from mainland Spain were relaxed. In time, he would amass a fortune there.[14] Contacts increased in subsequent years, although the volume of Barings' Cuban operations remained insignificant when compared to its transactions with the British West Indies, especially Jamaica, which produced 47% of the world's sugar by the end of the Napoleonic Wars, and was the main supplier of sugar to Britain.

As Cuban sugar production increased from 14,455 tonnes in 1792 to 56,150 in 1825, Baring Brothers augmented its Atlantic World connections, and channelled Caribbean exports to meet growing demand in Europe and North America.[15] An account was opened in 1822 for Francisco Fesser & Co., a prominent export house which became Barings' most important correspondent, and remained so until 1837.[16] In 1828 the turnover of Fesser's various accounts had risen to more than £50,000, and by 1830 to £100,000.[17] Cuban exports to Britain grew fast in those years, doubling from 4,887 tonnes in 1821–5 to 9,009 tonnes in 1836–40, while West Indian production declined in the wake of abolition.[18]

At the same time the number of foreign merchants settled in Cuban ports increased. From 1829 a new commercial code favoured their settlement by allowing them to become naturalised, and to engage in trade as local merchants. The British trading community remained, but in reduced numbers, and very little is known of its activities.[19] In 1846 British residents numbered no more than 474 men and 131 women, most of whom were living in Havana and the environs of Santiago de Cuba, where they were employed by British mining companies. According to the 1841 census, 22 firms out of 135 in Havana belonged to foreign traders, although few were British. Foreigners frequently associated with local merchants based in the country, as was the case with Mariátegui, Knight & Co., Barings' main agents in Cuba in the 1830s. No British commercial houses were registered in Puerto Rico at that time.[20]

Joshua Bates, a merchant from Boston, became a partner in Baring Brothers in 1828. For several years he had represented the interests of several North American shipowners in London.[21] Bates injected great dynamism into Barings' affairs as a trading house, and was responsible for maintaining the firm's role as the most important 'American house' in London. His work relegated to a position of secondary importance the strictly financial activity of the partnership, such as cash advances made to foreign governments and loan placements on the London market. To reinforce the position of the House in American trade, Thomas Baring went to the US in 1829. There he established contact with Thomas Wren Ward, another prominent Boston merchant, who became Barings' first agent and special representative in North America, a position he retained until the second half of the century.[22] Barings opened a branch in Liverpool in 1832, and was soon deeply involved in the New Orleans cotton trade, through advances on produce consigned to be sold by them in Britain. At the same time they specialised in sending other American products to England and Continental Europe. Consignments were received to sell in London, or to forward to other European markets. The proceeds from sales were collected, cargoes of British manufactures purchased and dispatched to North America and the West Indies, and British manufacturers' drafts on American houses paid. These changes left their mark on the sources of Barings' profits. Commission income, which averaged £26,463 annually between 1823 and 1827, increased to £40,855 during each of the following five years, and to £81,085 annually from 1838 to 1842.[23] By that time Barings had become the leading 'American House'.

The level of activity in Cuba increased as Bates implemented changes to the firm's management. The firm began to devote itself almost entirely to trade finance and the purchase and sale of sugar to meet demand in Europe and the US, supplemented by trade in coffee, and occasionally in tobacco. Its role as suppliers to the *Renta del Tabaco* in mainland Spain was exceptional; the tobacco leaf supplied to the *Renta's* contractors was shipped from Kentucky.[24] Like many other merchant bankers operating in the Atlantic Barings traded in merchandise on their own account, and made capital advances to exporters to induce them to hand over their consignments. They extended credit to certain (few and carefully selected) exporters, accepting them to draw sterling bills of exchange, generally for two-thirds of the probable market value of the sugar shipments consigned to Barings, and seldom for the full estimated selling price in payment of sugars shipped. The bills issued on Barings were usually covered by the proceeds of the sales of the goods, which were typically sold before the sterling bills drawn in Cuba had to be honoured (usually several months after the merchandise had been loaded onto ships). Agreed lines of credit were reinstated as soon as any outstanding bills had been met through remittances. Barings also took charge of insurance and shipping for merchandise consigned to them, which was generally sent first to Cowes, the main entrepôt port for sugars re-shipped to European ports of the Atlantic and the Baltic, then on to a market. Barings' profits came primarily from the interest on advances, from commissions on consigned produce, and from the acceptance of bills of exchange. Very rarely did the firm provide direct financial services to planters, as will be seen below.

Barings had no interest in placing loans on the London market for railroad projects in Cuba, as Robertson & Co. did in 1838 for the Havana–Güines railway, or like Schröders & Co., which in 1853 placed its first loan, worth one million pesos, for the Matanzas to Sabanillas railway,[25] or as Barings themselves did in the US, where the house was heavily involved in railway finance. Sugar transportation was effectively favoured by Barings over railway development in the mid-nineteenth century; despite various company launches in the railways sector, the bankers at Eight Bishopsgate generally turned down requests to place loans for such projects.[26]

Mariátegui, Knight & Co. of Havana

Mariátegui, Knight & Company was founded in Havana in the mid-1820s.[27] The two named partners were George Knight, a North American merchant who had been established in Cuba from at least 1816, and Juan José Mariátegui, a member of a prominent family from Guipúzcoa in Spain. One of José Mariátegui's brothers was Interior Minister in 1823 and a parliamentarian in 1834,[28] and he himself was a prominent figure in the business community, as indicated by his membership in 1823 of the Governing Council of the *Junta de Gobierno del Real Consulado de Comercio*, the Royal Consulate of Trade.[29] The third partner was Gonzalo Alfonso y Soler, a rich planter and slave-trader born in Havana in 1794. He was probably one of Cuba's wealthiest men, with ties of kinship to the Poey and Aldama families, who were also prominent sugar planters. The Alfonso family owned several plantations around Matanzas, including San Gonzalo, Acana, Triunvirato, and Concepción, and later the Compañía de Caminos de Hierro de la Habana railway, alongside important shares in other railroad companies and ports.

With its own warehouses and port installations in Matanzas, Mariátegui, Knight & Co. operated primarily in commission-based sugar, molasses, and coffee exporting. This business was balanced by the firm's importation of manufactures and dry goods from Glasgow and other British and European ports, and especially from the US. Its balance sheet from

September 1830 to August 1831 records the export of 67,528 cases of sugar, more than 105,000 quintals of coffee, and 2,346,112 gallons of molasses, plus some tobacco, all of which was valued at $2,733,840 – far from a trifling amount.[30] Cuba exported 2,433,423 cases of sugar in the years 1826 to 1830,[31] an annual average of 487,298 cases. Mariátegui, Knight & Co. were shipping around 15% of these Cuban sugars. In the same period they paid export duties to the value of $548,512. Receipts at all customs points in 1831 reached $4,798,405;[32] payments by Mariátegui represented 11% of the total. Net profits rose in the same period to $112,000, and to $315,000 in 1835,[33] which was not far from those of Drake Brothers & Co., which was considered one of the most important trading firms of the 1840s.[34] Although less well known, Mariátegui, Knight & Co. were also among the most prominent sugar merchants in Cuba. Perhaps because of this they soon became Barings' sole agents on the island.

Barings' first business ventures with Mariátegui go back to 1828, when they provided insurance for the ship *Colón* and its cargo.[35] In March 1829 Barings opened a credit of £10,000 in favour of Mariátegui with Goodhue & Co. of New York, to favour sugar exports to the North American market. Goodhue was outstanding among the firms dealing in the Russia trade, where Barings had important business through Stieglitz & Co., Goodhue's chief trading partner. Ten years later, Burnham & Co., at the time Barings' main agents in Cuba, was still sending sugar to Goodhue, which was closely tied to Baring Brothers.[36]

Barings' relationship with Mariátegui, Knight & Co. intensified following George Knight's visit to London in autumn 1831, when new credit facilities which had been placed at the firm's disposal allowed it to operate in the most active months, when harvests were exported.[37] A credit was opened in favour of Pasaverino & Co. of Buenos Aires, enabling them to issue bills on Barings for £2,000 a month against shipments of jerked beef consigned to Knight & Co.,[38] an arrangement which allowed the latter to participate in this lucrative trade. As trade in sugar expanded, and with the number of slaves doubling to reach 436,000 between 1817 and 1846, consumption rose from 500,000 arrobas in 1829 to 800,000.[39] A joint exchange account of £10,000 was opened, with shared profits and losses. Funds were to be drawn down when the exchange rate was high between the pound and the peso, and remitted when it was low. On these bases the turnover of the accounts of Baring Brothers and its trading partners rose from £14,533 in 1830 to £164,000 in 1832.[40] Confidence in Barings rose, and in March 1834 Mariátegui, Knight & Co. were appointed the bank's authorised agents in Cuba, replacing the Bostonian Thomas Ward, who had managed Barings' interests there, in the West Indies, and in the US since 1831.[41]

George Knight & Co.: from commission merchants to promoters of sugar estates

Juan José Mariátegui died in Montpellier in November 1835,[42] but his death did not interrupt the firm's business. In keeping with agreements made, his heirs were obliged to withdraw their funds in instalments, and the company continued to operate henceforth under the name George Knight & Co. When Gonzalo Alfonso abandoned it in 1837, Peter Lambert Fernández became a partner.[43] He was the owner of half of *Unión*, a beautiful estate immortalised by Eduardo Laplante in the lithographs of *Libro de los ingenios*.[44] In order not to harm Knight's interests, Alfonso agreed to recover his capital gradually from the company's annual profits. For example, that spring he asked Barings for a credit of £6,000 to acquire a new property in Cárdenas, to be paid back from the credit Alfonso had with Knight.[45] This was one of the few occasions when Barings advanced funds directly to a planter.

For a couple of years George Knight & Co. restricted its activities to purely mercantile business such as commission trading, freighting, insurance broking, and providing guarantees of receipt for products sent on commission to Barings. The company also bought sugars directly from planters on its own account to sell on to European and American importers. However, when sugar prices began to rise from 23 and 30 shillings a hundredweight to 38 and 45 in 1836,[46] Knight reoriented his business towards the provision of large-scale, short-term, high-interest loans to planters, with their crops as guarantees. Such loans were dispensed primarily to meet planters' increasing demand for capital to fund expansion. The firm, with an eye on the phenomenal dynamism of the sugar sector, followed a deliberate strategy of taking over estates. In a letter of 8 January 1836 Knight stated his amazement at the prices fetched by sugar, and assured Barings that Cuba would very soon become one of the wealthiest islands in the world. He concluded, 'If I could I would change my house into sugar estates.'[47] When the Boston trader Theodore Thorndike bought the *Santa Ana* estate, he instructed his agent in Cuba, Richard P. Dana, to negotiate with Knight for the funds necessary to acquire slaves and machinery. In exchange for further advances in 1838 Thorndike mortgaged the estate.[48] By 1840 George Knight had advanced Thorndike a total of $110,000; the assets securing his mortgages were sugar and coffee estates which together were worth $680,000 (Table 11.1).

Table 11.1 Property, claims and debts of George Knight & Co, 17 May 1840, pesos

Property	
half of the sugar estate ARROYO (177 slaves)	150,000
half of the sugar estate RECURSO, property of Eliseo Martín (160 slaves)	100,000
half of the coffee estate SANTA ISABEL (150 slaves and 360,000 plants)	100,000
half of the coffee estate CORNETA (40 slaves and 200,000 plants)	20,000
half of the sugar estate UNIÓN, property of L. Fernández (150 slaves and 3,000 boxes)	150,000
half of the coffee estate CAROLIN, property of L. Fernández (220,000 plants)	80,000
Other assets	20,000
Store and dwelling house in Matanzas	50,000
Refinery and other assets of his real estate property in Matanzas	100,000
Interests in Regla Bay (steamships and foundry)	120,000
Total property	**890,000**
Mortgage claims	
Sugar estate SANTA ANA and its fruits, for advances to Thorndike (110 slaves)	110,000
Sugar estates SONORA and ROBLE and coffee estates BÚFALO and INDUSTRIA, for advances to Theodore Phinney (450 slaves)	140,000
Coffee estate PALMA SOLA, for advances to B. Tales	70,000
Debt of Alex Taylor for advances on his half of the ARROYO	100,000
Debt of Eliseo Martín, for advances on his half of the RECURSO	80,000
DeConnick, balance purchase of sugar estate SANTA AMALIA	80,000
Advanced to T. Mason, secured by mortgage on crops	7,000
Advanced to Lewis de Mun, against his estate SANTA MARÍA	16,000
Debt of W. M. Kimon for the purchase of the coffee estate UNIDAD	27,000
Interests in molasses establishment in Cárdenas	50,000
Total mortgage claims	**650,000**
Other claims	
Advanced to Enrique Disdier on his sugar and coffee estates	72,400
Advanced to Robert Steel on his estate	16,000
Claims in the United Estates	60,000
Debts of several estates whose produce is consigned to Knight & Co. for sale	50,000
Debts to Knight & Co. in their current mercantile business in Havana	85,000
El Parisino (ship)	50,000
Probable proceeds of the Regla ferry and foundry	20,000
Total other claims	**383,400**
GRAND TOTAL	1,859,400

Source: BA.HC.4.6.2. One sterling pound equals five pesos ($).

Disaster struck when the crisis of 1837 devastated American trade. In anticipation of a financial catastrophe, which Barings seemed to be expecting, the company had restricted credit to its agents, but, in spite of their prudence, registered a loss of £168,000 at the end of 1837, and bad debts of almost £118,000 were written off its books in 1838.[49] Knight & Co. were also unable to avoid the effects of the crisis. In the absence of liquidity for non-payments by debtors, Knight began to exceed the credit limits set by Barings, drawing down amounts far greater than those agreed. The company's difficulties increased: imprudent management had led it into a 'mortgage trap', wherein large amounts of capital were locked up in illiquid assets. To exacerbate matters, adverse climatic conditions and a sharp fall in sugar prices on the world market paralysed demand, and extreme monetary strains did the rest, preventing the estate owners from meeting their obligations in time, even as they asked for more resources. Once monies were advanced to planters with produce as collateral, or on the back of mortgaged estates, the exporter had to maintain the planter through a continued supply of liquidity sufficient for the needs of the harvest, the sale of which was earmarked for repayment of loans.[50] Knight's expenses in 1838 included \$60,000 for advances to planters. At the same time falling sugar prices had led to large losses for Knight arising from shipments over the preceding two years. In 1838 and 1839 the company had consigned 8,000 and 11,251 cases valued at \$254,475 and \$380,000 respectively to Cramer Brothers, M. Brandt & Co., and John Thomas & Co. of St Petersburg, to sell in the Russian market, with proceeds subject to instructions from Barings.[51]

George Knight & Co. finally suspended payments on 15 May 1840. Two days later, local creditors met, and a liquidation panel was formed, which includied Gonzalo Alfonso, Luis de Mariátegui (nephew of the deceased Juan José), and Nicolás Domínguez, a notable Havana merchant. Knight wanted, at any cost, to avoid the fragmentation of his estates, and was anxious to reach an agreement with creditors, as he indicated to Barings two or three days before the suspension.[52] At the meeting on the seventeenth, it was agreed that George Knight & Co. would be dissolved, and the firm's rights and properties – and those belonging to Knight and to Peter L. Fernández individually – would be assigned to their creditors. The liquidators would receive all produce and goods, and would sell the merchandise and the ships. They would undertake the operation of the estates, and would distribute the proceeds among creditors. If prices did not vary, it was estimated that four or five years would suffice to repay the accumulated debt. Luis Mariátegui counselled Barings and other foreign creditors to accept the deal, in order to avoid the possible stagnation of the business amidst judicial proceedings. He knew the country and its infernal courts well enough to advise that they would bring 'anything but peace'.[53]

The Cuban creditors of George Knight & Co. were the heirs of Juan José Mariátegui ($90,000), Gonzalo Alfonso ($82,000), Julián Alfonso ($62,000), and Domingo de Aldama ($25,000). The remainder of the debt in Cuba was distributed among numerous merchants. The tax authority, *Intendencia de Hacienda,* was owed $23,212 for unpaid customs duties. Debts in Europe and the US amounted to $626,354. Sums were owed to Reid, Irving & Co., the leading merchant bankers in the West India trade,[54] while Fould Frères of Paris and Cramer & Son of St Petersburg were among other firms touched by the failure. Outstanding among the North American creditors were Holford & Co. of Boston and Mauran & Co. of New York (£40,000),[55] but Barings was the company most badly affected. Their credits with Knight had risen to £70,000, a considerable sum relative to their capital, which stood at £691,489 in 1839 and £501,944 in 1842.[56]

At the meeting of 17 May the liquidating committee estimated the value of the Knight & Co.'s assets at $1,866,000, against debts of $982,000.[57] In July the firm's assets subject to liquidation stood at $2,256,639, and the debts at $1,325,153. The real asset base was greater, however, because the properties' value was estimated very conservatively, and some assets had been hidden.[58] It was a relatively important estate. Barings believed it more convenient to sell the properties, should there be sufficient proceeds to satisfy all creditors, and if the planters were so disposed, because a Cuban law, the *privilegio de ingenios,* forbade the execution of mortgages against sugar estates or their slaves if the debt was not at least to the full value of the property. In any case Barings accepted the liquidators' proposal and named the Havana firm DeConnick, Spalding & Co. as its attorneys, instructing them to subscribe to the minutes signed on 17 May, and insisting that in the course of liquidation the estates should be administered only for the sole purpose of paying debts. No improvements or slave purchases were to be made. In any case, signing the deal would not affect Barings' claims against Gonzalo Alfonso for the loan made in 1837.[59]

Running sugar estates and becoming slave proprietors

Recovering its money turned out to be no easy task for Barings. Knight's liquidation became difficult when the economy took a negative turn, with a prolonged period of declining sugar prices, from 49 shillings per hundredweight in 1840 to 40 in 1841, 33 in 1844, and 22 in subsequent years.[60] The high prices of 1835 and 1836 had encouraged planters to buy slaves on credit and to invest in breaking new lands for cultivation, some at a considerable distance from the mills and ports, so transport absorbed an important part of the value of produce, and was impossible to cover unless prices were high.[61] The effects of the 1841 banking crisis in the US were keenly felt in Cuba, given the close commercial and financial ties between the two economies, to the point when, in spring 1842, after a prolonged visit to Europe, Mariátegui observed that 'things have changed so much here upon my return that I no longer seem to be in wealthy and prosperous Havana'.[62] There were several bankruptcies in June, and as he stated, it seemed that very few plantation owners were able to meet their obligations.[63] In the face of these adverse conditions in the sugar market, Knight's creditors modified the 1840 agreement. To obtain greater guarantees, they decided in June 1843 effectively to transfer the mortgages on the estates.[64] Existing law stipulated that mortgage creditors who had not assumed a mortgage could find themselves passed over in favour of those that had taken direct and absolute control of mortgaged property, in the manner stipulated by the law.[65] The transfer was delayed for the prior settlement of monies owed to the *Intendencia de Hacienda* (which as the tax authority was one of the privileged creditors in bankruptcies) through the transfer of the mortgage on a coffee plantation.

Soon after came the death of George Knight. Following a long negotiation, DeConnick, Spalding & Co. agreed with Robert Morrison, representing Reid, Irving & Co., that Barings would keep the mortgages on half of the *Arroyo* estate, half of *Santa Ana*, and on the whole of *Santa María* (valued together at $376,000). Reid Irving would take the mortgages on the *Sonora*, the property of Theodore Phinney, and the *Unidad*, a coffee estate owned by W.M. Kimon (Table 11.1).[66] In January 1844, Mathias Purton, Barings' agent in New Orleans, accepted on behalf of the firm the transfer of Knight's mortgage on the *Arroyo* and its 177 slaves. However, in September the bank suddenly backtracked, and the mortgage transfer was cancelled. Barings' advisers believed that holding mortgages on estates with slaves could be considered a contravention of an act passed on 24 August 1843. In September 1841 the Anti-Slavery Society had asked Parliament to act to prevent British firms operating in Cuba and Brazil from favouring the slave trade or slavery's expansion, as was occurring with mining companies operating in Santiago de Cuba and Brazil. The result was the Act for the more Effectual Suppression of the Slave Trade, which penalised slave ownership by British subjects 'wheresoever residing or being, and whether within the Dominions of the British Crown or of any foreign country'.

Instead mortgages on *Santa Ana* and *Arroyo* were written in the names of front men, and were to be administered by Storey, Spalding & Co. (successors to DeConnick, Spalding & Co.), a firm described by Samuel Bierly as a confidant of the Barings, and as having little capital, but excellent connections.[67] *Santa María* was handed to Mariátegui & Co., the company created after Knight's bankruptcy. DeConnick and Mariátegui supervised the operations of the mortgaged estates remaining in the hands of their proprietors, and provided intermittent reports of their progress to Barings. The bank in turn gave them precise instructions on managing the estates, and on the shipment of sugars sent by the owners to Barings' agents, to be sent to Western & Grey or Goodhue & Co. of New York, or to London, in keeping with prevailing market conditions.

The Cuban debt became a veritable nightmare for the bankers. A series of adverse conditions prolonged repayment of the debt for over 15 years. The period of depression that began in 1842 lasted until 1852.[68] In addition to serious social disruptions provoked by slave uprisings in certain estates in Matanzas and Cárdenas, there was an extended drought that year, followed by a cyclone that ruined the harvests. The crop of 1845 was around 98,000 tonnes, about half that of preceding years. In 1846 another cyclone hit the countryside, after which came the effects of the 1847 financial crisis expanding from Europe, and the revolutions of spring 1848, which paralysed sugar sales in Europe, causing prices to tumble there.

Desperation increased over time, as evidenced by the hundreds of letters sent to the estates' administrators. In July 1848 Spalding & Co. were told that Barings had decided to reduce the credit extended to Burnham and to be drawn against consignments, as the firm did not wish to be left with merchandise that could not be sold. In September Bates wrote to Spalding stating that he despaired of obtaining any repayment that year, and lamenting that *Santa Ana*'s production could not cover costs. *Arroyo*'s accounts yielded equally poor results.[69] In December Spalding announced that nothing could be repaid. Prices had fallen so much and demand was so limited that the Barings advised him to limit the sugar cane zone, and to extend areas for foodstuffs, since it was possible to buy all output only when market prices were high.

The 1850s were no better. At the start of the decade a string of calamities unrelated to the world of trade and finance continued to obstruct recovery of the debt. In 1851 *Arroyo* suffered a devastating fire and lost various plots of cane. It also became necessary to import new machinery from Scotland to replace existing equipment,[70] while a number of slaves were lost that year due to a cholera epidemic. Corresponding with its agents, Barings lamented the elevated exploitation costs, a circumstance shared with other planters who were no longer raising the revenues achieved in earlier years. In a state of complete desperation, Barings proposed to Burnham in March 1850 that he seek a buyer disposed to acquire the mortgage on *Santa María*, which stood at £4,142. The bank wished to cede its rights with a discount of 20% to one of the owners, Edmund de Pestre.[71] Even that did not prove easy: although de Pestre wanted to cancel the mortgage, and Santiago Bayley from Matanzas appeared ready to finance the operation,[72] the transaction failed under the tight monetary conditions prevailing after the landing in Cárdenas of an invading force led by Narciso López, which caused a public commotion. Facing insecurity, those with cash preferred to invest in the US. Under these circumstances the transfer of the Baring mortgage on the estate was delayed until January 1851, after Burnham received authority to move it.[73] More than ten years had passed since the $16,000 loan had ended up in the hands of Barings.

The cholera epidemic that broke out late in 1851 extended over the final two months of the year. In *Arroyo*'s neighbouring estates 'negros... died by the dozen', but 'order, improvements in the negros' nutrition, and less exhausting work days' prevented the same occurring at *Arroyo*. Burnham insisted that the administrators absolutely did not permit slaves to be overworked, but this obliged them to hire more hands, with a consequent rise in costs. He sought Barings' permission to buy more slaves and to sign on Asians, the indentured labourers that had begun to arrive Cuba in 1847, to solve labour shortages on the estates. In July 1853 Burnham wrote, alarmed, that 'the *Álava*, the magnificent property of Zulueta, had lost 200 of its 1,000 slaves' when cholera had invaded the adjacent estate. 'We fear the *Arroyo* may not escape this time', he wrote.[74]

In 1843 Thorndike's debt reached $120,672. Barings did not receive a single penny against it between 31 December 1850 and 4 March 1853, but the bank may have managed to recover its credits against *Arroyo* and *Santa Ana* from the elevated prices realised for the abundant harvests of 1856 and 1857. From spring 1854 developments in the sugar market began to astonish Burnham. The business being done was quite astonishing, as if 'fortune were to be made all at once'. Prices escalated to unimagined levels up to June 1857, an increase in circulating money reduced the interest rate from 18% to six per cent, and foreign payments for exports led to a continuous inflow of gold.[75] The estate owners apparently made good use of this moment to pay outstanding debts and free themselves from their creditors. In any case, the relief was short-lived. The 1857 financial crisis and its effects, which were to be felt for several years, were not long in coming. Freed from the *privilegio de ingenios* that had protected planters, and which began to be eased in 1852 following a decree which allowed owners of sugar estates to renounce their rights, many merchants were now able to take over debtors' estates. The fate of Barings' mortgages on the properties is unknown, however, as is the timing of the plantation owners' debt clearances, because all reference to them disappears in Burnham's letters at this time.

The Barings case was not unique; other creditors of Knight, such as Drake Brothers & Co., suffered similar penalties. Reid, Irving & Co. had complications with the *Sonora* estate owned by Theodore Phinney, a North American from Rhode Island. Following the conspiracy of *La Escalera* in 1844 the military authorities dealt ruthlessly with his slaves. Some were murdered and many tortured, at a time when England was trying to impose on Spain the implementation of the anti-trafficking treaty of 1835. To that end in 1842 it appointed the abolitionist David Turnbull, British consul since late 1840, as superintendent of freed African slaves. Turnbull was accused of inciting the slaves to revolt, while slaves and freed blacks were brutally suppressed.[76] Phinney died around 1852, and a little later his widow was asking Moses Taylor of New York for $50,000 to finance the harvest.[77] The estate had no mortgage at the time, which meant presumably that Phinney had cancelled the debt with Reid, Irving & Co.

Other bankers with business in Cuba, such as the Rothschilds, also experienced setbacks. In the spring of 1837 Charles C. Tolme,[78] a British trader of unquestioned solvency who acted as British consul and whom the Rothschilds had as their agent, suspended payments. Tolme's creditors agreed that he should continue operating his establishment in Cárdenas and the *St George* estate under the supervision of a liquidating committee, which would receive profits and pay dividends to creditors while the estate was not sold.[79] The merchant Francisco de Goyri y Beazcoechea, the Rothschilds' agent in Cuba since 1837, had been charged with the task of collecting in Havana bills issued by the Spanish government to Weisweller, Rothschild's agent in Madrid, to secure the bank's advances to the Spanish Treasury in the metropolitan capital. Goyri acted as the Rothchilds' attorney in the Tolme liquidation, and Tolme himself joined Goyri & Co., which assumed all its business, receiving all accounts and properties held from third parties. The firm pledged to repay its new partner's debts with the commissions produced by transactions undertaken on the Rothschilds' account. In 1840, that sum reached $6,828.[80] For years Goyri sent the corresponding dividends from the *St George* estate's production. By 1848, ten years after Tolme's suspension, the Rothschilds had yet to recover their money.[81]

Burnham & Co. and the loss of the old markets

L. Mariátegui & Co. was formed in 1840, following Knight's bankruptcy. All orders and merchandise consigned to him were transferred to the new firm. Among his partners was R. C. Hogan, a North American merchant who had been established in Matanzas for years, and who acted as a Barings agent before leaving his business to Mariátegui upon his retirement to the US, where he was a partner in the prominent import firm Hogan and Miln.[82] The second partner was Auguste Guillaumin, a brother-in-law of Knight, who had been his agent in Europe. The third was Gonzalo Alfonso y Poey, son of a former partner of Mariátegui, Knight & Co., who had contributed capital of $5,000 to the assets of the company, as well as the benefit of his close family connections. Luis Mariátegui had become one of the most respected merchants of Havana. He presided over the *Junta de Comercio*, the Trade Council, and was elected in 1845 as a member of the board of *Real Consulado y Junta de Fomento*, the Royal Consulate and Public Works Council.[83] He was then an estate and slave owner and a frequenter of Havana's elite social circles, and in 1854 he became a Member of Parliament in Madrid.

Eager to resume relations with Barings, Mariátegui wrote to the firm explaining that commission business was very lucrative at that moment, and that Knight's problems had been related to his estates and planters, rather than with his mercantile business.[84] L. Mariátegui & Co. enjoyed a wide range of facilities: a shipping credit to draw against consignments of a certain number of shillings per hundredweight, the rate fixed by Barings and variable in line with the conditions in the European market; a blank credit of £5,000 to be covered by consignments or bills of exchange, in the latter case paying the usual banking commission; a credit of £5,000 at the disposal of the partnership's agent in Europe against British manufactures consigned to Mariátegui; another of £3,000 at the disposal of the agent in Buenos Aires for shipments of jerked beef; and £2,000 for the agent in Hamburg against consignments of German goods for Mariátegui.

L. Mariátegui & Co. dissolved itself in the summer of 1845,[85] and was succeeded by James C. Burnham & Co. This partnership included Mariátegui and Guillaumin, who were joined by Burnham, who had made his career remarkably quickly in North America.[86] By 1843 he had already become a formidable rival to Drake, whose administrator observed that when it came to consignments, Burnham took everything presented to him. In autumn 1845 Mariátegui travelled to London to meet with Bates. He wished to raise the blank credit to £10,000, at least from January to August, which would enable the new firm to issue drafts during the shipping season, when demand was high.[87]

James C. Burnham & Co. must have conducted a very important volume of business.[88] In 1850 it had transactions with Moses Taylor, valued at $180,000. The company had doubled Drake's operations, and became the wealthy North American merchant's premier Cuban client.[89] It was to Burnham that the Baron Edmond de Rothschild referred in 1849, in the letter quoted above. Even if business with Barings is ignored, it was precisely from 1846 when, following the liberalisation of tariffs for sugar imports, that British imports from Cuba rose from 197,460 hundredweight in 1845, to 875,420 in 1847, and 946,826 in 1852.[90] In 1844 Parliament had passed a law increasing duties on slave-grown sugar and reducing them for sugars produced by free workers.[91] Free-trade arguments soon prevailed, however: under a law of August 1846 duties on slave-grown sugar were gradually and annually to decrease until 1851, when duties on all sugars were to be equalised, whatever their origin.[92] West Indian planters had lost their influence, and after abolition saw production fall from 195,000 tonnes in 1834 to 118,000 in 1840.

The partnership agreement ended in September 1850. Mariátegui retired to Europe; Guillaumin was already living in Paris.[93] A new commercial partnership was formed which continued as James C. Burnham & Co. at least until Burnham's death in February 1881. Burnham hastened to consolidate ties with Barings, assuring the firm that his capital was not reduced.[94] Barings maintained close ties, and entrusted the firm with all its interests in Cuba. In 1853 Burnham expended an open credit line of £60,000, with freedom to make use of £10,000 before sending any shipping documents, the remainder to be covered with shipments or remittances within 60 days.[95] In May 1861 the facility was extended to £100,000 against shipments consigned to Barings. Over all those years the prolific correspondence between the two houses refers to prices, shipments, sugar insurance policies, and exchange rates.

It was in those middle years of the century that Barings ventured into a limited relationship with Cuban railroads. The Havana–Güines railroad was built in 1838 under the auspices of the *Real Junta de Fomento y Agricultura*, the Royal Public Works and Agriculture Council, with a loan of £450,000 issued in London by Alexander Robertson. The funds were in part transferred to Cuba through Barings, thanks to the deep relationships the bankers had established in Cuba.[96] In 1842 a group of estate owners led by Gonzalo Alfonso bought the railway from the *Real Junta de Fomento*, and they created the *Compañía de Caminos de Hierro de la Habana*, which took charge of the British debt.[97] Alfonso proposed, and Barings agreed, that he would become the firm's agents in London in charge of dividend payments. Nonetheless a few years later the merchant bank refused to place a public loan for the railway company to cover the final instalment of the debt, which was due in January 1860. Instead Barings negotiated with bondholders the conversion of their six per cent instruments into new ones offering seven per cent return, and £107,150 of the debt of £123,000 was converted. The difference was anticipated by Barings, which undertook to service the new bonds for a commission of one per cent of the dividends paid, and 0.5% of the capital reimbursed. The advance would be repaid in April through an annual interest charge of seven per cent.[98]

In January 1868 Barings had ordered from Burnham 5,000 boxes of sugar on their joint account.[99] This is the last shipment recorded. Correspondence with Burnham, practically their sole agent since 1845, ceases at this time. Certainly the outbreak of the war of independence that began in Cuba at the end of that year created a difficult situation for Barings. Over the previous two years the bank had been providing the planter Miguel Aldama with credit against sugar cargoes.[100] In July 1867 it had loaned him £8,000 to buy a new sugar estate. A little later Aldama embraced the cause of independence, and early in 1869 colonial authorities seized all his properties. As creditors, Barings made a claim in Havana, but without much success. The matter was taken to the authorities in Madrid in November 1871, following a British minister's mediation, but in 1872 Barings had yet to recover the investment, and it may have taken several years more.[101]

Still, in the late 1860s it was not war that was responsible for the decline of the formerly intense relationships which had existed between Barings and the Cuban sugar merchants. Nor were Barings unique in this respect; the business of other British merchant bankers was also in decline.[102] For example, the number of acceptance credits for clients of Schroders fell from 53 in the period 1848–68 to 10 in 1869–94. Likewise the accounts that Kleinwort and Sons had maintained with Cuban exporters disappeared in the 1870s, in keeping with a drastic fall in Cuba's importance in the firm's commission business.[103] By this time European markets were completely dominated by sugar beet, and consumption of Cuban sugar had collapsed. In 1864 the Spanish colony exported to Britain 22.52% of its sugars; by 1877 the share had fallen to 4.4%. In those years North American imports rose from 35.94% to 82%.

The emergence of new commercial methods for buying sugar was another important factor in these developments. From the 1860s onward a string of changes fundamentally modified the nature of Cuban sugar export channels, and drastically reduced the demand for the services of firms that specialised in accepting bills of exchange. The extension of the transatlantic telegraph in 1886 proved to be a fatal blow to the traditional sales system based on consignments and advance payments. North American refiners used the telegraph to buy Cuban products, and sent their agents to the island. They came to direct agreements with the planters with firm offers, which reduced drastically the number of commission merchants and the consignment business.[104]

Barings, credit flows, and a century of Cuban sugar

Spanish Caribbean colonies gradually became integrated into Atlantic World commerce as the system of imperial protection linking trade traffic with Seville and other ports of mainland Spain gave way to the need to supply the colonies, and to the impossibility of metropolitan consumption and processing of their products. Trade with North America and Britain had effectively opened before the liberalisation decree of 1818.

Barings had begun with a hesitant involvement in the Cuban sugar trade in the early 1810s, when sugar production was expanding in Cuba and demand in both Europe and the US was increasing rapidly. It moved towards its most intensive period of activity in the middle decades of the century, until its connections with the island disappear entirely around 1870, amid changing conditions in the world sugar market, when beet was providing more than 36% of sugar consumed, effectively closing European markets to Cuban sugar. Throughout those years Barings acted as an acceptance house, such that its financial activity was mostly trade-related, and contributed, through the flow of trade credit it provided, to the opening up of markets for Cuban production. This was achieved through the creation of a solid network between the bank's Cuban agents, its North American correspondents, and others located in different parts of northern Europe and Russia. Merchants' short-term loans, secured by mortgages on land and crops, were the main methods used by planters to obtain liquidity, but Barings only occasionally loaned directly to planters, and then in only very particular cases, such as to Alfonso and Aldama.

In spite of the volume of trade undertaken in Cuba Barings operated through agents, and did not establish a branch.[105] Its correspondence reveals contacts with a very small number of local sugar merchants, in contrast to other bankers such as Schroders. The four key commercial firms with which Barings conducted most of its Cuban business were closely related, and had become intertwined over time due to continuity among some of the partners. Barings' correspondents were few, and all were people of proven reputation who attracted the highest confidence, based on very long relationships. To minimise agency risks, Barings sought to limit its links with other bankers, even if this was not always possible.

Providing short-term credit was obviously a complex business, and was not without risks. As a result of the 1837 crisis and the bankruptcy of one of its agents, Barings suffered considerable losses, and came into possession of plantations lodged as security for debtors' mortgages. This was no simple process, because they were slave estates at a time when slavery had been abolished in the West Indies, and the slave trade was the subject of a vigorous challenge by British abolitionists. Sugar cultivation under the administration of front men proved to be a complex task, and in many years the bank faced enormous difficulties in recovering debts. The story of Barings in Cuba, and its few but deep and important relationships with local business partners and others, shows clearly how the risks and challenges of the Atlantic World affected the business of all in the network, and how the bankers of Bishopsgate faced them, with their business partners and rivals, day to day.

Notes

1. Knight, Franklin W. and Peggy K. Liss (eds): *Atlantic port cities: economy, culture and society in the Atlantic World, 1650–1850*, Knoxville: The University of Tennessee Press, 1991.
2. Perkins, Edwin J.: 'The Anglo-American houses in nineteenth century', in Cassis, Youssef and Cottrell, Philip L. (eds): *The world of private banking*, Burlington: Ashgate, 2009, p. 112.
3. Platt, D.C.M.: *Latin America and British trade*, London: A. & C. Black Ltd., 1972; Jenks, Leland H.: *The migration of British capital to 1875*, London: Thomas Nelson and Sons, 1963. For an earlier period, Burne Goebel, Dorothy: 'British trade to the Spanish colonies', *The American Historical Review*, Vol. 43, No. 2 (1938), pp. 288–320.

4. Curry-Machado, Jonathan: 'Running from Albion: migration from the British isle in the nineteenth century', *International Journal of Cuban Studies*, Vol. 2, No. 1 (June, 2009), pp. 25–36; *Cuban sugar industry: transnational networks and engineering migrants in mid-nineteenth century Cuba*, New York: Palgrave Macmillan, 2011.
5. Platt, *Latin America*, p. 289. See Zanetti, Oscar and García, Alejandro: *Caminos para el azúcar*, Havana: Editorial de Ciencias Sociales, 1987; Roberts, Richard: *Schroders, merchants & bankers*, Basingstoke and London: Macmillan, 1992.
6. Evans, Chris: *Slave Wales: The Welsh and Atlantic slavery, 1660–1850*, Cardiff: University of Wales Press, 2010, pp. 115–29; Roldán de Montaud, Inés: 'The Cuban copper cycle in the nineteenth century, 1830–1868', in Claughton, Peter and Mills, Catherine (eds), *Mining perspectives: proceedings of the 8th Mining History Congress*, Truro: Cornwall and West Devon Mining Landscape World Heritage Site, Cornwall Council, 2011, pp. 203–16.
7. Stubbs, Jean: *Tobacco on the periphery: a case study in Cuban labour history, 1860–1958*, Cambridge: Cambridge University Press, 1985.
8. Platt, *Latin America*, p. 289; Cobiellas, Michael: 'British economic presence in Havana, 1900–1930', *International Journal of Cuban Studies*, vol. 2, no. 1 (2009), p. 49.
9. Rothschild Archive London (RA), T8/125, 14.01.1849.
10. Jenks, *The migration*, p. 33 ff.; Wechsberg, Joseph: *The merchant bankers*, Boston: Little, Brown 1967, cap. III; Hidy, Ralf W.: *The House of Baring in American trade and finance: English merchant bankers at work 1763–1865*, Cambridge: Harvard University Press, 1949.
11. Orbell, John: *Baring Brothers & Co. Limited: A History to 1939*, London: Baring Brothers & Co., 1985.
12. Labouchere, G: 'L'annexion de la Louisiane aux États-Unis et les maisons Hope et Baring', *Revue d'histoire diplomatique*, Vol. XXX (1916), pp. 223–450; Hidy, R.H.: *The House of Baring in American trade and finance: English merchant bankers at work 1763–1865*, Cambridge: Harvard University Press, 1949, pp. 33–4.
13. In the 1940s they acted as financial agents of the Spanish government, disputing the Rothschild monopoly. Roldán de Montaud, Inés: 'Los intereses de los banqueros británicos en España: La Banca Baring y las minas de Almadén', *Hispania*, Vol. LXIII/1, 210 (2003), pp. 255–93.
14. Baring Brothers Archive (London) (BA), *General Ledgers, American and Colonial* (GL) 18.323/20, p. 261. In ledger for 1811–13, no Cuban accounts are recorded. For the use of this source, Orbell, John: 'Private banks and international finance in the light of the archives of Baring Brothers', in Cassis and Cottrell, *The World*, p. 146; Diaper, Stephanie: *The history of Kleinwort, Sons & Co. in merchant banking, 1855–1961*, PhD thesis, University of Nottingham, 1983, p. 27.
15. Moreno Fraginals, Manuel: *El ingenio: complejo económico social cubano del azúcar*, Havana: Editorial de Ciencias Sociales, 1878, Vol. III, p. 43.
16. BA.GL.18.325/18, p. 210; Became agents of NMR & Sons in 1838, RA.XI/38/106.
17. BA.GL.18.325/31, pp. 207, 209, 211, 212; 18.325/33, pp. 263–4.
18. Moreno Fraginals: *El ingenio*, vol. II, p. 158.

19. British mercantile communities in other places are better known. See, Blinn Reber, Vera: *British mercantile houses in Buenos Aires: 1810–1880*, Cambridge: Harvard University Press, 1979.
20. Dávila Cox, Emma A.: *Este inmenso comercio: las relaciones mercantiles entre Puerto Rico y Gran Bretaña, 1844–1898*, San Juan: Instituto de Cultura Puertorriqueña, Universidad de Puerto Rico, 1996, pp. 160–7.
21. http://www.baringarchive.org.uk/barings_people/biographies/joshua_bates
22. Thomas Wren Ward's papers in the Massachusetts Historical Society, www.masshist.org/findingaids/doc.cfm?fa=fa0013&hi=on&within=1&query=baring&submit=Search#firstmatch
23. Orbell, *Baring*, pp. 30–5; Ziegler, Philip: *The sixth great* power, London: Collins, 1988.
24. Otazu, Alfonso de: *Los Rothschild y sus socios españoles*, Madrid: O. Hs Ediciones, 1987, p. 181; Roldán de Montaud, *Los intereses*, pp. 260–2.
25. Roberts, *Schroders*, p. 59.
26. In 1840 they refused to issue a loan for the group Aldama-Alfonso, and again in 1850, Zannetti, *Caminos*, p. 51; Hidy, *The House*, p. 413.
27. ANC, *Tribunal de Comercio*, 308/13. He had been a partner of Miralles, Knight & Co. in 1823, ANC, Escribanía de Daumy, 804/5.
28. ANC, Escribanía de Gobierno, 482/13, ab intestato of Juan José Mariátegui.
29. *Guía constitucional de forasteros de la isla de Cuba, y calendario manual para el año 1823*, Havana: Arazoza y Soler, 1823, p. 115.
30. BA.HC4.6.2, Balance sheet of Mariátegui, Knight & Co.
31. Humboldt, Alexander von: *The island of Cuba*, New York: Derby & Jackson, 1856, p. 253.
32. Madden, Richard R.: *The island of Cuba: its resources, progress, and prospects, considered in relation especially to the influence of its prosperity on the interests of the British West India Colonies*, London: Partridge and Oakey, 1853, p. 61.
33. BA.HC4.6.2, Knight to BB, 7.04.1835.
34. Ely, Roland T.: *Cuando reinaba su majestad el Azúcar*, La Habana: Imagen Contemporánea, 2001 (1963), pp. 277–8. Bergad, Laird W: *Cuban rural society in the nineteenth century: social and economic history of monoculture in Matanzas*, Princeton: Princeton University Press, 1990, p. 113.
35. BA.GL.18.325/31, p. 185. The correspondence of Knight to Bates goes back to January 1826, BA.HC5.1.31.
36. BA.HC4.6.2, Knight to BB, 20.11.1828; Kirchner, Walther: *Studies in Russian-American commerce: 1820–1860*, Leiden: Brill, 1975, pp. 189 and 221.
37. BA.HC4.6.2, Knight to Bates, 23.10.1831; BA.LB.18, pp. 757–8, 769, 844.
38. BA.HC4.6.2, 6.11.1833.
39. Turnbull, David: *Travels in the West. Cuba, with notices of Porto Rico and the slave trade*, reprinted, New York: Negro University Press, 1969, p. 154 [first ed. 1840].
40. BA.GL.18.325/34, p. 922, and 18325/36, pp. 757–8.
41. Archivo Nacional de Cuba (ANC), Tribunal de Comercio, 284/10.
42. RA.XI/112/36A, Goiry to Rothschild, 6.02.1836; BA.HC4.6.2, 6.02.1836; ANC, Escribanía de Gobierno, 482/13, ab intestato of Juan José Mariátegui.
43. BA.HC4.6.2, Knight to BB, 7.01.1837, and ANC, Escribanía de varios, 887/15707.
44. Cantero, J. Germán: *El libro de los ingenios: colección de vistas de los principales ingenios de azúcar de la isla de Cuba: dedicado a la Real Junta de Fomento*, Havana: Luis Marquier, 1857.

45. BA.HC3.51.2, Alfonso to BB, 7.06.1837.
46. Deerr, Noël: *The history of sugar*, London: Chapman and Hall, 1952, Vol. 2, p. 531.
47. BA.HC4.6.2, 8.01.1836.
48. ANC, Escribanía de Gobierno, 786/4, Knight & Co. in liquidation to Israel Thorndike.
49. Hidy, *The House*, p. 248; Orbell, *Baring*, p. 31.
50. BA.HC4.6.2, Estimate of income handed to M. Bates in 1838.
51. BA.HC4.6.2, Statements of sugars shipped by Messr[es] Geo. Knight.
52. BA.HC4.6.1, Knight to BB, 12.05.1840 and 22.08.1840.
53. BA.HC4.6.2, Mariátegui to BB, 26.05.1840.
54. Kynaston, David: *The City of London, vol. 1: A world of its own, 1815–1890*, London: Chatto & Windus, 1994, p. 157; Hidy, *The House*, p. 375. He led the opposition against the Whig policy of equalising duties on sugar.
55. BA.HC5.2.30, Ward to BB, 30.09.1840, and Knight to Ward, 22.08.1840; BB.HC4.6.2, 26.05.1840; ANC, Escribanía de varios, 887/15707 for Knight's bankruptcy and his probate.
56. Ziegler, *The sixth*, p. 375; Chapman, *The rise*, p. 41.
57. BA.HC4.6.2, Minutes of the meeting of 17 May 1840. On the bankruptcy also RA.XI/38/125A, Goyri to Rothschild, 18.05.1840; ANC, Escribanía de Hacienda, 269/9787, and Escribanía de Gobierno, 786/4–5.
58. Such as an important amount of shares of the Improvement Bank of New Orleans, BA.HC4.6.2, Mariátegui to BB, 26.05.1840.
59. BA.HC5.2.30, 25.09.1840, Ward to BB. DeConnick, Spalding & Co., was established in Matanzas by 1836, when they were correspondents of Rothschild, RA.XI/112/36A. They were like Fesser or Drake of best reputation, according to Barings informant Bierly.
60. Deerr, *The history*, vol. II, p. 531.
61. RA.XI/38/125A, Goyri to Rothschild, 22.08.1840.
62. BA.HC4.6.2, Mariátegui to BB, 30.04.1842.
63. RA.XI/38/125B, Goyri to Rothshild, 4.06.1840.
64. BA.LB, 12, BB to DeConnick, Spalding & Co., 19.12.1842.
65. Turnbull, *Travels*, p. 92.
66. BA.HC3.6.2, Reid Irving to BB, 28.04.1843; BA.HC5.7.8, Purton to BB, 05.04.1843.
67. The partners were Charles W. Storey, a merchant from Massachusetts, partner of James Drake in 1822, and Edward Spalding. When he died in 1845, the firm was succeeded by Adot, Spalding & Co., formed by Edward Spalding and José Vicente Adot. They were Barings agents, at least until Adot died in 1865. BB.HC.16/1845–1849, Reports on business houses, part II, Bierley to BB, May 1845. At the time they were the most important commercial firm in Havana, Ely, *Cuando reinaba*, pp. 209 and 343.
68. Riverend, Julio: *Historia económica de Cuba*, La Habana: Editorial de Ciencias Sociales, 1985, p. 391.
69. BA.LB.18, 1.09.1848; 01.01.1848.
70. BA.HC4.6.10, Storey to BB, 7.06.1851 and 26.04.1856.
71. BA.LB.20, 30.07.1850; LB.21, 1.11.1850, 16.12.1850, 1.03.1850 and 1.05.1850. Also HC4.6.8, Burnham to BB, 22.03.1850, 25.06.1850, 25.12.1850 and 22.01.1851. In March 1847 their credit against the estate rose to $32,690.

72. BA.HC4.6.10, Storey & Spalding to Baring, 30.09.1850, and BA.HC4.6.8, De Pestre to Burnham, 1.6.1850. On Bayley, see Bergad, *Cuban rural*, p. 171.
73. BA.HC4.6.8, Burnham to BB, 22.01.1851; De Pestre to Burnham, 15.09.1850.
74. BA.HC4.6.8, Burnham to BB, 14.12.1851, 26.02.1852, 29.11.1852.
75. BA.HC4.6.8, Burnham to BB, 7.05.1854 and 10.01.1857.
76. Bergad, *Cuban rural*, p. 242.
77. Ely, *Cuando reinaba*, pp. 186–7. Paquette, Robert L.: *Sugar is made with blood: the Conspiracy of la Escalera and the conflict between Empires over slavery in Cuba*, Middletown: Wesleyan University Press, 1990.
78. RA.XI/38/1250, circular de la casa Tolme, 30.10.1837. Martínez y Fernández, Luis: *Fighting slavery in the Caribbean: The life and times of a British family in the nineteenth century Havana*, New York: M. E. Sharpe, 1998, p. 27.
79. RA.XI/38/125A,Goyri to NMR, 18.12.1839; Circular of Tolme of Havana, 30.12.18380.
80. RA.XI/38/125A, Goyri to NMR, 30.0.1841.
81. RA.XI/38/125B, 10.07.1848. That year he remitted $3,789 from Tolme's estate.
82. Ely, *Cuando reinaba*, p. 145. On Luis Mariátegui & Co., ANC, Escribanías, 109/2; BAHC4.6.4, Hogan to Bates, 25.04.1830; Mariátegui and Guillaumin to BB, 26.05.1840, 21.06.1840 and 25.09.1840.
83. ANC, *Real Consulado y Junta de Fomento*, 10/702.
84. BA.HC4.6.2, Guillaumin to BB, 2.06.1840.
85. Archivo Histórico de Protocolos Notariales in Madrid, 25.567, Paris series; ANC, *Real Consulado y Junta de Fomento*, 10/702.
86. BA.HC4.6.2, Knight to Bates, 5.6.1832.
87. BBA.HC4.6.2, Mariátegui to BB, 8.10.1845.
88. Moreno Fraginals, *El ingenio*, Vol. III, p. 70.
89. Ely, *Cuando reinaba*, pp. 278–9.
90. *Return of the quantities of sugar imported and entered for home consumption, with rate of duty charged and revenue there, from 1793 to 1849, inclusive, ordered by the House of Commons*, 1854, pp. 20–1; Moreno Fraginals, *El ingenio*, Vol. II, pp. 160 and 209.
91. *Correspondence respecting the sugar of Cuba and Puerto Rico presented to both Houses of Parliament*, London: T.R. Harrison, 1845, n. p.
92. Deerr, *The history*, p. 431; Green, William A: *British slave emancipation: the sugar colonies and the great experiment, 1830–1865*, Oxford: Clarendon Press, 1976, p. 228.
93. BA.LB.21, BB to Spalding, 18.09.1850.
94. BA.HC4.6.8, Burnham to BB, 5.09.1850.
95. BA.HC4.6.8, Burnham to BB, 21.01.1853.
96. BA.Spanish Ledgers, 18.336/2; RAL.XI/38/125A.
97. Friedlaender, Heinrich: *Historia económica de Cuba*, Havana: Editorial de Ciencias Sociales, 1978, vol. I, p. 298.
98. BA.Save, No. 45, J.A. Echevarría to BB, 21.11.1859; BA.HC4.6.11, 25.11.1859.
99. BA.LB 40, BB to Burnham, 15.01.1868.
100. BA.LB.38, p. 427.
101. BA.HC4.6.8, Burnham to BB, 23.05.1867 and 6.07.867; Archivo Histórico Nacional (Madrid), Ultramar, 754/27, Minister of Great Britain to Secretary

of Estate, 5.10.1871; BA.HC5.2.30, Ward to BB, 24.12.1875, mentions Aldama's debt of $45,718.

102. Bergad, *Cuban rural society*, p. 170.
103. Roberts, *Schroders*, pp. 50 and 96.
104. Roberts, *Schroders*, p. 98; Diaper, *The history*, pp. 150–1 and 187; Brown, *Financing*, p. 69.
105. In Matanzas existed a firm named James Baring & Co. The main partner was a cousin of Thomas Baring, but they were never Baring's agents, BB.HC.4.6.12.

12

Circuits of Knowledge: Foreign Technology and Transnational Expertise in Nineteenth-century Cuba

David Pretel and Nadia Fernández-de-Pinedo[1]

European and American industrial nations and the slave plantations of the Spanish Caribbean were not opposing economic systems during the nineteenth century; rather they were deeply entwined: the 'sugar industrial revolution' occurring on Cuban plantations coincided with the industrialisation of the Atlantic World economy.[2] During the central decades of the century Cuba experienced a period in which agrarian capitalism expanded, and creole planters managed to transform their small-scale slave plantations into large agro-industrial complexes. Cuba emerged as an advanced agro-industrial region where planters, sugar masters, and prominent businessmen embraced the latest technical innovations and participated in transnational networks of commercial and knowledge exchange. It became the largest sugar producer in the Atlantic World.

This chapter treats the technological 'revolution' that occurred on Cuban plantations during the nineteenth century as essentially part of the systemic transformation of the Atlantic World economy. The primary motivation for adopting this macrohistorical perspective is to understand the interconnections between the Cuban economy and the historical processes of the greater Atlantic World.[3] The intensified integration, from the late eighteenth century, of Cuba into the Atlantic World economy can be explained by, on the one hand, the expansion of sugar consumption in the industrial world and, on the other, the slave revolution in Haiti that opened up opportunities for other countries to take over its position in the world sugar market.[4] This expansion of the sugar trade, along with the technological advancement in mechanical engineering and chemical improvements in sugar production, also led to the consolidation of transnational circuits for the transfer of sugar technologies to Cuba – circuits that involved the participation of a variety of intermediaries and legal and technical experts.

The first part of the chapter places the nineteenth-century Cuban plantation economy within the context of the technological globalisation that stimulated the circulation of technology and expertise in the Atlantic World economy. The second examines the activities of foreign firms and experts in nineteenth-century Cuba, in areas such as the commercialisation of technology and patenting. It presents some explicit examples of the circulation of technical expertise between Cuba and advanced industrial nations, particularly the United States, France, and Britain. The chapter ends with a detailed examination of the transnational operations of the French machinery manufacturer Derosne & Cail. It was one of the most innovative engineering firms of the mid-nineteenth century, and one of the first European companies to supply advanced technology to the Caribbean sugar industry. It examines the company's strategies for commercialising steam technologies (such as the use of foreign agents, legal and technical experts, and offices) for the Cuban sugar industry between 1812 and 1898, and its management of intellectual property rights vis-à-vis sugar technology, and the relationship of this firm to Cuban sugar planters.

The technological transformation of Cuban plantations

In contrast to the extensive literature on technology and colonialism in the nineteenth century in the British and French worlds, the networks and circuits of technological exchange in the Spanish Caribbean have been relatively little studied.[5] The scant literature on the technological changes within the nineteenth-century Cuban plantation economy has focused primarily on the relationship between technical improvements and slave labour. One of the most important recent developments in Caribbean history has been the recognition of the influential role played by Cuba's creole elite and the experts who migrated to the island in promoting the technological modernisation of the Cuban sugar industry. Among others, recent works by Jonathan Curry-Machado, Reinaldo Funes, Daniel Rood, and Dale Tomich have shown how modern machinery and organisational innovation were disseminated. These authors explored the implication of these factors for Cuba's role in the Atlantic economy and their scholarship has shown how European and American circuits of technological exchange expanded to include even the Spanish Caribbean.[6]

During the period 1815 to 1868 the Cuban plantation economy underwent its first remarkable transformation. In the words of Moreno Fraginals, a 'jump from manufacture to big industry', a sort of 'sugar industrial revolution' occurred.[7] The Cuban sugarcane industry became, from the mid-nineteenth century onward, a modern tropical enterprise. By 1868 Cuban sugar mills were producing nearly 30% of the total world market of this commodity.[8] The modernisation and industrialisation of sugar production cannot be explained solely by factors such as the expansion of the island's sugar frontier, its fertile soil, its or ideal climate.[9] Nor can these changes be explained by the sugar industry's extensive use of coercive labour until the abolition of slavery in Cuba in 1886. Even more important in the transformation of Cuba's plantation economy were the technological changes and organisational innovations introduced on the island in the nineteenth century. As Dale Tomich asserts, it is 'not an exaggeration to suggest that technical innovation was the condition for the expansion of sugar and slavery in Cuba'.[10]

Given that sugar was a uniquely tropical commodity, and that metropolitan Spain lacked refineries with which to process sugarcane, Cuban planters processed the cane at plantation sites on an industrial scale. Through the 1870s this on-site refining gave Cuban sugar producers competitive advantage in terms of both the quality of their product and the ease of distribution. The on-site processing may also explain why the colony's sugar industry and auxiliary sectors, such as railroads and steamships, experienced such a surge in innovation during this period.[11] By the mid-1860s Cuba had 1,200 kilometers of railways, and plantations on the western part of the island had become advanced sites of technological and chemical experimentation.[12] The new technologies were introduced on Cuban plantations during the 1840s and 1850s despite a lack of strong incentives for introducing labour-saving innovations. Mechanical and chemical innovations, along with the early introduction of the railway, radically transformed the production level, productivity, and scale of Cuba's sugar industry.

The architects of the first technological transformation of the Cuban sugar industry were the creole 'sugarocracy'. During this period these sugar planters acted as the chief promoters of technology transfer, establishing agreements and partnerships with foreign inventors and mechanical manufacturers. Cuba's mid-nineteenth-century creole elite relied on formal colonial institutions such as the *Real Consulado de Agricultura, Industria y Comercio*, the *Junta de Fomento*, and especially the *Sociedad Económica*. These institutions were devoted to fostering the modernisation of colonial industries.[13] Although still constrained by political and legal ties with a declining metropolis, these colonial institutions were already controlled by creole economic elites that used them as forums which favoured cooperative efforts among planters intended to develop Cuba's sugar industry.[14]

The support provided through this institutional infrastructure and the administrative networks set up by the creole elite were crucial in encouraging technology transfer to the island and its developing sugar industry. This support allowed sugar-mill owners to participate actively in the global circuits of technological exchange that smoothed away any obstacles presented by the Spanish metropolitan economy, and enabled Cuba to obtain the technology it needed. These institutions adopted a variety of policies to promote industrial advancement, such as the establishment of commissions to study foreign technological progress, the creation of research laboratories, the establishment of advanced botanical gardens, and the proliferation of scientific and technical societies. As early as 1794, the sugar planter Francisco de Arango y Parraño traveled to Europe to learn more about how the steam engine could enhance sugar production. Many other Cuban planters and scientists would tour Europe during the next decades in search of technological improvements.[15]

Planters' increasing propensity to invest in capital-intensive technology should also be seen as part of a broader trend of optimism about the virtues of technical advancement and industrial modernisation. To remain innovative, some planters went heavily into debt – not only because the new technology was expensive, but also because of the additional costs of maintenance, repairs, and the salaries of foreign machinists. For planters the investment in new technology was not only an economic opportunity, but also an indication of status and wealth.[16] Rich planters such as Julian de Zulueta and Antonio Parejo, who had grown wealthy through the slave trade, were among those who demonstrated the greatest willingness to invest in very expensive machinery such as the Derosne vacuum pan. Cultural motivations as well as economic ones lay behind planters' attraction to cutting-edge technology; it is not clear that they experienced strong pressure to introduce labour-saving innovations. As Franklin Knight puts it, cooperation among wealthy planters operated alongside 'friendly competition' which fostered positive attitudes toward technological progress.[17]

The dramatic expansion of Cuba's sugar industry also transformed the patterns of supply of credit, machinery, and labour to the island.[18] In this context the political enterprise of reshaping Cuba's territory into a productive sugar 'factory' became a central component of export-led Cuban economic development. Botanical and agricultural sciences were tools which enabled this transformation and commoditisation of the

Cuban landscape during the mid-nineteenth century.[19] However, not all planters were seized by the urge to innovate. Some were small-scale operators whose limited resources precluded innovation; others were dubious about the profitability of the new machinery and preferred to diversify than to reinvest their profits in sugar technology.[20] As the work of contemporaries Carlos Rebello, Alvaro Reynoso, and Justo Cantero made clear, what occurred was not the complete replacement of old technologies by new ones, but rather the coexistence of semi-mechanised and mechanised sugar mills.[21]

Cuba's technological requirements and position in the Atlantic World economy were different from those of mainland Spain. The Spanish colonies, as advanced sites of sugar production, required different policies than those of continental Spain. Like other colonial or postcolonial sugar producers, such as the British West Indies, Brazil, Hawaii, and Java, Cuba had to look abroad for its technology. However, a significant – albeit hardly surprising – contrast can be observed in Cuba's case. While in the other current and former colonies the metropolis supplied a significant part of the technology, as well as the capital and experts necessary for its introduction, in the case of Spain's Caribbean colonies the role of the metropolis was comparatively inconsequential, apart from its central role in fiscal and commercial policy.[22]

A study of Cuban trade figures from 1826 to 1863 reveals that the introduction into the colony of new sugar production and railway technologies in Cuba was characterised by well-defined patterns.[23] Figures 12.1 and 12.2 show a sharp increase in the imports of technology and coal, as well as increased sugar exports during the central decades of the century. Table 12.1 shows the countries of origin for the technology imported into Cuba. Sugar and railway technology originated not in Spain but in the more advanced industrialised nations. For example, according to trade figures for the year 1850 all the machinery, equipment, and tools that Cuba imported came from Britain and the US, and was shipped to the island by commercial firms located in these two countries.[24] The following passage from an 1851 edition of the patent journal *Scientific American* neatly sums up the situation: 'Cuba is almost wholly supplied with machinery from the United States. There is in nearly every plantation in Cuba a sugar mill driven by steam engines, built usually in New York or Boston.'[25]

One complementary way to study the degree of technological change in the Cuban plantation economy is through analysis of the available aggregated data on patenting activity. Patent data are only a partial technological proxy, however, and present important methodological

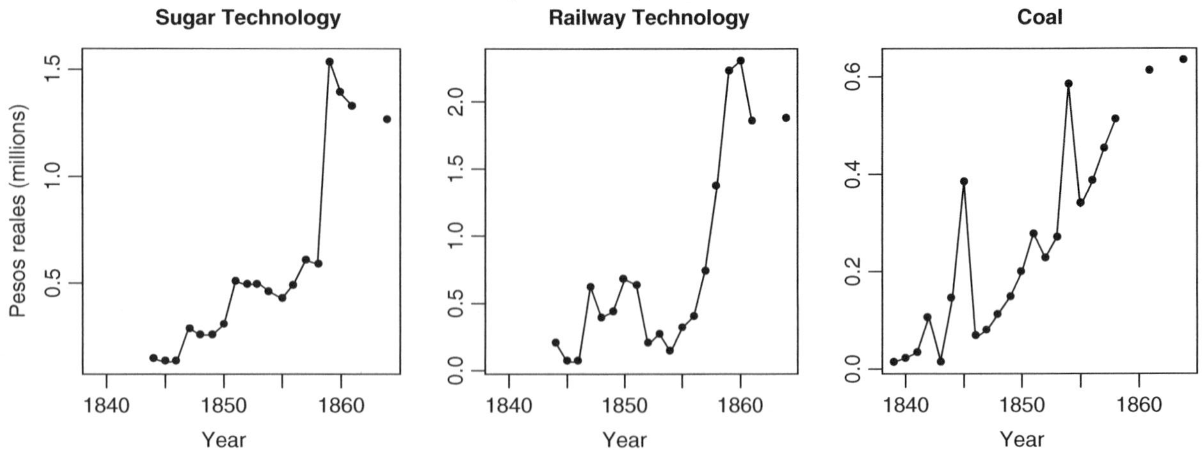

Figure 12.1 Cuban imports of technology and coal, 1839–1863

Table 12.1 Machinery imports for Cuban sugar mills, physical units and country of origin, 1844–1857

	Britain	*US*	*France*	*Germany*	*Denmark*	*Belgium*	*Holland*	*Spain*	*Warehouse*
Distillery (1)		8							
Trapiche (grinding mill) (2)	233	264	3						
Boiling house (3)	9,956	2,420		154	6		3	2	
Boiling house tools (4)	252,434	16,509		120	6,151				4,748
Steam machinery (5)	2,744	1,722	8	13	15	7			
Purging house (6)	166,003	45,318	154	2,069	1,650				23,472

Source: *Balanzas de la Isla de Cuba, 1844-1857.* (1) *aparatos para aguardiente* and *alambiques;* (2) *guijos, trapiches, trompos para guijos, coronas para trapiche, tambores* and *dados para guijos;* (3) *clarificadoras, condensadores, pailas, puertas de fornalla, refinadores, tachos, compuertas,* and *bocas y puertas de fornalla* (According to trade figures, 21,316 British pailas were imported in 1853. The authors think this is a mistake, and have have taken into account only 316 pailas for this year; (4) *repartideras, bombas, bombones, cachimbos, cubos* and *espumaderas;* (5) *calderas, chimeneas, cilindros para máquinas, máquinas de vapor* and *parrillas (juegos).* Parrillas have been counted by sets not by units, as follows: 2,505 sets were imported from Britain, 1,165 from USA, 11 from Germany and 14 from Denmark (6) *filtros, aparatos para purgar, tanques para meladura* and *hormas.*

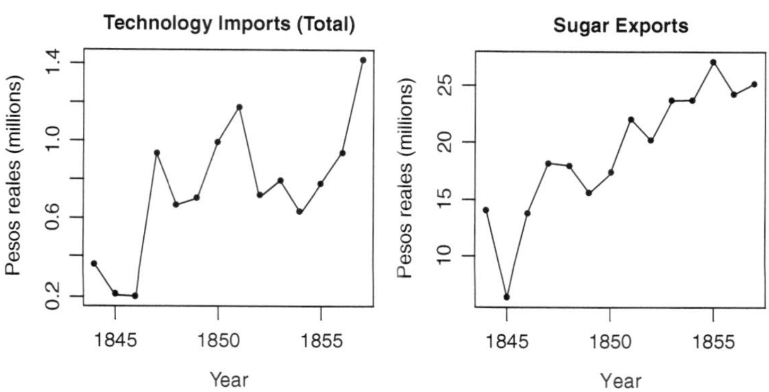

Figure 12.2 Cuban technology imports and sugar exports, 1839–1863

problems. Still, they can reveal degrees of technological change and dependency. It is clear that in the mid-nineteenth century it became routine for planters, machinists, and machinery manufacturers to register their inventions and technical improvements at Havana's patent office. From the 1820s onwards some of the most economically valuable technologies transferred from advanced economies to Cuba – ranging from chemical processes to capital-intensive vacuum pans – were channeled through the patent system.[26] These transfers were carried out through either the metropolitan office in peninsular Spain or, more frequently, the Cuban patent register. The technical information contained in the patents granted at the Cuban register was regularly published in *La Gaceta de La Habana,* a channel through which planters released key technical information.

At least 2185 patents were directly registered in Havana between 1826 and 1898.[27] It seems that technology transfer to Cuba through patenting was most prevalent between the late 1830s and 1868, the beginning of the Ten Years' War (1868–78). The main patenting sectors during this period were sugar improvements, combustibles, and steam technology. The 1850s was the decade with the highest number of patent applications – more than 830. A qualitative study of certain patent files reveals that in cases in which it was necessary to protect creole interests, Cuban administrators managed to deny foreign inventors' and firms' requests for patents on sugar technology. A tacit limitation was placed on the granting of patents of 'introduction' or importation to Cuban residents in an effort to avoid monopolies and foster the exchange of technological knowledge among planters.[28] This strategy, in conjunction with Spain's commercial policy, had the consequence of encouraging the importation of machinery into Cuba.

The last third of the nineteenth century saw the second great technological transformation of the Cuban sugar industry: the transition to the central factory system of production.[29] The new phase began in the 1870s when the Cuban sugar cane industry entered a crisis rooted in, among other things, increasing competition from beet sugar producers, the reduction of freight rates due to the construction of Suez Canal in 1869, improvements in packaging and storage, and the extension of sugar cane plantations to new regions in the Atlantic region and beyond.[30] According to Carlos Marichal, in 1878 world sugar prices began a downward trend, resulting in significant losses for producers.[31] Against this backdrop Cuban mills initiated a process of merging and modernisation. The total number of sugar estates declined significantly, and the Cuban mills became the largest in the world. The change in the size of the industry was closely associated with the introduction of technical and organisational innovations related to the so-called 'second industrial revolution', such as continuous process innovations.[32] The change in the scale of production also had consequences for the location of the industry, namely the expansion of the sugar industry to the eastern part of the island. Although from the 1880s incentives to patent modern technology related to sugar cane exploitation increased, no documentation in Cuba indicates that these incentives actually increased the number of patent applications. However, there was increasing activity surrounding the extension of corporate and 'elite'[33] patent rights from the metropolitan registers in Spain to Cuba during the two last decades of the century.[34]

Engineering firms and transnational expertise

Mid-nineteenth-century Spain could neither provide the necessary technology to Cuban plantations nor serve as a sufficient market for Cuban sugar. Moreover, Cuba did not have a sufficient industrial base from which to supply the technology necessary for the expansion of its plantation economy.[35] In this context foreign machinery manufacturers entered into collaborative relationships with Cuban creole elites, rather than facing competition from indigenous manufacturers. The political and economic elite offered these foreign firms advantageous commercial and material conditions, including institutional change, weak legal enforcement, and infrastructure, to help the imported technology reach Cuban plantations. The early introduction of railroads in Cuba in 1837 had the further effect of stimulating European and American manufacturers to commercialise their technologies on the island, thereby deepening Cuba's technological dependence.[36]

Cuban sugar plantations were one of the largest markets for these foreign machinery makers and engineering firms.[37] Companies based in New York, Paris, Liverpool, and Glasgow became closely interconnected with the sugar elites of the Spanish Caribbean, developing commercial relationships that favoured the transfer of technical innovations to the region. By the late 1830s American and British companies were introducing the overwhelming majority of machinery used on sugar plantations throughout the Spanish Caribbean. The development and growth of mechanical engineering in cities such as Glasgow and New York was directly stimulated by the mechanisation of Caribbean sugar plantations, which demanded a large quantity of tools, equipment, and machinery. Cuban plantations also became sites of incremental innovation, industrial experimentation, and trials for foreign firms.[38]

Engineering firms on the east coast of the US had the advantage of geographical proximity to Cuba, which reduced shipping costs. During the 1840s the New York firm West Point Foundry enjoyed a brief period of dominance on the island, which was soon lost to Merrick & Towne (Philadelphia) and Novelty Iron Works (New York). According to Roland T. Ely, records of the New York merchant banker Moses Taylor show that West Point Foundry lost its market share in sugar mill equipment because it lacked confidence in planters' solvency, and thus refused to accommodate their perpetual need for credit. Meanwhile its reputation suffered among planters who objected to its late deliveries, relatively high prices, and lack of professional services.[39]

In the 1840s steam-engineering companies, such as Britain's Fawcett Preston and the aforementioned Novelty Iron Works and West Point Foundry, were the most important suppliers of sugar machinery to Cuba. Only French firms, such as Derosne et Cail, managed to compete with the British and American companies. Other firms, such as the Glasgow-based Duncan Stewart & Co. and Babcock & Wilcox, along with the French firms Compagnie de Fives Lilles, Société Anonyme des Anciens Établissements Cail, and Frères Brissoneau et Compagnie, would take over during the last decades of the nineteenth century.[40]

Large machinery makers based in Glasgow, such as W. & W. McOnie and Mirless Watson & Co., found a primary market in Cuba, where they signed hundreds of export contracts in the mid-nineteenth century. The order books of Mirless Watson and MacOnie allow reconstruction of the process of international technology transfer to Cuba. These two firms not only sold machinery, tools, and equipment (steam engines, sugar mills, hammers, juice pumps, cane carriers, and so on) to the Cuban sugar plantations, but also provided planters with instructions on how to set up and operate the machinery, sketch plans indicating the arrangement of buildings and the position of machinery, repair services, and testing of new machinery once installed.[41] Some of the planters even asked Mirless Watson to cast the names of both the Cuban plantation and the machinists erecting the new equipment prominently into the fabric of the largest machines they sent to Cuba.

Mirless Watson carried out long-term maintenance of the machinery they sold to Cuban planters, and sent tools, duplicates, and spare parts with each order of equipment. This firm's order books reveal that Cuban plantations relied on the after-sales service provided by foreign suppliers, and on the assistance of foreign technicians to maintain and repair the machinery introduced. For example, in 1874 Mirless Watson sent to San Joaquin sugar estate in Matanzas – through the British mechanical intermediary William Ross & Co. – a new steam spindle that was a duplicate in every respect to the one sent in 1854, some twenty years earlier.[42] Similarly, in 1891 Mirless Watson sent to the Soledad estate in Cienfuegos a new steam cylinder which had been finished to fit the dimensions of the steam engine supplied by the company in 1857.[43]

The study of international technology transfer to Cuba requires attention not only to the activities of foreign machinery makers in the transmission of knowledge, but also to the actors involved in this process. Foreign engineering firms were connected to Cuban plantations through various transnational experts, ranging from machinists to intermediary agents. In the 1840s and 1850s machinery makers and Cuban plantations relied on a highly mobile group of practical experts, notably the migrant engineers and chemists who acted as intermediaries in the transfer of technology.[44] The careers and lives of some of these experts in machinery and sugar production had transatlantic dimensions. According to the registers of machinists that the *Capitanía General* maintained for Matanzas, Santiago, Cardenas, and Cienfuegos, it is clear that hundreds of foreign engineers worked on Cuban plantations in the 1850s, the vast majority from Britain, France, Ireland, and the east coast of the United States.[45] Some of them were to remain in Cuba for the rest of their lives – including the German Karl Rossum, who in 1867 began working at the Central Tinguaro in Matanzas, first as an administrator and then, from 1877 to 1899, as owner.[46]

The Cuban plantation provided many career opportunities for well-qualified individuals. Machinists attained their reputation as experts in sugar technology through personal and professional connections gained after many years of practical work experience in the Cuban sugarfields, as well as in European and American engineering firms. Two good examples of expert migration in the 1850s are the Scottish machinists Edward Beanes and William H. Ross, agents and technical experts for Mirless Watson & Co. and the Liverpool-based firm Fawcett Preston. The two of them had spent many years working on Cuban sugar plantations, finally returning to Liverpool where they acted as intermediary agents for British engineering firms, trading technology in Cuba, first as business partners and, from 1862, independently.[47] Beanes, who had spent more than 25 years working on Cuban plantations,[48] was also a prolific inventor who obtained several patents for technical improvements in the manufacture of sugar, such as a US Patent in 1865 for improvements in the refining of raw sugar by neutralising the acids of the cane juice.[49]

As Curry Machado has shown, these foreign technical experts and creole technocrats acted as 'sub-imperial' agents in the technological transformation of Cuba's sugar industry.[50] Herein the term 'sub-imperial' refers to the island's economic and technological liberation from metropolitan Spain, and its increasing ties with rival empires several decades before it attained political independence in 1898. Cuba was dependent not only on foreign technology, but also on foreign engineers and machinists. It was not until July 1845 that a school for machinists was set up in Havana with the support of the *Sociedad Económica*.[51] This school was established with the explicit objective of providing the country with enough skilled technical experts in steam technologies. The school's charter made clear that Cuba urgently needed machinists trained in theoretical and practical questions to work on the island's sugar plantations, as well as in its railway and steamship sectors.[52] However, the school lacked the modicum of resources it needed to carry out its activities.

The activities of the firm Babcock & Wilcox on the island in the late nineteenth century are noteworthy. From its inception in 1881 the company embraced a global strategy of commercialising technology and the management of property rights through foreign agents. According to Kristine Bruland, between 1881 and 1891 the European branch of this firm, established in Glasgow, sold and installed technology in 44 countries. Its third most important market, after France and Russia, was Cuba, where it signed 88 export contracts over those ten years. The importance of Cuba for the firm is clear. It set up an office in Havana in 1882, directed by Alberto Verastegui, a Spaniard who had originally migrated to New York, where he became director of exports for Babcock & Wilcox, and who subsequently moved to Cuba to work as a commercial agent for the firm.[53] Another example is the partnership between inventor Thomas Edison and Basque businessmen José Francisco Navarro who, in 1881, established in New York the Edison Spanish Colonial Light Company, later renamed The Havana Electric Light Company. This firm had as its declared purpose to 'own, manufacture, sell, operate and licence' technology patented in Cuba.[54]

The role of transnational networks of experts was crucial in colonial patenting. When the transfer of capital-intensive central sugar technology to Cuba began in the 1880s, the number of patent agents and other intermediaries transferring inventions to the island increased decisively. The extension of patent rights granted in metropolitan Spain to the colonial territories was a lucrative activity. Foreign machine and engine manufacturers required agents who were experts in the particularities of Spanish patent regulations and administrative procedures in the colonies. During the last decades of the century, the extension of patents and trademarks issued in metropolitan Spain to *Ultramar* (colonial Spain) was an important activity undertaken by the leading Spanish patent agencies. The three agencies that monopolised this activity in Cuba – Elzaburu-Vizcarrondo, Centro Auxiliar de la Industria and Clarke, Modet & Co. – also happened to be the most active agencies in international patenting in mainland Spain.[55] Machinery manufacturing companies, such as the Glasgow-based Duncan Stewart & Co. and the French firms Compagnie de Fives Lilles, Société Anonyme des Anciens Établissements Cail, and Frères Brissoneau et Compagnie, made extensive use of these agencies to secure their colonial patents. Moreover, many of the foreign machinery firms active in the Spanish patent system, such as Babcock & Wilcox and Société des Constructions Mécaniques Saint Quentin, had branches in Havana during the last decades of the century.[56]

Spanish and Cuban lawyers, businessmen, and planters guided and assisted foreign corporations in registering, publicising, and commercialising their inventions in Cuba. Experts' assistance in preparing patent applications, including mechanical drawings, had become essential as early as 1850. One patent practitioner, who worked intensively for foreign sugar machinery manufacturers including Duncan Stewart and Fives Lille, was the Puerto Rican Julio Vizcarrondo, the first professional agent to work intensively in Spain for foreign firms and inventors such as Thomas Edison and the German steelmaker Krupp. A renowned lawyer based in Madrid from 1863, Vizcarrondo was a prominent liberal politician and journalist, and leader of the Spanish abolitionist movement. Vizcarrondo was educated in San Juan, Paris, the US, and Madrid, where he obtained his law degree from the Universidad Central. Because of his political stand against slavery on Spain's Caribbean plantations, Vizcarrondo was exiled in 1850 to New York City, where he met and married Harriet Brewster, daughter of the American abolitionist Henry Brewster Stanton. Vizcarrondo's international education and four years of exile in New York gave him the ideal personal and professional background to build, late in his life, a sizeable international industrial property agency in Madrid: the Anglo-Spanish General Agency and Commission House, established in 1865.[57] In 1854 he had returned to Puerto Rico where, in addition to his political activities, he worked as an attorney and helped to promote industrial activity on the island through the mercantile publication *El Mercurio*, which he founded in 1857, then edited. Before moving to Madrid in 1863 Vizcarrondo worked mainly as a representative of American and British firms in Puerto Rico and Cuba.[58]

The patenting activity of the firm Duncan Stewart and Co. provides a good example of Vizcarrondo's roles as a transnational expert in 'colonial patents' and an intermediary in technology transfer to Cuba. This machinery manufacturer, based in Glasgow, used the services of Vizcarrondo's agency in several of its patent applications under the Spanish patent system. For instance, in April 1887, Vizcarrondo presented an application – in the Madrid Register – for a patent of introduction for an improved sugar mill.[59] Vizcarrondo supported Duncan Stewart in the patent application process, translated the technical memorandum, and arranged the necessary mechanical drawing services. A year later this agent would also assist Duncan Stewart in officially certifying that the new invention was being used in Cuba in accordance with the legal requirements of the 1878 Spanish patent law, which had been extended to Cuba in 1880.[60] The new mill was set up on the Soledad sugar estate, a large modern 'sugar central' owned by the Boston firm E. Atkins and Company, and one of the first major direct investments by an American firm on the island.[61]

Derosne & Cail in Cuba

An excellent example of the transnational operations of a large engineering firm in Cuba in the mid-nineteenth century is Derosne & Cail. This firm pursued a global strategy through the use of foreign agents and the management of patent rights in different countries. Its roots date back to 1812, when the prominent French chemist and pharmacist Charles Derosne established Ateliers de Constructions Mécaniques in Paris – a company that by 1818 had evolved into Derosne et Cie. In 1836, when the boilermaker Jean François Cail became Derosne's business partner, the firm was incorporated as Derosne & Cail. Cail had started working at Derosne et Cie. in 1824, where, like many other employees, he learned his trade as an apprentice in the workshop. After the death of Charles Derosne in 1846 the activities of Derosne & Cail would be continued by Cail, resulting in the creation of Société Cail & Cie., a firm that by the late 1860s had around 5,000 employees. In 1883 it became Société Anonyme des Anciens Établissements Cail, and in 1898 Société Française de Constructions Mécaniques – until its merger with Fives-Lilles in 1958.[62]

By the middle of the nineteenth century the firm had become one of the world's foremost sugar machinery manufacturers, and the second-largest firm in France after Schneider. Interestingly, it not only had factories in France, but also pursued a global strategy of production and commercialisation of technology, as well as international management of property rights through foreign agents and offices. Derosne & Cail had its headquarters in Paris (Chaillot), and soon set up factories in Grenelle and Denain. At this time it became a multinational corporation with an extensive network of factories, representatives, agents, and branches throughout Europe, Russia, Southern Asia, Latin America, and the Caribbean. In France, the firm pursued business partnerships with other inventors and mechanical engineers, simultaneously reducing its competition and increasing economies of scale. In the case of the French Indies, Derosne & Cail invested in several sugar mills and engaged in business partnerships with planters in Guadeloupe and Martinique.[63]

From the 1820s Derosne et Cie. had extended its network to the sugar-producing islands of Cuba, Puerto Rico, Guadeloupe, Martinique, Mauritius, Java, and Bourbon.[64] Contact between Derosne & Cail and Cuban sugar planters began in the late 1830s, with an attempt to introduce Derosne's vacuum pan to Cuban plantations.[65] The first refining system sold and installed by the company in Cuba was a vacuum pan erected in 1841 on *La Mella,* a sugar estate owned by the wealthy planter Wenceslao Villaurrutia. It was the inventor himself, Charles Derosne, who provided all the machinery and supervised the assembly of the new system on Villaurrutia's plantation.[66] The May 1843 crop was the first one to be processed entirely with the new apparatus. According to a report by Villaurrutia on the performance of Derosne's new 'sugar machinery', the system of vacuum pan evaporation increased production, significantly saved labour, and reduced charcoal consumption.[67] Data collected by Ramón de la Sagra, Carlos Rebello, and Justo Cantero likewise indicated that the Derosne system was superior, and reduced consumption.[68] On the other hand, although the Derosne system reduced labour requirements in the refining process, it demanded large quantities of sugar cane, which in turn required more labour in the field. Moreover, the initial investment was considerably higher than had been required for the technically inferior vacuum boilers; Villaurrutia needed to borrow $9,000 from the *Junta de Fomento* to purchase Derosne's installation.[69]

In their 1844 treatise describing the new method – translated into Spanish by the Cuban chemist José Luis Casaseca – Derosne and Cail recognised that the new apparatus needed a skilled sugar master, as well as other semi-skilled labourers to operate it. They emphasised, however, that it actually simplified the tasks of unskilled slave labourers.[70] According to the United States Patent Office, Villaurrutia paid $32,000 for the Derosne installation – which, though it included steam clarifiers, filters, and charcoal reburners, was not a full Derosne refinery system. That would had cost almost $100,000.[71] Derosne himself trained Villaurrutia's machinists to use his innovation, and supervised installation on other Cuban plantations, such as Bolumbre[72] (also owned by Villaurutia), Güinía de Soto (owned by Justo Cantero), Flor de Cuba (owned by Joaquin de Arrieta), La Gran Azucarera (San Martino and Santa Susana), and the plantations of the Zulueta family (Habana, Vizcaya and Alava).[73]

As Carlos Rebello's statistics for the year 1859 confirm, use of Derosne's apparatus proliferated throughout the island during the 1840s and 1850s.[74] Although, according to Rebello, only 77 of the 1,300 Cuban sugar mills were using vacuum pans in 1860, these 77 included the majority of the largest Cuban sugar mills. Of the 77 plantations using this steam technology, at least 33 were using the Derosne technology, and 20 were using the Rillieux system, commercialised in Cuba by Merrick & Towne. However, according to several complementary sources (including mechanics journals, contemporary treatises, and documentation on specific sugar plantations obtained in Cuba's National Archive), it seems that the number of sugar mills employing some sort of Derosne technology was higher than indicated in Rebello's statistics.

Derosne & Cail employed several French machinists and engineers with experience in sugar technology to install its machinery in Cuba. Some of them, such as the *ouvriers mécanicien* Jean Baptiste Superville and Jules Rossignon, lived for long periods in Cuba. Among the various commercial agents of Derosne & Cail was the French lithographer Eduardo Laplante, who was well known for his detailed depictions of Cuban sugar mills.[75] Commission houses such as Société Durège, Ducrey et Cie – present in Havana between 1855 and 1870 – also acted as consultants and commercial agents for Derosne & Cail technology.[76] This company was created with the express purpose of selling Derosne & Cail machinery in Cuba. Société Durège was run by the French engineer Daniel Ducrey, who in the early 1850s had installed Derosne & Cail technology in the sugar estates of Antonio Parejo and Eusebio Alfonso, among others.[77]

The introduction of Derosne's vacuum technology was a massive investment for Cuban planters, who lacked easy access to credit due to the weak Cuban banking institutions of the time, and the inability of the Spanish metropolis to provide capital for the introduction of modern technology to the island.[78] Because planters needed loans to buy the technology, Derosne & Cail provided credit to planters unable to borrow money from Cuban institutions or foreign money lenders such as Moses Taylor, Baring Brothers, and Brown & Brown.[79] In the 1840s, to encourage planters to buy its apparatus, Derosne & Cail offered a 25% discount and payment in two installments – the second delayed a full year.[80] This was in significant contrast to firms such as West Point Foundry, which required payment in cash or the backing of a reputable banking house. The installation of Derosne technology and other similar vacuum pans left Cuban planters in considerable debt during the second half of the century, and many of them eventually fell into bankruptcy.[81]

After the successful introduction of the new vacuum pan in Cuba, Derosne & Cail tried to secure the property rights of their apparatus in the Cuban patent system. Derosne and his business partner Jean François Cail had already secured the patent rights of this invention in France and Britain, thereby amassing a small fortune from sales of the new invention. In June 1842, they applied to Havana's *Junta de Fomento* for a 15-year 'royal privilege of invention'. Their agent in Cuba was Joaquín de Arrieta, a sugar planter, who acted as an intermediary in the application process to obtain this patent.[82] In 1843 Arrieta went on to introduce Derosne's apparatus into his sugar mill *Flor de Cuba*. Soon after setting up the Derosne machinery, the *Junta de Fomento* organised a commission of sugar planters whose task was to visit Arrieta's sugar estate in order to evaluate the results of Derosne's new technology.[83]

The patent application was officially rejected by Havana's *Junta de Fomento y Agricultura*. The reasons offered were twofold. First it was argued that, according to Spanish law, the new technology had already been introduced into the island. Second, Cuban institutions controlled by the planters, such as the *Junta de Fomento* and *Real Sociedad Económica*, had already invested significant capital in the introduction of Derosne's invention.[84] Indeed, in 1843 the *Junta de Fomento* commissioned the chemist Jose Luis Casaseca to travel to Brussels to visit the Derosne & Cail factory and investigate the improvements that the firm had introduced in sugar technology.[85] A year later, in 1844, Casaseca translated Derosne's treatise with the economic support of the *Junta de Fomento*, distributing 500 copies among Cuban sugar planters and five copies to the island's public libraries.[86]

Although Derosne's patent application was rejected, the episode shows how the control and management of patented technology in colonial plantation economies became a commonplace among foreign firms in the mid-nineteenth century. However, it is difficult to know the extent to which patent rights were important to foreign firms such as Derosne & Cail and Babcock & Wilcox. It is striking that, despite the persistent corporate patenting activity occurring in Cuba and mainland Spain to protect inventions on the island during this period, sugar machinery manufacturers usually did not manufacture, or commercially exploit, the industrial property rights they had obtained in Cuba.[87] The Cuban patent system might have been used more as a marketing tool by foreign manufacturing firms, specifically as a means of advertising new technologies on the island, than as a means of exploiting industrial property rights. Although patent documentation serves as a good gauge for the dynamic of technological innovation and experimentation in Cuba, it seems that in the mid-nineteenth century the most salient transfer mechanisms to the island continued to be the migration of experts, the circulation of technical literature, and the direct trade of foreign machinery.

The activities of Derosne & Cail in the Caribbean make clear that the company based its development on an astute global strategy of diversification and internationalisation, effectively harnessing an international network of intermediaries and thereby successfully establishing itself as one of the pioneering European companies to supply steam technologies to the Cuban plantation economy.[88] Derosne & Cail's good reputation and extensive international network of agents and intermediaries allowed it to charge higher prices than its British and American competitors.[89]According to Ramón de la Sagra, agricultural botanist and director of the Havana Botanical Gardens in the 1830s, the spread of the Derosne system among Cuban plantations was attributable to Derosne's intelligence and perseverance, and the substantial personal and material resources of his firm.[90] Certainly Derosne & Cail can be regarded as an early example – significantly predating the proliferation of multinationals – of a transnational corporation transferring knowledge and technology to emerging markets. The firm was a key actor behind the reorganisation of the world sugar cane industry, with its ability both to adapt innovations to sugar cane production, and to open up an international market for its equipment and technology.[91]

The introduction of Derosne's vacuum pan in Cuba also provides an understanding of how mid-nineteenth-century sugar planters exchanged and shared technological information (rather than competing among themselves) through the institutions they controlled, such as the *Junta de Fomento*. This successful strategy of cooperative efforts among planters promoted innovation and satisfied the demand for technology and expertise in Cuba. In the last decades of the nineteenth century, creole planters would become organised in associations which shared technological knowledge. One such organisation was *El Círculo de Hacendados*, a corporation set up in the late 1870s to disseminate agricultural developments and foreign sugar technologies in Cuba's sugar plantations through the exchange of knowledge and information.[92] The activities of *La Junta de Fomento* and *El Círculo de Hacendados* show how Cuban creole elites used their political power to further their corporative interests in the sugar industry. It is not clear, however, to what extent planters' political regulations and investments in technology succeeded in modernising the Cuban economy, and to what extent they were merely a manifestation of a rent-seeking impulse to secure their own economic enterprise.

From macrohistories to broad understandings

This chapter has adopted a macrohistorical approach to the study of technological modernisation on Cuban sugar plantations in the nineteenth century. It has argued that these changes can best be understood in the long run, and in the broader context of the Atlantic economy, a region that constituted an interdependent economic system. Building on work by Sidney Mintz and Dale Tomich, among others, this chapter has placed the Caribbean at the centre of the history of nineteenth-century industrial capitalism, focusing on the study of international technology transfer and expert migration.[93] It has considered the interplay between local conditions and global developments by delineating the means and actors by which technology was transferred from Europe and the United States to Cuba. In doing so, it illustrates the role of indigenous and foreign agents in the diffusion of technology during this period, uncovering the rent-seeking and cooperative behaviours of the Cuban creole elites in fostering technology transfer and investment by foreign firms.

The mid-nineteenth-century Cuban sugar industry is one of the earliest examples of industrial agriculture in the tropics, being highly intensive in both machinery and capital. A study of Cuban trade figures reveals that metropolitan Spain was unable to provide the technological innovations required by the industry. The evidence shows that the technology used on Cuban plantations was neither imported from the Spanish metropolis, nor produced in Cuba. On the contrary, interimperial and intercolonial technological exchange was far more important. In developing its sugar industry, Cuba became dependent on materials, equipment, and durable goods from American, British, and French engineering firms. Although Cuba maintained its political and fiscal ties with the declining Spanish metropolis, it established technological links with foreign manufacturers from the most industrially advanced Atlantic nations. In essence, for the technological transformation of

Cuban sugar plantations, economic links assumed greater importance than political control.

Demand for technology and agro-industrial experimentation on the Cuban plantations were central to the development of some of the largest mechanical engineering firms in Britain, France, and the US. In the mid-nineteenth century foreign engineering firms from the most industrialised nations began to dominate the trade to Cuba in modern sugar-related technologies. Many of these foreign manufacturers had branches and agents in Havana. The increasing number of commercial prospects in the Cuban economy in the mid-nineteenth century led western manufacturers of refining equipment to patent their innovations in Cuba. The expansion of Cuban plantations, in turn, was made possible by the technology sold by these firms, and by the engineers who linked the plantations with the centers of production in Europe and the US, while adapting the technology to local requirements. The case of Derosne & Cail highlights the global strategy of a firm that made an important contribution to the introduction of machinery and equipment that would be employed in Caribbean plantations.

The Cuban creole elite functioned as the main promoter of the modernisation of industry in the mid-nineteenth century through the diffusion of foreign technology and the attraction of foreign experts, who were critical social actors in the rapid technological change that occurred on Cuban plantations in the mid-nineteenth century. Active transfer agents, from machinists to intermediaries, not only carried technological information to Cuba, but also assisted machinery manufacturers in the commercialisation of their technology on the island. During these years not only did Cuba receive experts from abroad, but Cuban professionals themselves travelled to America and Europe to learn about new techniques and industrial methods.

It is important to note that the focus cannot be placed solely on the transfer of knowledge and information from industrialised nations to Cuba. Also relevant were the processes of experimentation and incremental innovation that occurred on the Cuban plantations – processes that would contribute to the development of mechanical engineering and chemical process improvements in Europe and the US. Innovation in mechanical engineering and chemicals was driven not just by the development of general purpose technologies such as the steam engine and railways in European or American factories, but also by the demands of sugar plantations, which required the adaptation of these technologies to sugar production and Cuban conditions. In other words, this is not a story of technologies brought from the industrial world to Cuba, but one of interaction, exchange, and appropriation that occurred in both directions, mediated by a variety of intermediaries and consultants.

Cuban agro-industrial complexes became a highly dynamic space open to foreign intervention at many levels during the mid-nineteenth century. Innovative activity and property rights management in Cuba during this period were rooted in overlapping long-term and transnational legal, technical, and economic networks. An extensive and interconnected group of machinists, chemists, consultants, merchants, and lawyers not only enabled Cuban planters to introduce new technology, but also reduced the risks and uncertainty that foreign machinery makers had encountered in the international transfer of technology and expertise. This intense transfer can be explained only within the larger, more complex Atlantic World economic system in which rival Atlantic powers – and their firms and experts – acted as a kind of shadow economic metropolis of Cuba.

Notes

1. The authors gratefully acknowledge financial and institutional support from the Spanish Ministry of Economy and Competitiveness (ref. HAR2012-35965 and ref. HAR2013-40760-R), the Max Weber Programme at the European University Institute, the Universidad Autónoma de Madrid, and the Economic History Society.
2. It seems that the term 'Sugar Industrial Revolution' was first used by the nineteenth century Cuban chemist José Luis Casaseca and has continued to be used since then. See, for instance, Moreno Fraginals, Manuel: *The sugarmill: the socieconomic complex of sugar in Cuba 1760–1860*, New York and London: Monthly Review Press, 1976, and and Schnakenbourg, Christian: 'From sugar state to central factory: the industrial revolution in the Caribbean (1840–1905)', in Albert, Bill and Graves, Adrian (eds), *Crisis and change in the international sugar economy 1860–1914*, Norwich: ISC Press, 1985.
3. See Tilly, Charles: *Big structures, large processes, huge comparisons*, New York: Russell Sage, 1989, pp. 74–6; Tomich, Dale: 'World slavery and Caribbean capitalism: the Cuban sugar industry, 1760–1868', *Theory and Society* 20(3) (1991), pp. 113–15; Tomich, Dale and Zeuske, Michael: 'The second slavery: mass slavery, world economy and comparative microhistories', *Review: A Journal of the Fernand Braudel Center*, XXXI, no. 3 (2008), pp. 91–100.
4. Tomich, Dale W.: *Through the prism of slavery: labor, capital, and world economy*, Lanham, MD: Rowman & Littlefield, 2004, pp. 123–33; Mintz, Sidney, W.: *Sweetness and power: the place of sugar in modern history*, New York: Viking Penguin, 1985, pp. 60–73 and 127–50.
5. Elena, A. and Ordóñez, J.: 'Science, technology, and the Spanish colonial experience in the nineteenth century', in MacLeod, Roy (ed.): *Nature and*

empire: science and the colonial enterprise (Osiris 15), Chicago: University of Chicago Press, 2000, pp. 70–82.

6. Rood, Daniel: *Plantation technocrats: a social history of knowledge in the slave-holding Atlantic World, 1830–1865*, Irvine: unpublished PhD Dissertation, University of California, 2010; Curry-Machado, Jonathan: *Cuban sugar industry: transnational networks and engineering migrants in mid nineteenth century Cuba*, London: Palgrave Macmillan, 2011; Funes, Reinaldo and Tomich, Dale W.: 'Naturaleza, tecnología y esclavitud en Cuba: frontera azucarera y revolución industrial, 1815–1870', in Piqueras, José Antonio (ed.): *Trabajo libre y coactivo en sociedades de plantación*, Madrid: Siglo XXI, 2009; Tomich, *Through the prism*. See also Tomich, this volume.
7. Moreno Fraginals, *The sugarmill*, pp. 113, 141, 142.
8. Tomich, *Through the prism*, p. 129. In 1875 Cuba produced 800,000,000 kilogrammes of sugar, according to Gautier, L. and von Wagner, R.: *Nouveau traité de chimie industrielle 'a l'usage des chimistes*, Paris: F. Savy, 1879, p. 78.
9. Balboa, Imilcy: *De los dominios del rey al imperio de la propiedad privada. Estructura y tenencia de la tierra en Cuba (siglos XVI–XIX)*, Madrid: CSIC, 2014; Ortega, Jose Guadalupe: *The Cuban sugar complex in the age of revolution, 1789-1844*, Unpublished PhD Dissertation, UCLA, 2007.
10. Tomich, *Through the Prism*, p. 130. See also Moreno Fraginals, *The Sugarmill*. Although for Moreno technology pushed slavery only in the first stage, being the principal cause of its abolition in the long run.
11. Zanetti, Oscar and Alejandro García: *Sugar and railroads: a Cuba history, 1837–1959*, Chapel Hill: University of North Carolina Press, 1998.
12. Zanetti and García, *Sugar and railroads*, p. 78; Rood, *Plantation technocrats*.
13. Knight, Franklin W.: 'Origins of wealth and the sugar revolution in Cuba, 1750–1850', *Hispanic American Historical Review*, Vol. 57, No. 2 (1977), pp. 231–53.
14. Goncalvès, Dominique: *Le planteur et le roi: l'aristocratie havanaise et la couronne d'Espagne (1763–1838)*, Madrid: Casa Velazquez, 2008; Piqueras, J.A.: *Sociedad civil y poder en Cuba: colonia y poscolonia*, Madrid: Siglo XXI, 2005; Barcia, Manuel: 'Sugar, slavery and bourgeoisie: the emergence of the Cuban sugar industry' in Bosma, Ulbe *et al.* (eds), *Sugarlandia revisited: sugar and colonialism in Asia and the Americas, 1800 to 1940*, New York: Berghahn Books, 2007; Bahamonde, Ángel and Cayuela, José: *Hacer las Américas: las élites coloniales españolas en el siglo XIX*, Madrid: Alianza, 1992.
15. Moreno Fraginals, *The sugarmill*, pp. 31–3, 98–9; Zogbaum, Heidi: 'The steam engine in Cuba's sugar industry, 1794–1860', *Journal of Iberian and Latin American Research*, 8 (2) (2002), pp. 37–60.
16. Knight, *Origins of wealth*, pp. 249–50; Zanetti and García, *Sugar and railroads*, pp. 57–77.
17. Knight, *Origins of Wealth*, p. 250.
18. Moreno Fraginals, *The sugarmill*; Tomich, *Through the prism*; and Iglesias García, Fé: *Del ingenio al central*, San Juan: Editorial de la Universidad de Puerto Rico, 1998.
19. Funes, Reinaldo: *From rainforest to cane field in Cuba: an environmental history since 1492*, Chapel Hill: University of North Carolina Press, 2008, pp. 179–215; McCook, S.: *States of nature: science, agriculture and environment in the Spanish Caribbean, 1760–1940*, University of Texas Press, 2002, pp. 11–25.

20. Rossignon, Jules: *Manual del cultivo de la caña de azúcar*, Paris, 1859, p. 183; Piqueras, J. A.: 'Capitales en el azúcar: los hacendados cubanos ante la rentabilidad económica y la oportunidad de inversión (1878–1895)', in *Revista de Indias*, 1998, no. 212, pp. 163–93; McCook, *States of Nature*, pp. 23–4.
21. Rebello, Carlos: *Estados relativos a la producción azucarera en la isla de Cuba*, Havana: Imprenta del Gobierno, 1860; Cantero, Justo G.: *Los ingenios: colección de visitas a los principales ingenios de azúcar de la isla de Cuba*, García, Luis Miguel and Santamaría Aranjuez, Antonio: Doce Calles: CEHOPU, CSIC y Mapfre 2005 (1857); Reynoso, Álvaro: *Ensayo sobre la caña de azúcar*, Havana: Ministerio de Industrias, 1963 (1862); Moreno Fraginals, *The sugarmill*, p. 85.
22. Fernández de Pinedo, Nadia, *Comercio y Fiscalidad: Cuba 1794–1860*, Bilbao: UPV, 2002; Roldán de Montaud Inés (ed.), *Las haciendas publicas en el Caribe Hispano durante el siglo XIX*, Madrid: Colección América, 2008. The French Consul in Havana was very explicit depicting the Spanish Antilles as dependent on foreign interests. See Archives Nationale, Affaires Étrangères, Correspondances Consulaires, La Havane, Vol. 13, f. 190, 1842.
23. See 'Balanza General de Comercio de la Isla de Cuba' for the years 1826–63, Biblioteca Nacional (henceforth BN) and Archivo Nacional de Cuba (henceforth ANC). On the dependence of Cuba on foreign railroad technology and technical experts in the nineteenth century, see Zanetti and García, *Sugar and railroads*.
24. It is important to underline that trade figures provide information only on the last port of shipping. It is not possible to know precisely the country of production of the technology imported to Cuba.
25. *Scientific American*, vol. 7, no. 8, 1851, p. 59. A similar reference of 1852 is given in the same publication: 'Very few persons are aware of the large amount of machinery that is usually shipped from the United States, particularly from Boston, to Cuba. On account of the large crops which have been produced there in the last few years, and the large demand for molasses and sugar, machinery has been, and is now, in constant demand, for the manufacture of it'. 'Machinery for Cuba', *Scientific American*, vol. 8, no. 9, 1852, p. 69.
26. Fernández de Pinedo, Nadia, Pretel, David, and Sáiz, Patricio: 'Patents, sugar technology and sub-imperial institutions in nineteenth-century Cuba', *History of Technology*, 30 (2010), pp. 46–62.
27. ANC (Patent documentation at the following collections: Gobierno General; Gobierno Superior Civil; Real Consulado y Junta de Fomento) and Archivo de la Oficina Cubana de la Propiedad Industrial (henceforth AOCPI). This patent data should be regarded as an estimation of the total of patent applications. It is difficult to assess if there were more patent applications in Cuba, given that the records are spread over different collections of the ANC and the AOCPI.
28. Patents of 'introduction' conferred exclusive use of inventions to individuals who were not the inventors, with the aim of stimulating national industry. From its establishment in 1826, the Spanish system of industrial property granted patents of introduction to importers of overseas inventions and not just to the 'first' and 'true' inventor.
29. Iglesias García, *Del ingenio al central*; Tomich, *Through the prism*, pp. 125–6; Schnakenbourg, *From sugar state*; Ayala, Cesar J.: 'Social and economic

aspects of sugar production in Cuba, 1880–1930', *Latin American Research Review*, 30, no. 1 (1995), pp. 95–9; Funes, *From rainforest*, ch. 5.

30. *Círculo de Hacendados de la Isla de Cuba. Exposición del Círculo de Hacendados, la Junta General de Comercio, la Real Sociedad Económica y la Sección de Agricultura de la misma, autorizada por sus respectivos presidentes en 4 de setiembre de 1878*, Habana: La Propaganda Literaria, 1878, p. 8.
31. Marichal, Carlos: La crisis mundial de 1873 y su impacto en América Latina, *ISTOR Revista de Historia Internacional*, IX, no. 36, 2009, p. 41.
32. Dye, Alan: *Cuban sugar in the age of mass production: technology and the economics of the sugar central, 1899–1929*, Stanford: Stanford University Press, 1998, pp. 10–14.
33. 'Elite patents' refers to the small groups of the most valuable patents. The elite status is measured according to the duration for which patentees maintain a monopoly and the number of assignments and licenses made. Foreign patents lodged at a higher cost in Cuba might also be considered 'elite' patenting activity. For in-depth insights into the idea of 'elite' patents and the value of patent rights, see Inkster, Ian: 'Patents as indicators of technological change and innovation – an historical analysis of the patent data 1830–1914', *Transactions of the Newcomen Society*, 73(2), 2003, pp. 179–208.
34. Patent documentation at ANC, AHOEPM and Archivo Histórico Nacional (Ultramar collection), henceforth AHN.
35. Moreno Fraginals, *The sugarmill*, pp. 101–27; Iglesias García, *Del ingenio al central;* Marqués Dolz, María Antonia: *Las industrias menores: empresarios y empresas en Cuba (1880–1920)*, La Habana: Ciencias Sociales, 2006.
36. Zanneti and García, *Sugar and railroads*, pp. 39–56.
37. Moreno Fraginals, *The sugarmill*, ch. 4, Iglesias García, *Del ingenio al central;* Ely, *Cuando reinaba*, p. 517.
38. Rood, Daniel, *Plantation technocrats.*
39. Ely, *Cuando reinaba*, pp. 516–17.
40. Matei, Ramos: 'The role of Scottish sugar machinery manufacturers in the Puerto Rican plantation system, 1842–1909', *Scottish Industrial History* 8(1) (1985), pp. 20–30; Curry-Machado, *Cuban sugar industry*; Iglesias García, *Del ingenio al central.*
41. Glasgow University Archives (henceforth GUAS). Ref. UGD/118.
42. GUAS. Ref. UGD/118/2/4/2, order book year 1873.
43. GUAS. Ref. UGD/118/2/4/10, order book year 1891.
44. Jules Rossignon, *Manual del cultivo*, p. 86.
45. See ANC, Gobierno Superior Civil, legajo no. 1284 (exp. no. 50242, no. 50244, no. 50245, no. 50250) and Legajo 1104 (exp. no. 40622).
46. Jiménez, G.: *Las empresas de Cuba 1958*, La Habana: Editorial de Ciencias Sociales, 2004, pp. 186–7.
47. Mirless Watson & Co. Order Books. GUAS. Ref. UGD/ 118/ 2/4.
48. *Chemical News and Journal of Industrial Science*, 20/01/1865, p. 35. See also Curry-Machado, *Cuban Sugar Industry.*
49. For examples of patents obtained by Edward Beanes in the US and UK, see *Scientific American*, vol. 1007, iss. 8, Aug. 23 (1862); vol. 1012, iss. 13, March 25 (1865); and *London Gazette*, 13 January 1863.

50. Curry-Machado, Jonathan: 'Rich flames and hired tears': sugar, sub-imperial agents and the Cuban phoenix of empire', *Journal of Global History* 4(1) (2009), pp. 33–56.
51. 'Cuba no es un estado cliente', *Cuba Contemporánea*, no. 150 (1925), pp. 129–30.
52. *Reglamentos para las Escuelas especiales de Maquinaria, Telegrafía, Agrimensores y Maestros de obras,* Habana: Imprenta del Gobierno Capitanía General por S.M., 1856, pp. 3, 5.
53. Bruland, Kristine: 'The Babcock & Wilcox Company: strategic alliance, technology development, and enterprise control, circa 1860–1900', in Brulan, Kristine and O'Brien, Patrick (eds): *From family firms to corporate capitalism,* Oxford: Oxford University Press, 1998; Bruland, Kristine: 'The management of intellectual property at home and abroad: Babcock & Wilcox 1850–1910', *History of Technology* 24 (2004), pp. 151–71.
54. Document of Incorporation of Edison's Spanish Colonial Company (5/09/1881), The Thomas Alva Edison Papers, Rutgers, ref. XX19.
55. See, for instance, the hundreds of files of patent rights extended to the colonies in AHN, Ultramar colection, Ministerio de Ultramar Remite Patentes de Invención, nos. 127, 155, 156, 131 for the years 1888, 1889, and 1891–93, and AHN, Ultramar, Convalidación de Patentes Dadas en la Peninsula, nos 178, 180, 182, 184, for the years 1893–96.
56. ANC, Gobierno General, 1885, Leg. 455, no. 22246; *Revista de Agricultura,* año XII, no. 4 (24 January 1892, p.48) and *Revista de Agricultura,* año XII, no. 11 (13 March 1892, p. 13).
57. Register of Patent Operations and Business Diaries, *Elzaburu Agency Private Records*, Madrid, and the original powers of attorney kept in the patent documentation of the AHOEPM for the period 1826–1903.
58. Historical Dossier, Elzaburu Industrial Property Agency (Madrid, 2009).
59. AHOEPM, priv. no. 6,915.
60. Julio Vicarrondo's Business Diaries, Elzaburu Private Records, Register of Patent Operations for the Year 1887.
61. Dye, *Cuban Sugar*, pp. 58–60; Davis-Oyesanya, Denise S.: *Atkins and Company and the sugar trade in Cuba*, PhD Dissertation, University of Oklahoma, 1996.
62. For the history of the firm Derosne & Cail see Chadeau, Emmanuel: *L'économie du risque: les entrepreneurs 1850–1980,* Paris: O. Orban, 1988; Thomas, Jean-Louis: *Jean-François Cail: un acteur majeur de la première révolution industrielle,* Chef-Boutonne: Editions de l'Association CAIL, 2004; Gaillard, Jeanne: 'Les usines cail et les ouvriers métallurgistes de grenelle', in *Le Mouvement Social,* no. 33/34 (1960), pp. 35–53; 'Constructeurs de machines: Maison Cail et Cie', *L'exposition populaire illustrée,* 1867, p. 77; Turgan, J.: *Les grandes usines, études industrielles en France et à l'étranger,* vol. 2, Paris: Michel Lévy frères, 1865; Barbier, Frédéric: *La patronat du Nord sous le Second Empire: une aproche prosopographique,* Genève: Librairie Droz, 1989, p. 107.
63. Thomas, *Jean-François Cail*, pp. 59, 167–205.
64. Chadeau, *L'économie du risque*, pp. 25, 28, 29; Bosma, Ulbe: *The sugar plantation in India and Indonesia: industrial production, 1770–2010*, Cambridge: Cambridge University Press, 2013, p. 108; Leon, J.A.: *The sugar question: on the sugar cultivation in the West Indies*, London, 1848. The firm set up agencies

in Cuba, Guadeloupe, Martinique, Bourbon, Maurice, and Java. Thomas, *Jean-François Cail*, p. 75.
65. Turgan, *Les grandes usines*, p. 25.
66. Bachiller, Antonio: 'Breve ojeada sobre los progresos de la agricultura y su estado actual', *Memorias de la sociedad económica de la Habana*, Havana, 1856; Leon, *Sugar question*, pp. 19–25; Funes and Tomich, *Naturaleza, tecnología*, pp. 108–9; Moreno Fraginals, *The sugarmill*, pp. 111–12.
67. Leon, *Sugar question*, pp. 19–20.
68. Ramón de la Sagra, *Historia física, política y natural de la isla de Cuba*, Tomo XIII, 1861, pp. 82–3; de la Sagra, *Cuba en 1860*, p. 82; Rebello, Carlos: *Estados relativos a la produccion azucarera de la isla de Cuba: formados competentemente y con autorizacion de la Intendencia de ejercito y hacienda*, Havana: 1860; Cantero, *Los ingenios*.
69. ANC. Junta de Fomento. Negociado de Cuentas. Legajo 41, expediente 1824 (1842) and expediente 4086 (1843); ANC, Real Consulado de agricultura, industria y comercio, Legajo 41, expediente 2576 (1843).
70. Derosne, C. and Cail, J.L.: *De la Elaboración del Azúcar y de los Nuevos Aparatos Destinados a Mejorarla*, La Habana, 1844, pp. 15–22.
71. Annual Report of the Commissioner of Patents for the Year 1848, *United States Patent Office*, Washington, 1849, p. 328.
72. Guicharnaud-Tollis, Michèle: *Le sucre dans l'espace caraïbe hispanophone aux XIXe et XXe siècles: stratégies et représentations*, Paris: L'Harmattan, 1998, p. 131.
73. *El Mundo Universal*, p. 101. Correspondance Letter, La Habana, 4/7/1847. Affaires Etrangères, Paris, Correspondance Consulaire, La Havane, 1847, vol. 15, p. 99.
74. Rebello, *Estados Relativos*.
75. Cantero, *Los ingenios*, p. 65.
76. Ibid., *Los ingenios*, pp. 191, 195, 287.
77. *Gazette de l'Industrie, et du Commerce*, 26/08/1855, serie B-N°25, Paris, p. 5.
78. Roldán de Montaud, I.: *La banca de emisión en Cuba (1856–1898)*, Madrid: Banco de España, 2004. Calavera Vayá, A. M.: 'El sistema crediticio español del siglo XIX y su reflejo en Cuba: los comerciantes banqueros', in Naranjo C. and Mallo, T.: *Cuba la perla de las Antillas*, Madrid: Doce Calles, 1996, pp. 335–44; Collazo Pérez, E: 'Crédito y proyectos bancarios en Cuba durante el siglo XIX', *Boletín del Archivo Nacional de Cuba*, no. 3 (1989), pp. 1–40.
79. Thomas, *Jean-François Cail*.
80. *El Heraldo de Madrid*, no. 384, 14/7/1849, p. 4.
81. Moreno Fraginal, *The sugarmill*, p. 73; Iglesias García, Fé.: 'Azúcar y crédito durante la segunda mitad del siglo XIX', Santiago, no. 52 (1983). Zogbaum, *The Steam Engine*. Similarly, in the French West Indies most of the planters fell into considerable debt through purchasing sugar technology. See Tomich, Dale: 'Sugar technology and slave labor in Martinique, 1830–1848', *Nieuwe West-Indische Gids* 63, nos. 1&2 (1989): p. 121. For a different approach, see Piqueras, *Capitales en el azúcar*. See also Ines Roldan, this volume.
82. ANC, Gobierno Superior Civil, Legajo 1476, no. 58365 (1842); ANC, Junta de Fomento, Legajo 205, no.9115 (1842).
83. ANC, Real Consulado y Junta de Fomento, Legajo 95, no. 4,006 (1843).

84. ANC, Gobierno Superior Civil, Legajo 1476, no. 58365 (1842).
85. ANC, Real Consulado y Junta de Fomento, Legajo 95, no. 3996 (1842) and no. 4004 (1843). See also Casaseca, José Luis: *De la necesidad de mejorar la elaboración del azúcar en la Isla de Cuba y de las mejoras de que es susceptible esta fabricación*, Habana: Imprenta del Gobierno y Capitanía General, 1843.
86. *Informe presentado a la Real Junta de Fomento, de Agricultura y Comercio de la Isla de Cuba, por el Sr. D. Wenceslao de Villa Urrutia sobre Los Resultados de la Zafra que este año ha hecho su ingenio en un tren Derosne*, Habana: Oficina del Faro Industrial, 1843.
87. See Sáiz, Patricio and Pretel, D.: 'Why Did Multinationals Patent in Spain? Several Historical Inquiries', in Donzé, P. and Nishimura, S. (eds): *Organizing global technology flows*, London: Routledge International Studies in Business History, 2013, pp. 39–59.
88. Chadeau, *L'économie du risque*, pp. 28–9; Smith, Michael S.: *The Emergence of Modern Business Enterprise in France, 1800–1930*, Cambridge, MA: Harvard University Press, 2006, p. 210.
89. Turgan, *Les grandes usines*, p. 25; Agents diplomatiques, *Bulletin consulaire français: recueil des rapports commerciaux adressés au Ministère des affaires étrangères par les agents diplomatiques de France à l'étranger*, Paris: Imprimerie nationale, 1885, p. 135; Ampère, Jacques: *Promenade en Amérique: États-Unis, Cuba, Mexique*, vol. 2, Paris: Michel Lévy frères, 1856, p. 141.
90. Ramón de la Sagra, *Historia física, política y natural de la isla de Cuba*, Tomo XIII, 1861, pp. 82–3.
91. Galloway, J. H.: *The sugar cane industry: an historical geography from its origins to 1914*, Cambridge: Cambridge University Press, 2005, p. 157. For Moreno Fraginals, Derosne & Cail 'acted like any big modern enterprise'. Moreno Fraginal., *The sugarmill.*
92. See *Reglamento del Círculo de Hacendados de la Isla de Cuba*, La Habana: Establecimiento Tipográfico O'Reilly, 9, 1889: Capítulo 1, Artículo 1, p. 4.
93. Mintz, Sidney W.: *Sweetness and power*; and Tomich, *World Slavery*.

13

Afterword: Mercantilism and the Caribbean Atlantic World Economy

Martin Daunton

This collection of essays widens the analysis of the Atlantic World in at least two very useful ways. On the one hand, it extends the time horizon by moving well into the nineteenth century, beyond the usual concentration by historians of the Atlantic World on the long eighteenth. On the other hand, it extends the geographical reach beyond the British empire to include France, the Dutch Republic, Spain, and the United States in a trans-imperial understanding of the complex web of connections on which the Atlantic economy rested. A shortcoming of much of the literature on the Atlantic World is that it focuses on only one empire, at most considering conflict with other imperial systems. One of the strengths of this volume is that it is concerned with the interconnections between different empires, and pays attention to their differences of structure and policies. These two themes of time and space are closely connected, for the extension of the time period draws our attention to a geographical shift, with the remarkable expansion of the Cuban sugar industry in the nineteenth century. The Spanish empire is brought into the picture alongside the apparently more dynamic and dominant British empire – with some surprising results.

The extension of the time horizon and geographic coverage connect with a third theme in this collection of essays: the nature of economic policy and the role of the state in the Atlantic World economy. How did the various empires respond to each other in competition for markets within a shifting pattern of comparative advantage in production of goods? This consideration raises the vexed question of mercantilism, a term that is used by many contributors, though without a clear definition or assessment of its nature.

Mercantilism

In David Ormrod's chapter, mercantilism is presented as a developmental strategy. He complains that in the writings of many historians 'state power in general is assumed to be antithetical to sustained economic growth, leaving little place for mercantilist regulation in shaping trade patterns, encouraging strategic industries, supporting grain markets and so on' (p. 16). He is not alone in claiming a positive role for mercantilism, and rejecting neoliberal assumptions that economic growth rested on the free market. Patrick O'Brien has argued that Britain's economic growth was at least in part the result of successful, aggressive mercantilism based upon the hugely expensive Royal Navy, itself sustained by high, politically legitimate taxes and loans.[1]

Other contributors disagree, and argue that the use of force in the pursuit of trade was wasteful. Nuala Zahadieh claims that the War of the Spanish Succession was not self-financing, and did not open new trades. She generalises to other wars of the long eighteenth century, arguing that the English did not learn the lesson of the War of Spanish Succession, and continued 'to interrupt their constructive commercial expansion with debilitating conflicts' (p. 80). Indeed, Knick Harley suggests that the British victories in the West Indies were a poor deal for British consumers of sugar. He asserts in his essay that the British sugar islands were less productive than the French, so that the prosperity of British planters rested on mercantile protection to keep cheaper French sugar out of the domestic market. The profitability of the sugar islands relied not on their efficiency, therefore, but on the ability of West Indian planters to buy seats in Parliament and shape policy to their own ends. As a result, he argues, British taxpayers suffered twice over: they paid for the wars to secure colonies that produced expensive commodities, and then paid more for their sugar. Harley remarks: 'In effect, British consumers subsidised British slave plantations' (p. 161). In Zahadieh's opinion, Richard Cobden and the Manchester School were right: military might was not the way to generate trade. Her approach has more in common with the Washington Consensus and the interpretation of Douglass North and his followers. In their view, British economic growth did not rest on successful mercantilism, but on the creation of secure property rights and the credibility of the state's commitment to loans as a result of the constitutional settlement of 1688.[2]

Mercantilism makes an appearance in another way, through reflections on the extent of interconnection between imperial systems.

A common underlying assumption in the literature (and many of the contributions here) is that empires in the long eighteenth century tried to maintain closed systems. In the words of Gert Oostindie, 'Much of the emergence of an Atlantic World took place in the context of state policies intended to hold a mercantilist grip on as large a share as possible of the newly explored overseas territories and trade routes, and minimise the impact of outsiders' (p. 108). However, a number of contributions show how these closed systems developed leakages, either by smuggling or evasion, or by deliberate policy. The illegitimate but often essential role of illicit trade is only mentioned in passing, though it merits more consideration. Merchants and planters in the colonies might collude in evading regulations, so posing the local representatives of the imperial state a dilemma, caught between the edicts of the metropole and the wishes of the locals. For example, the loss of the Thirteen Colonies in 1783 posed a serious threat to the British colonies in the West Indies, which relied on North America for commodities.

The same issue arose for the French Caribbean after the loss of Canada and Louisiana, as explained in the contribution of Manuel Covo. In theory, metropolitan France was meant to supply the needs of the colonies, but in practice they came to rely on the United States, by means of regulated and licensed trade through Baltimore. The French subsistence crisis of 1789 led to the collapse of colonial regulation, and Covo argues that French mercantilism came to an end by 1793. Similarly, Oostindie sees the Dutch Republic as consciously benefitting from the closed mercantilist systems of other empires by adopting anti-mercantilist policies. It remained open in a way that enriched itself and 'lubricated the predominantly mercantile Atlantic economy', thereby paradoxically prolonging the mercantilist system of others by its openness (p. 132). As a result, the Dutch West India Company (*Geoctroyeerde Westindische Compagnie*, GWC) did not secure support for its mercantilist ambitions in the metropole.

Mercantilism is, of course, a notoriously slippery concept, one which was a product of later debates over the virtues of free trade. Mercantilism was a construct of free trade supporters, the antithesis of their preference for a multilateral world economy. In the historiography of the 1930s the concept had a historical and contemporary resonance, both as a description and critique of what came before the triumph of Richard Cobden in the second quarter of the nineteenth century, and as an attack on the world of beggar-thy-neighbour policies of bilateralism and autarky, represented above all by the 'Schachtian' policies of the Third Reich.[3] The concept emerged from later political controversies, and gave a dubious precision and ideological coherence to something that did not exist as a systematic and coherent policy in the seventeenth and eighteenth centuries. When we turn to the actual policies pursued at different times and in different places, a much more complex picture emerges. In reality, mercantilism was a 'jumble of devices' designed to meet particular circumstances – whether state finance, economic depression, sectional interests, or power politics.

For Adam Smith, the defining feature of what he called the 'mercantile system' was the desire to accumulate bullion, based on the belief that 'all wealth consists in gold and silver, and that to multiply those metals is the great object of national industry and commerce'.[4] In his view, trade policy rested on prohibiting the export of gold and silver and attracting bullion by means of a balance of payments surplus, created by export bounties, import restrictions, the exploitation of the colonies as captive markets, and the exclusion of ships and merchants from other nations. But was the concern as irrational as he and later free traders imagined? By the time Smith was writing, Britain had a sophisticated monetary and credit system, but a concern with bullion was arguably more rational in the seventeenth century, in order to settle adverse trade balances with the Baltic and East Indies. Furthermore, an inflow of gold could hold up prices, reduce interest rates, and increase domestic prosperity. Bullionism was not necessarily a fetish, and Keynes had a point in dismissing the classical economists for their 'presumptuous error in treating as a puerile obsession what for centuries had been a prime object of practical statecraft'.[5]

Similarly, the Navigation Acts of the seventeenth century had a practical aim in capturing the carrying and entrepôt trade of the Dutch, so securing income from freights, encouraging ship building, creating profits from marine insurance, exploiting the processing of imported commodities, and developing warehouses. The policy was remarkably successful, and contributed to the rise of London as the largest city in Europe. As Ormrod points out, English economic policy rested on two complementary approaches. On the one hand, internal corporate monopolies were dismantled, giving more freedom and allowing foreign merchants to settle in London, where they engaged in the trade with Europe. On the other hand, British merchants were able to concentrate on the extra-European and regulated markets. Internal free trade was complemented by external regulation (pp. 24–25).

Trade regulation both generated revenue for the government and protected 'infant industries', whether as a conscious policy or an unintended consequence. British policy become more protective in the course of the eighteenth century, largely as a result of a search for revenue, with the precise pattern of duties shaped by an interplay between governments and interest groups. A good example is the imposition of duties on cotton and silk textiles from India and China in 1690, followed by a complete ban on silk cloth and printed and dyed calicoes in 1701. Such a policy was detrimental to the English East India Company (EIC), which argued for free trade in textiles (at the same time as it sustained its own monopoly in tea). The duties and regulations were designed to protect the woollen textile industry – the largest industrial employer in the country – for a depression in the sector would have had a devastating impact on employment, and hence on the costs of poor relief (one of the largest items of public spending) and social stability (at a time when there was no police force or standing army).

The full title of the 1701 Act made the motivation clear: *An act for the more effective employing the poor by encouraging the manufactures of the kingdom*. The ability to import plain calicoes led to the development of a printing and dyeing industry in London, in contrast to the situation in the Dutch Republic, where the Dutch East India Company (*Vereenigde Oostindische Compangie*, VOC) successfully prevented a ban on the import of finished cloth, so contributing to the demise of the previously flourishing finishing trades. The British woollen industry pressed for stricter controls on imported textiles, and did secure a total ban on the import of calico in 1721, and even on the wearing of cotton cloth. The result was to encourage the growth of domestic competition to woollen textiles. The government was anxious to develop a domestic linen industry, not least in the troublesome areas of Ireland and Scotland, as an antidote to rebellion and Jacobitism. This new industry started to produce a new fustian cloth of linen and cotton, providing the basis for the emergence of the cotton textile industry. Contrary to the claims of free traders, the cotton industry emerged within a context of state regulation – an ironic point that was lost on the 'Manchester School' of free trade based in its home town.[6]

This example of the British textile trades indicates that the balance of trade and the pursuit of bullionism were not the only or even the main considerations. Although the purchase of goods from the East did lead to a drain of silver that worried the government, it was equally concerned by questions involving population and labour. The case study also suggests that the situation in the Dutch Republic was more complex than the pursuit of an open, anti-mercantilist policy. The ambition of the GWC was defeated, and Dutch planters did not secure a protected home market, but the VOC was successful, to the detriment of domestic textile interests. Above all, what is needed is a close analysis of the dynamics of policy formation in the metropolis, whether Britain, the Dutch Republic, France, or Spain – to indicate the complexities that are masked by the blunt concept of mercantilism.

In the case of Britain, other voices from the middle of the seventeenth century were arguing that growth was not a zero-sum game of appropriating trade from other countries, but that the economy could grow through improvements in productivity.[7] However, calls for free trade must be read with care, for the outcome could be ambiguous and complex. A good example is the opposition of 'separate' or independent slave traders to the monopoly of the Royal African Company (RAC) in the slave trade. The monopoly was ended in 1698, and an attempt at restoration in 1712 failed. Free trade seemed to have triumphed, but the story is in fact somewhat more complicated. The Act of 1698 could be interpreted as a partial success for the RAC, which had effectively lost its monopoly of the slave trade a decade earlier. In admitting the loss, the Company was able to require the separate traders to make a contribution to the infrastructure of forts in West Africa, which were also supported by the government in the middle of the eighteenth century. Free trade took place within a context of Company and state power. When the Company was finally dissolved in 1752, it received generous compensation from the state.

Much the same ambiguity of the definition of free trade applies to the Petition of the London Merchants of 1820, which can be interpreted as a manifesto for free trade involving classical economists such as Thomas Tooke. In reality, many merchant signatories were concerned with their particular interests, just like the separate slave traders. Hence a merchant trading with the Baltic resented preferences given to colonial timber from Canada: free trade was a vested interest as much any other policy.[8] As Perry Gauci has commented, policy was driven by economy politics more than by conceptions of political economy; the state was not simply a reactive player, but an active participant, concerned for issues of revenue, welfare, and public order. The notion of a monolithic mercantilism misses the point of how policy was made, and for what reasons.

Matters of policy

This definition of mercantilism means that we need to know just how policy was formulated in different imperial systems. Gauci remarks that 'British exceptionalism lay far more with the general freedom of the political process, rather than the liberalism of its economic outlook'. The making of national policy was debated in a wide range of public spheres, which extended far beyond parliament.[9] Douglass North's stress on the importance of Parliament after 1688 in creating secure property rights misses far too much about this wider public sphere. Many of the significant changes in policy, such as the emergence of the excise, predated 1688, as did the origins of economic growth and structural transformation, so that too much weight is placed in the existence of annual parliaments.[10]

What about other countries? The essays in this collection hint at ways in which our understanding of the Atlantic World could be extended through closer scrutiny of the domestic politics of the other empires, by examining their political structures or economic motivations. Britain was always the major market for the produce of her empire, in a way that was not true of Spain, which did not provide the market, capital, or technology for the development of Cuban sugar production (Pretel and Fernández de Pinedo, pp. 266–269). Just how did the economics of empire play out within the very different polycentric imperial structure of Spain? Oostindie claims that the Dutch differed from other states, since it did not direct its empire from the metropole. He characterises the Dutch state as decentralised and frail. But was there any more interest in central, metropolitan direction in Spain, with its very different state structure? One feature of the English economy, we have noted, was the combination of internal freedom with external regulation: what was the outcome in other states, where internal regulation might have lasted longer? More remains to be done on the domestic politics of empire and the formation of policy.

Property rights

The work of North, and his protégés, such as Daren Acemoglu, stresses the emergence of secure property rights and inclusive institutions in Britain and its empire as critical to economic growth, in comparison with other parts of the world.[11] This interpretation is far too simple, for the definition of what a property right was lay at the heart of politics in the long eighteenth century. The story is arguably one of the radical insecurity of property rights. The proprietors of the RAC were compensated for the loss of their property – as were the slave owners in the British West Indies – on a massive scale, about £20 million. But the owners of common lands or property often failed in their claims for compensation or recognition.[12] Secure property rights for settlers in North America meant loss or redefinition in a new mental framework of the property rights of native Americans. William Penn argued against his more acquisitive fellow colonists that the king's patent did not allow them to appropriate Indian land, which had to be purchased. He was not successful – and even if land were purchased rather than appropriated, what did 'purchase' and 'ownership' mean?

Different notions of property came into collision. English settlers assumed a fixity of possession, a Lockean notion that ownership rested on mixing labour with the land. By contrast, Native American conceptions of ownership were more flexible, resting on seasonal movements for fishing, hunting and gathering, as well as growing crops, within communal or corporate ownership by a *sachem*. To the settlers, the natives were wandering on the land, and not using it in a properly productive manner. As a result, they could declare Indian land as *vacuum domicilium*, empty of habitation. Even if not expropriated, a land market on the English style could soon lead to the liquidation of native ownership, substituting English conceptions of property as individualised and transferable.

A court case in Massachusetts in 1718 turned on the issue. Jacob Seeknout, the *sachem* of Chappaquiddick, impounded a flock of sheep that another member of the community, in defiance of his authority, had allowed settlers to run on the land. Rather than a direct expropriation of the natives, the settlers were creating disputes over authority within the community. Was Seeknout sustaining his legitimate rights, or was he stealing the sheep? His English lawyer, Benjamin Hawes, defended Seeknout's political legitimacy, and developed an argument about 'custom'. He won the case, but within English legal and societal definitions – and by himself acquiring the land in payment.[13] The arrival of English settlers destabilised native society, creating tensions over the appropriate response.

To take one example, between 1770 and 1775 the Cherokee ceded vast tracts in Virginia and Kentucky to traders in return for goods. The explanation is not merely fraud, misunderstanding, or superior force; it was also about a strategy pursued by the elders to prevent destructive warfare. They were influenced by the memory of war and the need to establish order. By ceding land, they saw the opportunity to secure goods to placate the younger men and stop raiding, and to establish a clear boundary between native and settler land. The strategy failed. The younger men were threatened by the continued encroachment of settlers on their hunting grounds, and they embarked on warfare in 1776. Their effort led merely to defeat and the loss of land.[14] Such issues were fought out throughout the empire, and also within the British Isles over the nature of customary tenure, the definition of clan lands in Scotland, or Celtic land tenure in Ireland.[15] Secure property rights to one was expropriation to another. It would be interesting to pursue these points further, by comparing the property regimes of the French and Spanish empires as they collided with native Americans who could sometimes play empires against each other through strategic alliances.[16] The Europeans brought their own distinctive property regimes to the new world, and had to negotiate with local custom and practice. The economics of empire was not just about factor costs of abundant land and scarce labour. It was also about the acquisition of land on new principles, and the displacement of existing rights.

Demography and resources

The population of the new world comprised indigenous peoples, slaves shipped from Africa, free blacks, and settlers from Europe, in different proportions. Britain differed from the other empires in possessing temperate settler colonies in North America for a longer time and on a larger scale, providing a market for British manufactures. The North American colonies supplied the slave economies of the West Indies with food and shipping, thereby producing more revenue to buy commodities from Britain – although Harley argues that internal growth in North America was more significant. The crucial factor was the settlement of the northern colonies by emigrants from Britain, which raises a point that is not considered here: the demographic structure of the Atlantic World. In their classic account of English population history, Tony Wrigley and Roger Schofield argued that England avoided a Malthusian crisis, and, in contrast to much of Europe, had a 'low pressure' demographic regime. Their main explanation is that England had a relatively late age of marriage, which reduced the number of births, but there was another consideration: migration. Between 1640 and 1699, emigration to North America might have absorbed as much as 69 per cent of the natural increase of the population. Over the entire eighteenth century from 1695 to 1801, about 20 per cent of the natural population increase of England was absorbed by migration, above all to North America.

A dual migration occurred. The largest component was made up of indentured servants, predominantly single, young men with some skill who agreed to work for a period of about four years in return for their passage and board and lodgings. The second group consisted of families, mostly farmers, who sold their assets to start a new life in the colonies. What was the impact of this migration across the Atlantic? Historians of the later nineteenth century have considered the impact on wage differentials both across the Atlantic and within the new and old worlds. In the land-abundant new world, wages were higher and rents lower, so equality was greater than in the old world, with its abundant labour and scarce land. A flow of migrants would narrow the wage differential across the Atlantic, and at the same time increase inequality in the new world and reduce it in the old world.[17] Did the movement of labour across the Atlantic in the long eighteenth century have a similar effect, and did it contribute to the distinctive demography of England? It would be worth pursuing these demographic factors. What proportion of population increase in other countries was absorbed by empires? What proportion of the indigenous population died from illness or violence?

England's low-pressure demographic regime in part rested on moderating the growth of population, and also on the injection of more resources. This point connects with the role of trade in general, and the Atlantic World in particular. Wrigley emphasises the role of domestic coal in providing a large injection of energy – a stock replacing the earlier flow from organic energy.[18] However, there were two other sources of energy. One came from the exploitation of the colonies, which Kenneth Pomeranz argues can explain the 'great divergence' between the West and Asia. In his view, the 'extraordinary ecological bounty of the new world' released the previous limits to the supply of land and energy, and was as important as coal.[19] But is this plausible for the late seventeenth and eighteenth centuries, when most of the imports from the new world were subtropical groceries and semi-luxuries (Ormrod p. 30)? Their impact was more likely to be indirect, through creating a demand for commodities from East and West – tea, coffee, chocolate, sugar – that could be purchased only by cash. Here was a motivation to restructure the household economy, in order to produce cash incomes through specialisation, and by becoming more industrious.[20] In terms of energy input, Ormrod argues that coal was more important, and that it was supported by iron and timber from the Baltic, which both injected energy into the economy, and generated trading profits for use in the Atlantic. 'The momentum behind the rise of the Atlantic World depended to a large extent upon change and adaptation within Europe itself' (p. 31).

A comparison of European empires would be of interest. Spain did not have temperate settler colonies, and depended initially on a flow of treasure from Latin America. It could not absorb the output of sugar from its colonies in the West Indies, which did not provide markets for domestic manufactures. The French did have settlements in North

America, most of which were lost in conflict with the British. In each case, the relationship of metropole with empire in the Atlantic was radically different, and had important consequences for one of the major issues in the history of empires: did they pay, and, if so, to whom? If they did not, who bore the burdens?

Transfer payments

The argument over the costs and benefits of empire goes back to eighteenth-century debates over the extent to which the benefits of North America for trade could be secured without the costs of imperialism – and the argument has run on through contemporary debates to modern history, without a clear resolution.[21] We have noted Zahediah's acceptance of Cobden's belief that the benefits of trade could be secured without the costs of warfare, and we have seen that Harley claims that the British sugar islands produced costly sugar at the expense of the British consumer. It would be interesting to pursue these points in more detail. Just how expensive was sugar from the British colonies, when compared with the French, and how much more productive were the French plantations? Harley does not provide any data, so it is impossible to know the scale of the transfer payment from consumer to West Indian planter.

As Harley points out, the transfer became more apparent in the early nineteenth century – a political point that merits more attention. The abolition of slavery in 1833 meant the payment of compensation to the planters, yet a few years later the equalisation of sugar duties on imperial and foreign sugar meant that domestic consumers were given priority over colonial producers. The Sugar Duties Act of 1846 was passed in the same session as the repeal of the Corn Laws, and led to a political campaign by the planters to reverse the decision. In the financial crisis of 1847, many West Indian merchant houses failed in London and Liverpool, and in Jamaica and British Guiana the local assemblies were in conflict with the governors, refusing to pay taxes and making draconian cuts in public spending. The British government was more concerned about retrenching at home to placate taxpayers than helping the merchants and planters by reversing the cut in sugar duties, or offering more compensation, as in 1833. The Sugar Act of 1848 only postponed, from 1851 to 1854, the introduction of the lower, equalised duties.

The planters claimed that the change in duties led to their ruin, but this is not entirely true. There were other reasons for problems in the West Indies, above all the difficulties of introducing technical change, when compared with their competitors elsewhere in the West Indies (above all in Cuba), and the problems of adjusting to free labour. The world market for sugar increased considerably, with world production of sugar rising by 98.2 per cent between 1839/46 and 1857/66. The West Indies and British Guiana accounted for a smaller share of the rising output, so that they produced 45.4 per cent more sugar over the period. The major problem was in Jamaica, where output *fell* by 24.7 per cent. The explanation for the 'ruin' of Jamaica was not the Sugar Duties Act, for other parts of the British empire were able to compete. Rather, it was the inability to adjust to new conditions. Political and social unrest in Jamaica continued, culminating in the notorious episode in 1865 when Governor Edward Eyre used martial law to put down what he considered to be an insurrection at Morant Bay. The event provoked a 'great moral earthquake of Victorian public life', as public intellectuals debated the rule of law, the limits of violence, and the character of the freed slaves.[22] One of the merits of this collection is that it focuses on the emergence of Cuba as a major sugar producer from as early as the 1790s, well before the equalisation of the sugar duties. As Dale Tomich shows, its extensive sugar frontier allowed it to expand and overtake Jamaica and British Guiana by the early 1820s, accounting for 30 per cent of world markets by 1868. It forms the obverse of Jamaica: a society which was never dominated by slavery, where a local creole elite could adopt capital-intensive technology, and in turn invest its accumulated profits in Europe, and above all Barcelona (Rodrigo y Alharilla; Pretel and Fernández de Pinedo).

Future directions

This collection of essays opens up many topics that can be pursued in the future, and that can be extended still further in both time and space. We could move beyond 1914, as migration was reversed from the Caribbean to the metropolis, and as European integration posed challenges for the former colonies. Cuban 'independence' from Spain led to a new form of economic dependence on American capital, with the consequences of default and intervention. Space could be extended by incorporating Portugal, Brazil, and Angola in another circuit in the southern Atlantic, into which Castro's Cuba made an intervention in support of the independence movement in Angola. New circuits of trade, money and knowledge emerged after 1914, and need their own collection of essays to continue the work of this thought-provoking and stimulating collection.

Notes

1. See review by O'Brien, P.K.: 'The contributions of warfare with revolutionary and Napoleonic France to the consolidation and progress of the British industrial revolution', London School of Economics Working Papers No. 150/11, 2011. He makes the same point in his review of Allen, Robert C: *The British industrial revolution in global perspective,* Cambridge: Cambridge University Press, 2009. For an account stressing the developmental role of the state, see Ashworth, William J.: *Customs and excise: trade, production, and consumption in England, 1640–1845,* Cambridge: Cambridge University Press, 2003.
2. The key statement of the position is North, Douglass C. and Weingast, Barry R.: 'Constitutions and commitment: the evolution of institutions governing public choice in seventeenth-century England', *Journal of Economic History* 49 (1989).
3. The classic statement of mercantilism was Hecksher, Eli F.: *Mercantilism,* 2 vols, English translation, London: George Allen and Unwin, 1935. It first appeared in Swedish in 1931.
4. Smith, Adam: *An Inquiry into the Nature and Causes of the Wealth of Nations,* Campbell, R.H. and Skinner, A.S. (eds), Oxford: Oxford University Press, 1976, vol. I, p. 450.
5. Keynes, John Maynard: *The general theory of employment, interest and prices: the collective writings of John Maynard Keynes, Vol. VII* (first edition 1936; this edition Macmillan and Cambridge University Press for the Royal Economic Society, 1973), p. 339.
6. O'Brien, P.K., Griffiths, T., and Hunt, P.: 'Political components of the industrial revolution: Parliament and the English cotton textile industry, 1660–1774', *Economic History Review* 2nd ser 44 (1991); Harte, N.B.: 'The rise of protection and the English linen industry, 1690–1790', in Harte, N.B. and Ponting, K.G.: (eds), *Textile history and economic history,* Manchester: Manchester University Press, 1973.
7. Comment of Julian Hoppit, 'A label and a pie', *Times Literary Supplement,* 8 August 2014, p. 23. See Innes, Joanna: *Inferior politics: social problems and social policies in eighteenth-century Britain,* Oxford: Oxford University Press, 2009, p. 116, where she refers to the 'heady vision of possible "improvement"' based on utopianism, science, conquest, trade, technical improvements, and the formation of a skilled and disciplined workforce in the mid- and late seventeenth century. See also Webster, Charles: *The great instauration: science, medicine and reform, 1626–1660,* London: Duckworth, 1975; Slack, Paul: *From reformation to improvement: public welfare in early modern England,* Oxford: Oxford University Press, 1998; Thirsk, Joan: *Economic policy and projects: the development of a consumer society in early modern England,* Oxford: Clarendon Press, 1978.
8. My comments are based on my *Progress and poverty: an economic and social history of Britain, 1700–1850,* Oxford: Oxford University Press, 1995, pp. 535–40; Hoppit, *A label and a pie,* p. 23; Stern, Philip J. and Wennerlind, Carl (eds): *Mercantilism reimagined: political economy in early modern Britain,* New York: Oxford University Press, 2014; Pettigrew, William: *Freedom's debt: the Royal Africa Company and the politics of the Atlantic slave trade 1672–1752,* Chapel Hill: University of North Carolina Press, 2013; Hoppit, J.: 'Bounties,

the economy and the state in Britain, 1689–1800', in Gauci, Perry (ed.), *Regulating the British economy 1660–1850,* Farnham: Ashgate, 2011.

9. Gauci, Perry: 'Introduction' in Gauci, *Regulating,* p. 19.
10. On the longer-term shifts in occupational structure, see Leigh Shaw-Taylor's project outlined at http://www.geog.cam.ac.uk/research/centres/campop/occupations. On the changes in excise prior to 1688, see Coffman, D'Maris: *Excise taxation and the origins of public debt,* Basingstoke: Palgrave Macmillan, 2013.
11. See Acemoglu, Daron and Robinson, James R.: *Economic origins of dictatorship and democracy,* New York: Cambridge University Press, 2006; idem., *Why nations fail: the origins of power, prosperity and poverty,* London: Profile Books, 2012, starts from a contrast between the British colonies of North American and the Spanish of Latin America, the one leading to inclusive and the other to extractive institutions.
12. The complex legal conflicts over property rights in different sectors is shown by Getzler, Joshua: *A history of water rights at common law,* Oxford: Oxford University Press, 2004, and Kostal, R.W.: *Law and English railway capitalism, 1825–1875,* Oxford: Oxford University Press, 1994. For an excellent and wide-ranging survey of what property rights were recognised and compensated, see Hoppit, Julian: 'Compulsion, compensation and property rights in Britain, 1660–1833', *Past and Present,* 210 (2011). The compensation of slave owners is analysed in Draper, Nicholas: *The price of emancipation: slave-ownership, compensation and British society at the end of slavery,* Cambridge: Cambridge University Press, 2010.
13. O'Brien, Jean: 'They are so frequently shifting their place of residence: land and the construction of social place of Indians in colonial Massachusetts', in Daunton, Martin and Halpern, Rick (eds), *Empire and others: British encounters with indigenous peoples, 1600–1850,* London: UCL Press, 1999, pp. 206–9; Plane, Ann-Marie: 'Legitimacies, Indian identities and the law: the politics of sex and creation of history in colonial New England', in Daunton and Halpern, *Empire and others;* Karsten, Peter: *Between law and custom: 'high' and 'low' legal cultures in the lands of the British Diaspora – The United States, Canada, Australia and New Zealand 1600–1900,* Cambridge: Cambridge University Press, 2002, ch. 1.
14. Bayly, C.A.: 'The British and indigenous peoples, 1760–1860: power, perception and identity', in Daunton and Halpern, *Empire and others,* pp. 32–3; Sheidley, Nathaniel: 'Hunting and the politics of masculinity in Cherokee treaty-making, 1763–75', in Daunton and Halpern, *Empire and others.*
15. For example, Searle, C.E.: 'The Cumbrian customary economy in the eighteenth century', *Past and Present* 110 (1986); Hoyle, R.W.: 'Tenure and the land market in early modern England: or a late contribution to the Brenner debate', *Economic History Review* 2nd ser 43 (1990); Devine, T.M.: 'Social responses to agrarian "improvement": the Highland and Lowland clearances in Scotland', in Houston, R.A. and Whyte, I.D. (eds): *Scottish Society, 1500–1800,* Cambridge: Cambridge University Press, 1989.
16. For example, White, Richard: *The middle ground: Indians, empires, and republics in the Great Lakes region, 1650–1815,* Cambridge: Cambridge University Press, 1991.
17. Bailyn, Bernard: *Voyagers to the West: a passage in the peopling of America on the eve of the revolution,* New York: Knopf, 1986; Wrigley, E.A. and Schofield,

R.S: *The population history of England, 1541–1871: a reconstruction*, Cambridge: Cambridge University Press, 1981, p. 175; Galenson, D.: *White servitude in colonial America: an economic analysis*, Cambridge: Cambridge University Press, 1981; O'Rourke, Kevin and Williamson, Jeffrey G.: *Globalization and history: the evolution of a nineteenth-century Atlantic economy*, Cambridge, MA: MIT Press, 1997; Aghion, Philippe and Williamson, Jeffrey G.: *Growth inequality and globalization: theory, history and policy*, Cambridge: Cambridge University Press, 1998.

18. Wrigley, E.A.: *Continuity, chance and change: the character of the industrial revolution in England*, Cambridge: Cambridge University Press, 1988.
19. Pomeranz, Kenneth: *The great divergence: China, Europe and the making of the modern world economy*, Princeton: Princeton University Press, 2000, pp. 11, 23.
20. de Vries, Jan: *The industrious revolution: consumer behavior and the household economy, 1650 to the present*, New York: Cambridge University Press, 2008.
21. As in Davis, Lance E. and Huttenback, Robert A.: *Mammon and the pursuit of empire: the economics of British imperialism*, abridged edition, Cambridge: Cambridge University Press, 1988; O'Brien, Patrick K.: 'The costs and benefits of British imperialism', *Past and Present* 125 (1989); Offer, Avner: 'The British empire, 1870-1914: a waste of money?', *Economic History Review* ns 46 (1993).
22. Curtin, P.D.: 'The British sugar duties and West Indian prosperity', *Journal of Economic History* 14 (1954), 157–64; Taylor, Miles: 'The 1848 revolutions and the British empire', *Past and Present* 166 (2000), 152, 158, 160–2. On the Eyre controversy, see Kostal, R.W.: *A jurisprudence of power: Victorian empire and the rule of law*, Oxford: Oxford University Press, 2005. On J.S. Mill, Hall, Catherine: 'Competing masculinities: Thomas Carylye, John Stuart Mill and the case of Governor Eyre', in *White, male and middle class: explorations in feminism and history*, Cambridge: Polity Press, 1992; Collini, Stefan: *Public moralists: political thought and intellectual life in Britain 1850–1930*, Oxford: Oxford University Press, 1991, pp. 144–62.

Index

Note: Page numbers in **bold** refer to Tables; those in *italics* refer to Figures

Printed and bound in the United States of America